ACTIVATED SLUDGE AND NUTRIENT REMOVAL

WEF Manual of Practice OM-9

Fourth Edition

Edited by Robert Otey, PE, PACP, and Mazen Kawasmi, PE, PACP, CFM

2025

Water Environment Federation
601 Wythe Street
Alexandria, VA 22314–1994 USA
http://www.wef.org

ISBN 978-1-57278-483-3

About WEF

The Water Environment Federation (WEF) is a not-for-profit technical and educational organization of more than 30,000 individual members and 75 affiliated Member Associations representing water quality professionals around the world. Since 1928, WEF and its members have protected public health and the environment. WEF and its global network of members and member associations provide water quality professionals around the world with the latest in water quality education, training, and business opportunities. WEF's diverse membership includes scientists, engineers, regulators, academics, utility managers, plant operators, and other professionals. WEF uses this collective knowledge to further a shared goal of improving water quality around the world. To learn more, visit www.wef.org.

Fourth edition edited by Robert Otey, PE, PACP, and Mazen Kawasmi, PE, PACP, CFM

With contributions by

Rachel Adkisson, PE
Payam Bahadorani, PE
Christopher Callahan, PE
Shay Coombs, PE
Thomas Brady Cross, PE, DBIA
Viraj deSilva, PhD, PE, BCEE
Coby C. Gee, PE
David R. Jackson, PE, BCEE
Nick Landes, PhD, PE
John L. Lopez, PE
David Munn, PE
Olivia Overton, MS
Tony Pevec, PE, BCEE
Lindsey Reid, PE
Rishi Saladi, PhD
Chad Simmons, PE, BCEE
Juan C. Vilca, PE
Mavis Wong, PhD, PE, BCEE

Review provided by

Frances Amoye, PhD
Natalie Beach, PhD, PE
Bryan D. Coday, PhD, PE
Christian Davies-Venn, PhD, PE, BCEE, BC.WRE, ASCE Fellow, WEF Fellow
Rajesh Doppalapudi, PE, BCEE
Dana Gonzalez, PhD, PE
Elaine Hung
Ishin Kaya
Paniteja Madala, MS, M.Tech
Michelle Young, PhD
Harry Zhang, PhD, PE, EWRI Fellow

Text originally prepared by the **Activated Sludge and Nutrient Removal Task Force** of the **Water Environment Federation**

Barton Jones, PE, *Chair*

Rich Atoulikian
Evangelina Belia, PhD, P.Eng
Marie Burbano, PhD, PE, BCEE
Paul Dombrowski
Patrick Dunlap
Richard Finger
Terry Gellner
James Goodley
Sarah Hooper
Sidney Innerebner, PhD, PE, PO
Saeid Khodaei, EIT
Adam Kohout
Steve Leach
Chris Maher
Henryk Melcer
Peter Petersen
Joel Rife
Adam Rogensues, PE
Bikram Sabherwal
Kenneth Schnaars, PE
Ron Schuyler
Andrew Shaw
Art Umble
David Wagoner
Paul Wood
Qingzhong Wu

Under the Direction of the **Plant Operation and Maintenance Subcommittee** of the **Technical Practice Committee**

2018

Water Environment Federation
601 Wythe Street
Alexandria, VA 22314-1994 USA
http://www.wef.org

Manuals of Practice of the Water Environment Federation

The WEF Technical Practice and Publications Advisory Panel (formerly the Technical Practice Committee and the Committee on Sewage and Industrial Wastes Practice of the Federation of Sewage and Industrial Wastes Associations) was created by the Federation Board of Control on October 11, 1941. The primary function of the Committee is to originate and produce, through appropriate subcommittees, publications dealing with technical aspects of the broad interests of the Federation. These publications are intended to provide background information through a review of technical practices and detailed procedures that research and experience have shown to be functional and practical.

Contents

List of Figures

List of Tables

Preface

This publication is an update of WEF's *Activated Sludge and Nutrient Removal, Third Edition*, published in 2018. This extensively revised fourth edition reflects the latest advances and industry best practices in the activated sludge process and nutrient removal. Revised content includes new aeration and process technologies, adding references to recent developments in attached growth augmentation and aerobic granular sludge; the addition of process control with nutrient probes and monitoring per- and polyfluoroalkyl substances (PFAS) in wastewater; a new section on high-purity oxygen systems; a new section that describes low dissolved oxygen/suboxic aeration, anammox, partial nitrification and denitrification coupled with anammox, denitrifying polyphosphate-accumulating organisms, and sidestream enhanced biological phosphorus removal configurations; and recommendations for monitoring of emerging contaminants such as PFAS. Practice questions are included at the end of each chapter to help the reader, whether a novice or experienced professional, enhance their understanding of the material and retain information vital to solving performance problems and improving operations. For related reference material, the following WEF manuals are recommended:

1. *Standards For Automation of Water Resource Recovery Facilities* (WEF 21)
2. *Standards for Operation of Nutrient Removal Facilities* (WEF 37)
3. *Standards for Design of Water Resource Recovery Facilities* (WEF 8)
4. *Energy in Water Resource Recovery Facilities* (Manual of Practice No. 32)
5. *Information Technology for Water and Wastewater Utilities* (Manual of Practice No. 33)
6. *Safety, Health, and Security Standards for Water Resource Recovery* (Manual of Practice No. 1)
7. *Planning, Design, and Implementation for New and Upgraded Water Resource Recovery Facilities* (Manual of Practice No. 28)
8. *Membrane Bioreactors* (Manual of Practice No. 36)
9. *Industrial Wastewater Management, Treatment and Disposal* (Manual of Practice FD-3)

10. *Nutrient Removal* (Manual of Practice No. 34)
11. *Odor Emissions and Control for Collection Systems and Water Resource Recovery Facilities* (Manual of Practice No. 25)
12. *Shortcut Nitrogen Removal—Nitrite Shunt and Deammonification*
13. *Solids Process Design and Management*
14. *Sustainability and Energy Management for Water Resource Recovery Facilities* (Manual of Practice No. 38)
15. *Ultraviolet Disinfection for Wastewater—Low-Dose Application Guidance for Secondary and Tertiary Discharges*
16. *Wastewater Biology: The Microlife*
17. *Wastewater Engineering: Design of Water Resource Recovery Facilities* (MOP 8 Series)
18. *Wastewater Solids Incineration Systems* (Manual of Practice No. 30)
19. *Wastewater Treatment Process Modeling* (Manual of Practice No.31)
20. *Water and Wastewater Laboratory Techniques*

The fourth edition of this Manual of Practice was edited under the direction of Robert Otey, PE, PACP, and Mazen Kawasmi, PE, PACP, CFM.

Authors' and reviewers' efforts were supported by the following organizations:

Carollo Engineers
Donohue & Associates
Freese & Nichols
GHD, Ltd.
Water Research Foundation

1

Introduction

1.0 INTRODUCTION

The activated sludge process is the most widely used biological treatment process for reducing the concentration of organic pollutants in wastewater. Well-established design and operational standards based on empirical data and scientific bases have evolved over the years. As a result, our understanding of the process has advanced from a system originally designed simply for the removal of solids and organic material to one that now removes nutrients such as nitrogen and phosphorus. New process configurations and technologies continue to evolve. Despite these advances, poor process performance can still present problems for many water resource recovery facilities (WRRFs).

The objective of this manual of practice is to help operators and other wastewater treatment professionals acquire a greater understanding of the activated sludge process, solve performance problems, and improve operations. Water treatment professionals operating potable reuse systems that use both the activated sludge process and drinking water treatment processes in conjunction will also benefit from this knowledge. The manual is intended as both a training and reference tool for newcomers and experienced professionals alike.

2.0 ORGANIZATION

This manual of practice is organized into the following chapters:

- Chapter 2 begins with activated sludge process fundamentals and a mechanical explanation of activated sludge microbiology. This chapter includes a brief overview of the activated sludge process, basic components and equipment, activated sludge terminology, an introduction to activated sludge process control, and a brief overview of process modifications and variations.
- Chapter 3 delves into activated sludge microbiology in more detail. It provides an overview of the composition of bacterial cells, their living processes, and growth patterns. It provides a description of the various types of microorganisms typically found in wastewater treatment, the conditions that foster their growth and the significance of their presence. It also covers both the contrast and required balance between floc-forming bacteria and filamentous bacteria. The chapter rounds out by presenting a method of assessing sludge quality via microscopic examination and the concept of using this assessment to aid in process control.
- Chapter 4 delves deeper into activated sludge process control by focusing on maintaining sludge quality through management of sludge age. It includes discussion of commonly used approaches to process control, how sludge age affects process variables such as microbiology, food-to-microorganism ratio and mixed liquor volatile suspended solids to mixed liquor suspended solids ratios (MLVSS/MLSS), what dissolved oxygen concentrations should be for different operating conditions, and what constitutes an appropriate sludge-wasting regime. This chapter will help operators answer the age-old questions of how to select the "right" sludge age and process control method for their treatment system based on system configuration, treatment goals, and operating conditions.
- Chapter 5 reviews process control for secondary clarifiers and introduces state point analysis, how to optimize return activated sludge (RAS) pumping rate, and how to predict the RAS concentration from other operating data.

- Chapter 6 includes descriptions of laboratory test methods used in activated sludge process control, with detailed procedures for some methods as well as guidance on how to use and interpret results. It also includes discussion of online process control monitoring tools and recommendations for monitoring emerging contaminants such as per- and polyfluoroalkyl substances.
- Chapter 7 presents troubleshooting procedures for solving common operating problems, including inadequate mixing and aeration, formation of foams, sludge bulking, and poor sludge quality.
- Chapter 8 describes the most common chemical and biological methods for nutrient removal. Biological processes covered include nitrification, denitrification, and biological phosphorus removal, with a focus on process control variable and operational considerations. In addition, chemical phosphorus removal is discussed.
- Chapter 9 includes an introduction to using biological process models for design and operational considerations. Best practices for modeling are described and several case studies illustrate the application of process modeling.

3.0 WHAT IS IN A NAME?

When the Clean Water Act (CWA) was passed in 1972 (originally known as the Federal Water Pollution Control Act Amendments of 1972), it referred specifically to publicly owned treatment works (POTWs). The term *POTW* did not just include the treatment facility, but also all the upstream infrastructure necessary to convey wastewater to the facility, including collection system pipes and lift stations. Both the CWA and legislation in many states still use the terms *POTW* and *sewage*. Over the years, as industry has worked to convey to the public the valuable services provided and help recognize that there are valuable resources in wastewater that can be recovered for the benefit of society, the preferred term for a treatment facility became *sewage* evolved into *wastewater* and the name for a POTW evolved to *wastewater treatment plant,* or *wastewater treatment facility.* Most recently the term *water resource recovery facility (WRRF)* has been more commonly applied to reflect the value of resources that can be recovered in a well-run treatment process. "WRRF" better reflects our goals as a profession—to maximize and optimize the recovery of water and other resources in wastewater while protecting the water environment using sound and disciplined science and engineering.

Indeed, we no longer simply treat water to remove pollutants. A well-designed and operated WRRF may also produce reuse water for irrigation and industry, biosolids for beneficial reuse, methane gas for heat and power, and nutrients for fertilizer. Protecting public health and the environment, while recovering valuable resources, are the goals of a modern WRRF. Water resource recovery facilities strive to be sustainable and energy neutral where possible. Water resource recovery facilities are increasingly incorporating advanced technologies, such as artificial intelligence and the Internet of Things, to optimize operations and resource recovery. Evolving environmental regulations and climate change adaptation measures are pushing WRRFs to meet higher sustainability and resource recovery standards. Public outreach and education efforts are essential to fostering community support for these facilities' environmental and resource conservation roles.

4.0 OVERVIEW OF WATER RESOURCE RECOVERY

Most WRRFs have both a liquid stream and a solids-handling stream (Figure 1.1). The liquid stream accepts raw influent wastewater, which is ultimately discharged as clean, treated effluent to the receiving body of water or recycled water system. In most facilities, solids removed from the wastewater and newly generated biological solids are transferred from the liquid stream

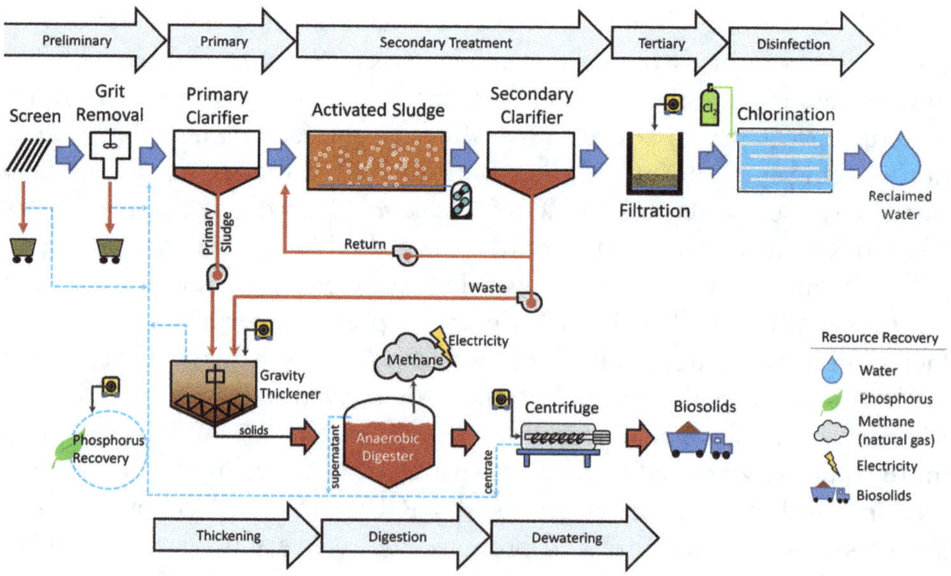

FIGURE 1.1 Treatment Process Overview (Reprinted with permission from Indigo Water Group)

to the solids stream for additional processing. The liquid and solids-handling sides of a WRRF interact as solids and liquids are recycled between them. In some facilities, solids removed may be sent to another WRRF for processing. More details can be found in WEF 23-11, *Standards for the Operation of Water Resource Recovery Facilities* (Water Environment Federation, 2023b) and *Operation of Water Resource Recovery Facilities Study Guide* (Water Environment Federation, 2018). A brief overview of the liquids stream is described in this section.

Water resource recovery facilities may contain preliminary treatment, primary treatment, secondary treatment, tertiary treatment, and/or disinfection, as shown in the activated sludge facility portrayed in Figure 1.1. A particular WRRF may have one or more of these stages of treatment. There is significant variation in the processes used across WRRFs, partly because they have been constructed over more than a century and due to changes in available technologies, vision for the utility, and differing permit requirements based on the facility's location. For example, a WRRF that discharges a large volume of treated effluent to a small, pristine trout stream will likely have more stringent ammonia limits in its discharge permit than a similarly sized facility that discharges through an ocean outfall or to a large river.

The purpose of preliminary treatment is to protect downstream equipment such as pumps from damage and/or blockages and to prevent inert material from settling and taking up valuable space in aeration basins and digesters. Preliminary treatment takes place in the WRRF headworks. Preliminary treatment typically uses screens as a first step to remove large materials such as sticks, rags, plastics, nondispersibles and other debris. Some facilities use two or three screening steps, with the initial screens used to remove the largest debris and subsequent fine screens used to remove smaller inert materials. Screens are typically followed by grit-removal facilities. Here, heavier particles like sand, grit, eggshells, and some heavier organic particles are removed while lighter particles pass on to the next process. Screenings and grit are often rinsed to return additional organic material to the liquid treatment stream for further processing. Screenings may also be compacted to reduce their volume before disposal. Screenings and grit are typically sent to a landfill for disposal. Reducing moisture and organic material content in grit and screenings is often required before the landfill will accept them. The volume of screenings and grit collected is highly dependent on the community served and the type of screening equipment used. Headworks may also include flow measurement, flow equalization, pumping, and/or odor control.

Typically, primary treatment follows preliminary treatment. Most facilities with primary treatment use primary clarifiers to slow down the water further so that heavier organic material and other particles can be removed by gravity. Grease and other floatable material will collect on the water surface where they are removed by a rotating skimmer arm. Wastewater remains in the primary clarifier for approximately 2 hours (Water Environment Federation, 2023a). It is important to note that a primary clarifier can only remove settleable and floatable material, which means removal rates are limited by the characteristics of incoming wastewater. For domestic wastewater primarily from residential and light commercial activities such as schools and stores, a primary clarifier can be expected to remove between 20% and 35% of the influent organic matter and as much as 60% to 75% of the influent suspended solids (Water Environment Federation, 2023a). Primary clarifiers do not, however, remove colloidal solids, dissolved solids, soluble organics, soluble phosphorus, or ammonia.

The purpose of the primary clarifier is to reduce the amount of organic and solids loading going to the secondary treatment process. Reducing the load to the secondary treatment process reduces its overall size and the amount of energy needed for process operation. For facilities that do not have primary clarification, the settleable solids, scum, and floatables will pass directly into the secondary treatment process.

By the time wastewater reaches the secondary treatment process, many of the larger particles that can settle on their own have already been removed by screening, grit removal, and/or primary clarification. Regardless of whether a facility has primary clarifiers, many of the particles that remain in the wastewater when it reaches the activated sludge process will not settle quickly on their own. For treatment to continue, the size of the remaining particles must be increased so they can be efficiently removed.

The activated sludge process and other biological treatment processes excel at converting smaller particles and soluble, biodegradable organic material into larger, heavier particles through bioflocculation. *Flocculation* refers to growing larger particles through collisions that help smaller particles stick together. Bioflocculation uses a combination of flocculation and biological conversion and growth to agglomerate smaller particles into larger, heavier ones. In the activated sludge basin, incoming wastewater is fed to a complex mixture of bacteria and other microorganisms known as the MLSS. Operators often refer to MLSS as simply the "bugs," or "inventory," which is a shorthand reference to bacteria in the process. The organic material in the influent wastewater becomes their food. As bacteria consume the available food, they form large colonies called *flocs*, which will, under the right conditions, grow large enough and heavy enough that they can be separated from treated wastewater by gravity. Flocs are kept suspended in the activated sludge process by either mixers, aeration, or a combination of mechanical mixing and aeration. In addition, nonbiodegradable solids in raw wastewater become part of the floc particles through bioflocculation.

Individual floc particles have a life cycle of initial formation and growth. As the floc particles age, they accumulate dead bacteria and other inert material and increase in size. As floc particles grow larger, it becomes increasingly challenging for bacteria at the center of the particle to rid themselves of waste products and gain access to nutrients and oxygen. Eventually, the floc will break into smaller flocs and the cycle will begin anew.

For most activated sludge processes, MLSS will be conveyed to a separate secondary clarifier and allowed to settle. For the separation step to be efficient, floc particles grown in the aeration basin must be large and dense. Process control aims to produce an MLSS that flocculates, settles, compacts, and complies with effluent discharge permit limits for organics, solids, ammonia, and other parameters (Wahlberg, 2016). Treated, clarified wastewater flows out through the top of the clarifier while MLSS settles to the bottom to form a sludge layer called the *blanket*. Most of the settled MLSS is returned to the activated sludge process, where it will be used to treat more influent wastewater. This is the RAS. Excess MLSS is removed from the process entirely as waste activated sludge. In a sequencing batch reactor activated sludge process, treatment and clarification take place in the same basin.

5.0 INFLUENT CHARACTERISTICS

Before going further, some terms that are used to describe the components of influent wastewater should be defined. Additional information on these and other parameters can be found in Chapter 6.

5.1 Biochemical Oxygen Demand

Biochemical oxygen demand (BOD) is a rough measure of how much oxygen is needed for bacteria to consume the biodegradable organic material in wastewater. Most WRRFs have a rated capacity for both flow and BOD. Discharge permits include limitations on the number of liters (gallons) of flow and kilograms (pounds) of BOD that a WRRF is allowed to receive and limitations on the concentration of BOD that may be discharged in the final effluent. Secondary treatment standards, which are part of the CWA, require WRRFs to remove at least 85% of influent BOD and total suspended solids (TSS). The CWA also limits the concentration of BOD and TSS that can be discharged to a maximum of 30 mg/L for a 30-day average and 45 mg/L for a 7-day average. Actual discharge parameters for BOD, TSS, and other constituents will vary depending on the permit requirements of the specific WRRF and, typically, are lower than these maximums.

Biochemical oxygen demand testing is conducted by taking a sample of wastewater, diluting it if necessary to bring it within the measurable range of the test, measuring the initial dissolved oxygen concentration, incubating the sample at 20 °C for a fixed period of time in the dark, and then measuring the final dissolved oxygen concentration. The BOD test is typically conducted over a 5-day period (5-day BOD [BOD_5]). The more organic material that the wastewater contains, the more oxygen bacteria will use to consume and stabilize it. By definition 1 kg of BOD will consume 1 kg of oxygen (1 lb of BOD consumes 1 lb of oxygen). Biochemical oxygen demand is thought of in terms of organic strength, but the test really measures how much oxygen is needed to treat or stabilize the wastewater. It is not possible to measure all of the organic compounds that might be in wastewater, so the BOD test is used to give a bulk estimate instead. It is worth noting that a 100-mL sample containing 1 g of glucose will have a different BOD than another 100-mL sample containing 1 g of starch. Each of these compounds is organic, but they have different numbers and types of chemical bonds. Bacteria use oxygen to break down organic material for energy and growth. The more complex the organic material, the more oxygen will be needed.

Biochemical oxygen demand can be further divided into carbonaceous biochemical oxygen demand (CBOD) and nitrogenous oxygen demand (NOD), as shown in the equation that follows. The amount of oxygen used in the BOD test that goes toward the breakdown of organic material is the CBOD. Additional oxygen may be used by a special group of bacteria, the nitrifying bacteria, to convert ammonia to nitrate. When this reaction takes place in the BOD test, the additional oxygen demand is the NOD. If an analyst wants to measure only the CBOD, a substance that is toxic to nitrifying bacteria will be added to the BOD test. For raw domestic wastewater (influent), typically, the CBOD is approximately 85% of the total BOD (Water Environment Federation, 2022).

$$BOD_{total} = CBOD + NOD$$

Biochemical oxygen demand consists of both soluble (dissolved) and particulate fractions. If a wastewater sample is filtered through a piece of filter paper, some of the organic material will pass through the filter and some will not be able to pass through. The organic material that passes through the filter is the dissolved or soluble BOD. The soluble fraction is consumed rapidly once it comes into contact with MLSS. The organic material that remains on the filter paper is the particulate BOD. If a facility produces primary sludge, a majority of the particulate BOD is captured before the activated sludge process. The particulate fraction may sorb rapidly to the biomass and degrade at a rate that depends on its composition. A can of soda, for example, contains a lot of sugar that is organic and able to pass through a piece of filter paper because it is dissolved. The percentage of BOD that is soluble depends on the size of the collection system and the types of customers served. For domestic wastewater, including light commercial users such as stores and schools, typically, soluble BOD is between 20% and 40% of total BOD (Water Environment Federation, 2022).

5.2 Chemical Oxygen Demand

Wastewater contains additional substances that may consume oxygen in a standard laboratory test and in the treatment process, but, at the same time, are nonbiodegradable. Heavy metals, hydrogen sulfide, and organic material that cannot be consumed easily by bacteria in the BOD_5 test are some of the substances that may consume additional oxygen during treatment or in the environment. This is the chemical oxygen demand (COD). The COD test does not measure oxygen demand from ammonia or organic nitrogen. The COD test is performed by taking a small sample and cooking it at high temperature with sulfuric acid and potassium dichromate. A color change indicates the amount of COD present. The test measures all of the biodegradable and nonbiodegradable substances in the wastewater. The COD result must always be equal to or greater than the BOD result for the same sample. Chemical oxygen demand is used in facility design, process modeling, and process control, and is the amount of oxygen required for the chemical oxidation of compounds in water. Carbonaceous biochemical oxygen demand is also used in facility design, process modeling, and process control and measures the oxygen needed by microorganisms to decompose only the carbonaceous (organic) matter in wastewater, excluding nitrogenous compounds. For domestic wastewater, including light commercial users such as stores and schools, a typical COD-to-CBOD ratio is between 1.9 and 2.2 (Water Environment Federation, 2023a).

5.3 Solids

Solids are classified according to their size, whether they can pass through a 1.5-mm filter, and whether they are organic or inorganic. Organic solids can be combusted (volatilized) in a furnace at 550 °C. Inorganic solids, or ash, are the residue remaining after a sample has been volatilized. Commonly used terms when discussing solids in wastewater are listed in Table 1.1.

Solids may be separated in the laboratory into different components according to Figure 1.2. If a sample of wastewater is measured and then dried at 103 °C in a preweighed dish, the water will evaporate and leave the solids behind. This is analogous to boiling a pot of water until dry, leaving behind a residue. This residue is the total solids.

TABLE 1.1 Types of Solids in Wastewater (ASTM International, 2025) (Reprinted with permission from Indigo Water Group)

Term	Definition
Total suspended solids	Solids that cannot pass through a 1.5-um filter
Total volatile suspended solids	Solids that cannot pass through a 1.5-um filter and are burned away when placed in a furnace at 550 °C
Total dissolved solids (TDS)	Solids that are small enough to pass through a 1.5-um filter; the sample must be dried completely before the dissolved solids can be seen with the naked eye
Total solids	All of the solid material in a sample; this includes both organic and inorganic solids Total solids = TSS + TDS
Total volatile solids	All of the solids in a sample that are burned away when placed in a furnace at 550 °C

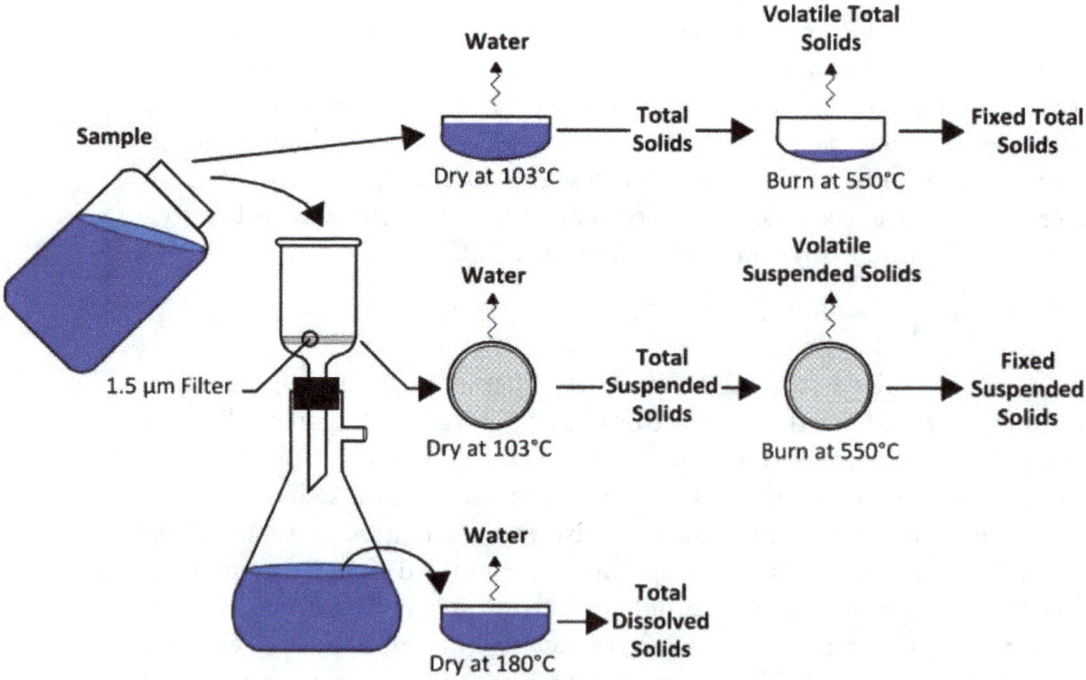

FIGURE 1.2 Classification of Influent Solids

If the wastewater sample is filtered through a piece of filter paper, the dissolved salts, minerals, sugars, fats, and other organic substances will pass through the filter paper. These are the dissolved solids. Think about dissolving sugar into coffee or tea. The solids that cannot pass through the filter paper are called the TSS or residue.

Total solids and TSS can be further broken down into inorganic and organic solids, typically called fixed and volatile solids, respectively. Think about the wastewater that comes into the WRRF. Sand, grit, and eggshells are solid, but they cannot be eaten by bacteria in the aeration basin and thus are fixed (inorganic). They are both insoluble and fixed solids. If a mixture of salt and sugar dissolved in water is allowed to dry in a dish, the residue left will be total solids. If that residue is heated to 550 °C, the sugar will volatilize but the salt will remain. The sugar is both soluble and a volatile solid, whereas salt is both soluble and fixed.

5.4 Nutrients

Influent wastewater also contains nitrogen, phosphorus, and other trace nutrients and minerals. Nitrogen is present in several compounds in influent wastewater, including organically bound nitrogen (proteins and other compounds) and ammonia. Organically bound nitrogen can be soluble or particulate, whereas all ammonia is soluble. This is why a primary clarifier cannot remove ammonia. Phosphorus can also be particulate or dissolved. Phosphorus can be organically bound (proteins and other compounds), chained together with other phosphate molecules to form condensed phosphates (detergents and antiscaling agents), or be present as orthophosphate. Orthophosphate has a single phosphorus atom bound to four oxygen atoms. Orthophosphate is also called *reactive phosphorus* because it reacts easily with other chemical compounds. All three forms of phosphorus can be either particulate or dissolved. Most of the phosphorus entering the treatment process will be dissolved orthophosphate.

Nitrogen compounds are expressed as "mg/L as N," for example, 1 mg/L of ammonia-nitrogen. This method of expression converts all of the different nitrogen compounds into the same currency: nitrogen. Organic nitrogen, ammonia, nitrite, and nitrate all contain other elements in addition to nitrogen, including hydrogen, oxygen, and carbon. Nitrogen is the element of concern from an environmental perspective. Expressing all the nitrogen compounds as nitrogen focuses the attention on the primary pollutant. It also allows the different nitrogen compounds to be added and subtracted to find total nitrogen, total Kjeldahl nitrogen, or total inorganic nitrogen. Total Kjeldahl nitrogen includes organic nitrogen and ammonia. Total inorganic nitrogen includes ammonia, nitrite, and nitrate.

Expressing nitrogen compounds as nitrogen requires some knowledge of chemistry. The periodic table contains the atomic weights of different elements. Adding atomic weights for a particular compound like nitrate together gives the formula weight.

Example: How much nitrate as nitrogen (NO_3-N) is in 20 mg/L of nitrate?

Step 1: Look up the atomic weights for nitrogen and oxygen. A mole of nitrogen atoms weighs 14 g/mol. A mole of oxygen atoms weighs 16 g/mol.

Step 2: Add the atomic weights together to find the formula weight:
1 nitrogen = (1)(14) = 14 g/mol
3 oxygen = (3)(16) = 48 g/mol
Total formula weight = 62 g/mol

Step 3: Use dimensional analysis to convert from milligrams per liter of nitrate to milligrams per liter of nitrate-nitrogen:

$$\frac{20 \text{ mg NO}_3}{\text{L}}\left|\frac{1 \text{ mol NO}_3}{62 \text{ g NO}_3}\right|\left|\frac{1 \text{ mol N}}{1 \text{ mol NO}_3}\right|\left|\frac{14 \text{ g N}}{1 \text{ mol N}}\right| = 4.5 \text{ mg/L NO}_3 - \text{N}$$

Phosphorus may be expressed as either phosphate (PO_4) or phosphate phosphorus (PO_4-P). Testing laboratories tend to report all nitrogen compounds as nitrogen but are not as consistent when reporting phosphorus compounds. It is crucial that operators be able to convert between milligrams per liter of PO_4 and milligrams per liter of PO_4-P.

Example: How much phosphate as phosphorus (PO_4-P) is in 3 mg/L of PO_4?

Step 1: Look up the atomic weights for phosphorus and oxygen. A mole of phosphorus atoms weighs 31 g/mol. A mole of oxygen atoms weighs 16 g/mol.

Step 2: Add the atomic weights together to find the formula weight:
1 phosphorus = (1)(31) = 31 g/mol
4 oxygen = (4)(16) = 64 g/mol
Total formula weight = 95 g/mol

Step 3: Use dimensional analysis to convert from PO_4 to PO_4-P:

$$\frac{3 \text{ mg PO}_4}{\text{L}}\left|\frac{1 \text{ mol PO}_4}{95 \text{ g PO}_4}\right|\left|\frac{1 \text{ mol P}}{1 \text{ mol PO}_4}\right|\left|\frac{31 \text{ g P}}{1 \text{ mol P}}\right| = 0.98 \text{ mg/L PO}_4 - \text{P}$$

Table 1.2 shows typical ranges for various influent parameters for domestic wastewater.

TABLE 1.2 Typical Composition of Raw Municipal Wastewater With Minor Contributions of Industrial Wastewater (g/m^3) (Eveline et al., 2023) (CC-BY-NC-ND 4.0)

Parameter	Low	Medium	High
COD total	500	750	1,200
COD soluble	200	300	480
COD suspended	300	450	720
BOD	230	350	560
VFAs (as acetate)	10	30	80
N total	30	60	100
Ammonia N	20	45	75
Nitrate + Nitrite N	0.1	0.2	0.5
Organic N	10	15	25
P total	6	15	25
Ortho-P	4	10	15
Organic P	2	5	10
Sulphate	24	36	72
TSS	250	400	600
VSS	200	320	480

Note. VFAs = volatile fatty acids; VSS = volatile suspended solids

5.5 Relationships Between Influent Parameters

A brief discussion of the relationship between COD, BOD, and TSS is needed to understand what happens to each component during treatment. A more detailed discussion can be found in *Basic Laboratory Procedures for the Operator-Analyst* (Water Environment Federation, 2022). Figure 1.3 shows a graphic representation of the relationships between influent solids and influent organic matter. For domestic wastewater that does not have a large industrial component, the COD-to-CBOD ratio, typically, will be between 1.9 and 2.2 (Water Environment Federation, 2023a). In Figure 1.3, COD is shown as being made up of two smaller parts: biodegradable COD, which is the same as CBOD, and nonbiodegradable COD. Both fractions can be

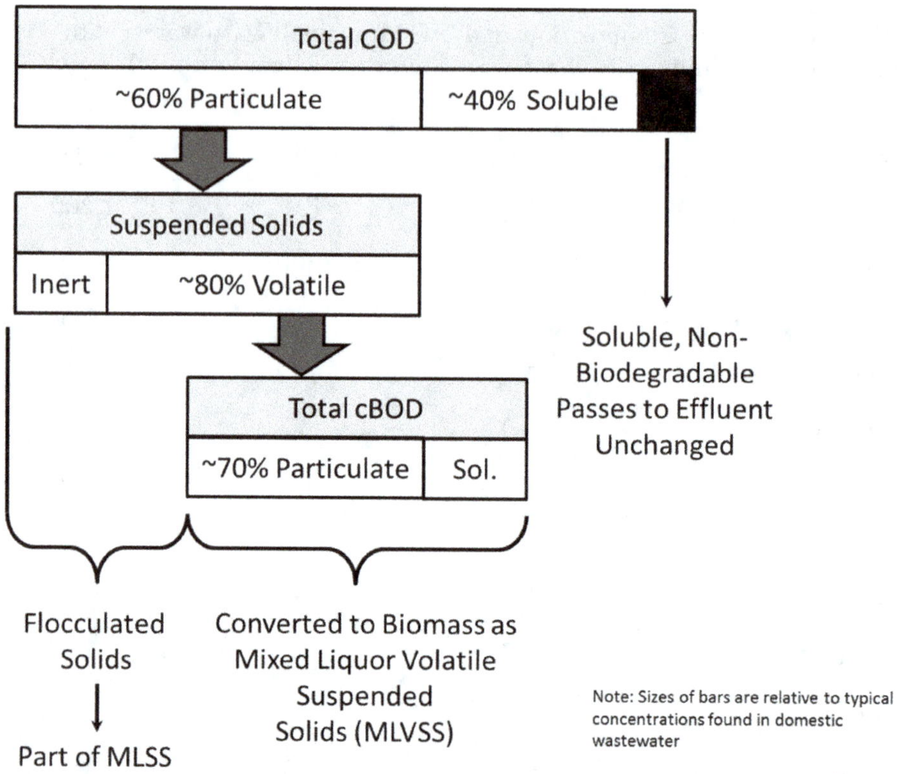

FIGURE 1.3 Relationships Between Influent Parameters (Reprinted with permission from Indigo Water Group)

further subdivided into particulate and soluble organic matter. For domestic wastewater, between 20% and 40% of influent CBOD will be soluble (Water Environment Federation, 2023a). Similarly, between 30% and 50% of influent COD will be soluble. Most of the TSS entering the WRRF will be organic, but between 15% and 25% is nonvolatile (inorganic) material such as sand, grit, and eggshells (Water Environment Federation, 2023a). Looking closely at Figure 1.3, it should be evident that organic solids are roughly equivalent to particulate CBOD and that the amount of soluble BOD in domestic wastewater is roughly equivalent to the amount of inert solids. Understanding these relationships can help an activated sludge process operator understand what happens to each fraction during treatment.

Organic material that is either soluble or particulate will be converted by bacteria into more bacteria, or MLVSS. Inert solids that are not removed during preliminary and primary treatment cannot be broken down by bacteria and will become part of the floc particles through bioflocculation. This is also true for particulate, nonbiodegradable COD. There is a small fraction of COD that is both nonbiodegradable and soluble. Bacteria cannot consume this fraction. It cannot be agglomerated through bioflocculation. This small fraction of the total COD passes through the WRRF unchanged and ends up in the final effluent.

6.0 PURPOSE OF BIOLOGICAL TREATMENT

The activated sludge process may be designed and operated to remove CBOD, convert ammonia to nitrate, remove nitrogen compounds, and/or remove phosphorus. Details of the different biological reactions are discussed in other chapters. The design of the system must provide for adequate basin size, oxygenation capacity, separation facilities, and other separation techniques to achieve target effluent requirements. The design of these systems is beyond the scope of this manual but may be found in *Standards for Design of Water Resource Recovery Facilities* (Water Environment Federation, 2023a). The following sections will address some of the process goals for each of the treatment objectives previously cited.

6.1 Conversion of Biodegradable Organic Material to Biomass

The purpose of biological treatment is to convert biodegradable organic material in influent wastewater to MLVSS and flocculate and settle nonbiodegradable particulate material. The conversion of biodegradable organic material is primarily through fast-growing heterotrophic bacteria. The goal of activated sludge processes is to grow larger, heavier colonies of bacteria that can then be separated efficiently from the treated water by gravity. The flocculated mass of bacteria and nonbiodegradable solids is collectively called MLSS. Uptake and conversion of CBOD is relatively rapid and is often complete within 90 minutes of influent wastewater entering the activated sludge basin. The importance of rapid CBOD uptake in systems with short hydraulic retention times, such as membrane bioreactors, is increasingly used in advanced municipal treatment. Bioflocculation takes much longer and requires that the solids (MLSS) remain in the process for many days.

With typical municipal wastewater, a well-designed and operated activated sludge system should achieve a CBOD effluent quality of 5 to 15 mg/L. Typically, effluent suspended solids should be less than 15 mg/L. To achieve consistent BOD and TSS concentrations less than 5 mg/L, some type of tertiary treatment may be required.

6.2 Conversion of Ammonia to Nitrate

Ammonia can be toxic to aquatic life when the concentration of ammonia in the stream or lake is high enough. For this reason, facilities that do not have enough in-stream dilution are required to remove ammonia. Adoption of anammox-based technologies has been used for more energy-efficient and cost-effective ammonia removal in high-ammonia wastewaters. The conversion of ammonia to nitrate is primarily carried out by slow-growing autotrophic bacteria in a two-step process. Autotrophic bacteria cannot use BOD and, instead, derive their carbon from inorganic carbon compounds, such as carbonate ions (CO_3^{-2}) and bicarbonate (HCO_3^{-}). Carbonate and bicarbonate are two components of alkalinity. Ammonia-oxidizing bacteria obtain their energy by oxidizing ammonia to nitrite and nitrite-oxidizing bacteria obtain their energy by oxidizing nitrite to nitrate. Nitrification processes may be designed as a combined system, in which both CBOD removal and ammonia oxidation can take place, or in two-stage systems in which CBOD removal is achieved in the first stage and nitrification is achieved in the second stage. There are advantages and disadvantages to each approach, which will be discussed in later chapters. Readers may also refer to *Nutrient Control Design Manual* (U.S. Environmental Protection Agency, 2010) and Chapter 8 of this book for more information on nitrification processes. Nitrification systems for municipal wastewater can achieve greater than 90% removal of ammonia, producing ammonia concentrations less than 1 mg N/L (U.S. Environmental Protection Agency, 2010).

6.3 Conversion of Nitrate to Nitrogen Gas

Conversion of ammonia to nitrate mitigates the toxicity issue for aquatic life but does not reduce the amount of nitrogen going to the receiving stream or lake. Excess nitrogen can result in algal blooms that can lead to eutrophication, disrupting aquatic ecosystems. Even without algal blooms, adding nitrogen to a natural system affects and modifies the quantity and variety of organisms that can thrive in it. Nitrate concentrations of 10 mg/L as nitrogen in drinking water supplies can cause methemoglobinemia, also known as blue baby syndrome, in babies, the elderly, and persons with blood disorders. To better protect streams, lakes, and downstream water users, total nitrogen removal may be necessary. Nitrate can be converted to nitrogen gas through denitrification by many different bacteria in the activated sludge process. The nitrogen gas produced is not very soluble in water. The gas leaves the wastewater and goes into the atmosphere, which is approximately 78% nitrogen.

Biological denitrification requires a source of carbon and the absence of oxygen. Ideally, the source of carbon will be CBOD from the influent; however, almost any biodegradable carbon source may be used, including methanol, waste beer, molasses, and other food wastes. In denitrification, bacteria consume CBOD using nitrate in place of oxygen. Theoretically, 2.86 kg of oxygen demand is satisfied per kilogram (2.86 lb/lb) of nitrate-nitrogen reduced to nitrogen gas (U.S. Environmental Protection Agency, 2010). Additionally, biological denitrification results in the release of approximately one-half of the alkalinity consumed in the nitrification process. Separate-stage and single-sludge denitrification processes can both achieve high nitrogen removal, on the order of 85% to 95% for municipal wastewater.

6.4 Biological Phosphorus Removal

Phosphorus is often the growth-limiting nutrient in the environment. Concentrations as low as 0.1 mg/L as phosphorus can cause algal blooms in streams and lakes. Wastewater effluent is a significant source of phosphorus in many watersheds. A typical activated sludge process may have an influent total phosphorus of 7 mg/L and an effluent total phosphorus of 3 mg/L. Water resource recovery facilities discharging into highly sensitive water bodies may be required to achieve total phosphorus removal between 1 and 0.1 mg/L in the final effluent, depending on permitted discharge limits.

In an activated sludge process, live bacteria uptake both nitrogen and phosphorus for growth and metabolism. Resulting MLSS will contain approximately 12% nitrogen and 2% phosphorus on a dry weight basis. Phosphorus cannot be transformed to a volatile gas like nitrogen and be released to the atmosphere; therefore, the only way for phosphorus to exit the treatment process is either through waste sludge or in the final effluent. Simply removing excess MLSS from an activated sludge process may result in 10% to 30% removal of phosphorus. Approximately 1 mg/L of phosphorus is removed for every 100 mg/L of $CBOD_5$ removed. The activated sludge process may be designed and operated to select for a population of microorganisms called *phosphate-accumulating organisms* that can store excessive quantities of phosphorus. This is called *enhanced biological phosphorus removal.* Phosphate-accumulating organisms can store up to 15% of their dry weight as phosphorus (Bond et al., 1999). The resulting MLSS may be 4% to 12% phosphorus (Liu et al., 1997). Removing this phosphorus-enriched sludge can result in effluent phosphorus concentrations below 1 mg/L.

7.0 PRACTICE QUESTIONS

1. The purpose of screening and grit removal is to
    a. Decrease loading to the secondary treatment process
    b. Prevent buildup of organic material in the clarifiers
    c. Protect downstream equipment
    d. Classify solids by their size and organic content

2. The purpose of primary treatment is to
   a. Reduce the organic loading rate to the secondary treatment process
   b. Remove settleable material from the influent wastewater
   c. Decrease overall WRRF energy use
   d. All the above

3. All of these are examples of secondary treatment EXCEPT
   a. Activated sludge
   b. Lagoons
   c. Trickling filters
   d. Primary clarifiers

4. The goal of activated sludge process control is to produce an MLSS that settles, flocculates, compacts, and meets the requirements of the discharge permit.
   a. True
   b. False

5. All the following influent components can be removed through bio-flocculation EXCEPT
   a. Particulate BOD
   b. Inert TSS
   c. Soluble, nonbiodegradable COD
   d. Particulate COD

6. When an activated sludge process is operated for total nitrogen removal, where does the nitrogen go?
   a. Into the atmosphere as nitrogen gas
   b. Into the waste sludge as nitrate
   c. Into the final effluent as organic nitrogen
   d. Into the digester as combined nitrogen

Answer key: 1. c, 2. d, 3. d, 4. a, 5. c, 6. a

8.0 REFERENCES

ASTM International. (2025, February 20). *Standard test methods for filterable matter (total dissolved solids) and nonfilterable matter (total suspended solids) in water* (ASTM D5907-18).

Bond, P. L., Keller, J., & Blackall, L. L. (1999). Anaerobic phosphate release from activated sludge with enhanced biological phosphorus removal. A possible mechanism of intracellular pH control. *Biotechnology and Bioengineering*, *63*(5), 507–515.

Eveline, I. P., Volcke, K. S., Comeau, Y., & Henze, M. (2023). Wastewater characteristics. In G. Chen, M. C. M. van Loosdrecht, G. A. Ekama, & D. Brdjanovic (Eds.), *Biological wastewater treatment: principles, modelling and design* (2nd ed.). IWA Publishing.

Liu, W.-T., Nakamura, K., Matuso, T., & Mino, T. (1997). Internal energy-based competition between polyphosphate- and glycogen-accumulating bacteria in biological phosphorus removal reactors—Effect of P/C feeding ratio. *Water Research*, *31*(6), 1430–1438.

U.S. Environmental Protection Agency. (2010). *Nutrient Control Design Manual* (EPA 600/R-10/100). Office of Research and Development, National Risk Management Research Laboratory—Water Supply and Water Resources Division.

Wahlberg, E. J. (2016). Personal communication.

Water Environment Federation. (2018). *Operation of water resource recovery facilities study guide.*

Water Environment Federation. (2022). *Basic laboratory procedures for the operator-analyst* (6th ed.).

Water Environment Federation. (2023a). *Standards for design of water resource recovery facilities* (WEF 8).

Water Environment Federation. (2023b). *Standards for operation of water resource recovery facilities* (WEF 11).

2

Activated Sludge Process Fundamentals

1.0 BACKGROUND AND PURPOSE

The purpose of this chapter is to give a broad overview of the activated sludge process and the biological processes that take place within it. The chapter will compare two groups of bacteria, the heterotrophs and the autotrophs, to combustion engines. Each group of bacteria has unique fuel and oxygen requirements that influence how they function under different environmental conditions. These basic principles will lay the foundation for more detailed and complex discussions in later chapters. The second part of the chapter presents examples of different activated sludge processes. For an in-depth discussion of process control and the interactions among process variables, refer to Chapters 4, 5, and 8.

2.0 BASIC SYSTEM COMPONENTS

2.1 Activated Sludge Basin

In its simplest form, the activated sludge process consists of an activated sludge basin where wastewater is combined with oxygen and a mixed population of bacteria and other microorganisms to consume organic material in wastewater (Figure 2.1). Activated sludge basins are the heart of the process. This is where treatment occurs. Microorganisms use organic material, a carbon and energy source, to grow, reproduce, and form larger, heavier particles called *flocs* through bioflocculation. Historically, activated sludge basins were referred to as *aeration basins* or *A-basins* because air is added to the process; however, not all activated sludge basins are aerated. Air, pure oxygen, or mechanical mixers may be used to mix the activated sludge basin, provide oxygen to bacteria, keep flocs in suspension, and blend microorganisms with influent wastewater.

The mixture of microorganisms and wastewater remains in the activated sludge basin for 4 to 36 hours, depending on the level of treatment required. Tanks are sized to provide sufficient hydraulic retention time (HRT) for

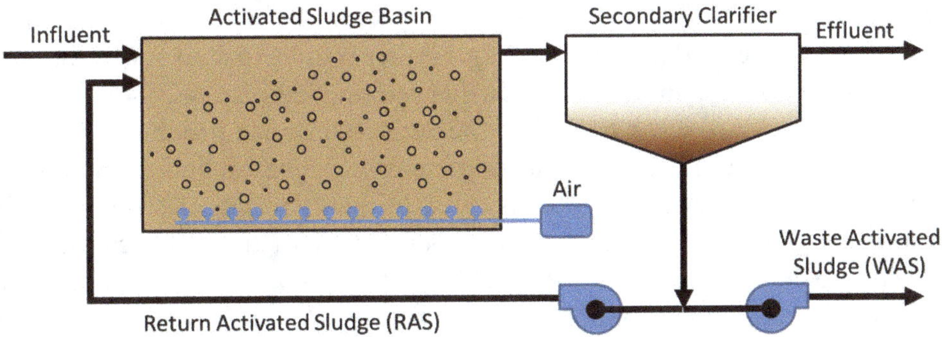

FIGURE 2.1 Activated Sludge Process (Reprinted with permission from Indigo Water Group)

proper flocculation and oxidation of the carbonaceous biological oxygen demand (CBOD) and ammonia, if nitrifying is a requirement. Depending on the requirements for effluent quality, the reactor may be subdivided or compartmentalized to achieve specific biochemical reactions.

The contents of the activated sludge basin are referred to as mixed liquor, which consists of wastewater; microorganisms; and biodegradable and nonbiodegradable suspended, colloidal, and soluble organic and inorganic matter. The solids are referred to as *mixed liquor suspended solids (MLSS)*, and the organic fraction is called *mixed liquor volatile suspended solids (MLVSS)*. If you looked at a sample of mixed liquor under the microscope, you would see a variety of different organisms living together in compact clumps suspended in the wastewater. In fact, mixed liquor borrows its name from the mining industry, where a liquor is any high suspended solids mixture or slurry. In activated sludge, it is a mixed liquor because of the variety of different organisms living in it. Microorganisms consist primarily of organic matter (70% to 80%) and are often measured as MLVSS; however, it should be noted that a portion of the MLVSS represents inert organic matter or nonviable biomass (i.e., not living and actively metabolizing) and the viable-to-nonviable biomass ratio is important for assessing the health of the microbial community.

2.2 Secondary Clarifier

While wastewater and biomass are blended into a mixed liquor in the activated sludge basin, downstream, the water must be separated from the biomass, typically in a secondary clarifier. The secondary clarifier is not aerated or mixed, allowing gravity to settle the solids. Floc particles produced in the activated sludge basin are heavier than water and, gradually, will sink to the bottom of the clarifier to form a layer called the *sludge blanket*. Clear, treated wastewater flows out over the top of the clarifier. This is the effluent. Typically, secondary clarifiers are smaller than activated sludge basins, with the water remaining in the clarifier for approximately 2 hours. Most of the settled solids are returned to the activated sludge process, where the bacteria will be used to treat more influent wastewater. This is the *return activated sludge (RAS)*. However, because microorganisms are continuously reproducing and increasing their numbers, some of the biomass must be continuously removed (i.e., wasted) from the activated sludge system to control the microorganism population. This is *waste activated sludge (WAS)*. Wasting is accomplished by removing a portion of the MLSS from the bottom of the clarifier, directly from the activated sludge basin, or from the RAS flow. Regardless of how WAS is removed from the activated sludge system, typically, it is transferred to solids-handling facilities for further processing. Operators determine how much WAS to remove each day by calculating the system sludge age and then adjusting the mass of WAS removed to maintain a target sludge age. Sludge age is discussed in detail in Chapter 4. This is the principal mechanism used for process control.

2.3 Return Activated Sludge

The aeration basin and secondary clarifier are connected by the RAS line. Typically, this pipeline runs from the bottom of the secondary clarifier to the front of the activated sludge basin, as shown in Figure 2.1. The RAS line is what makes activated sludge activated. Its purpose is to take the settled, active biological solids or MLSS from the bottom of the clarifier and return them to the treatment process.

The RAS line is what differentiates an activated sludge process from a lagoon treatment system. A *lagoon* is a natural treatment system consisting of one or more ponds that may or may not be mechanically aerated. Wastewater enters a lagoon system, bringing with it a population of microorganisms. As microorganisms consume organic material in wastewater, they reproduce and form heavier particles, similar to an activated sludge process. Typically, the last pond in a lagoon treatment process serves as a settling pond. Settling ponds, also called *polishing ponds*, perform the same function in lagoons as secondary clarifiers do for activated sludge processes. They separate treated wastewater from solids. With a lagoon, settled solids remain in the bottom of the settling pond, where they are broken down further by bacteria before being removed by the operator. They are not recycled. Because the lagoon is a pass-through system, wastewater and microorganisms spend the same amount of time in the process. The HRT and solids retention time (SRT) are the same. This limits the total number of microorganisms available for treatment to the number of microorganisms entering with the raw wastewater plus the number added through reproduction. Reproduction is limited by the amount of CBOD and oxygen available, water temperature, and the amount of time wastewater is allowed to spend within the lagoon. Other environmental factors can also affect reproduction. As a result, the concentration of microorganisms in a lagoon system is relatively low, which both extends the time required for treatment and limits the amount of biochemical oxygen demand (BOD) that can be processed.

Adding an RAS line allows the operator to keep solids in the activated sludge process longer than the water. Wastewater still moves through the system in a single pass, but solids are continuously recycled, which gives microorganisms in the process more time to grow and reproduce. As a result, the concentration of microorganisms in an activated sludge process is relatively high, which reduces the time required for treatment and increases the amount of BOD that can be processed. A typical, aerated lagoon system producing a final effluent with less than 30 mg/L of 5-day BOD (BOD_5) and total suspended solids (TSS) may need an HRT between 5 and 20 days to treat a BOD_5 load of 1682 kg/km^2·d (15 lb/ac/d). An activated sludge process typically has an HRT between 4 and 36 hours, which is much shorter, and is able to treat BOD_5 loads up to 641 kg/10^3 m^3·d (40 lb/10^3 cu ft/d) (U.S. Environmental Protection Agency, 1974, 1978). Reducing the HRT reduces the size of the treatment process. The fundamental difference between the two treatment systems is the RAS line. One specialized type of activated sludge system is a sequencing batch reactor (SBR). Sequencing batch reactors are batch processes that do not have separate secondary clarifiers or an RAS line. Different types of activated sludge systems will be discussed later in this chapter.

Return activated sludge, with influent flow, pushes MLSS from the activated sludge basin into the secondary clarifier. The mass of MLSS entering the secondary clarifier is the solids loading rate (SLR). The SLR may be calculated according to the following formula:

In International Standard units:

$$\text{Solids loading rate, kg/m}^2\cdot\text{d} = \frac{\left(Q_{\text{inf}} + Q_{\text{RAS}},\ \text{m}^3/\text{d}\right)\left(\text{Concentration, }\dfrac{\text{mg}}{\text{L}}\right)}{(1000)\left(\text{Clarifier surface area, m}^2\right)} \quad (2.1a)$$

In U.S. customary units:

$$\text{Solids loading rate, lb/sq ft}\cdot\text{d} = \frac{\left(Q_{\text{inf}} + Q_{\text{RAS}}\right)\left(\text{MLSS, }\dfrac{\text{mg}}{\text{L}}\right)\left(8.34\ \dfrac{\text{lb}}{\text{mil. gal}}\right)}{\text{Clarifier surface area, sq ft}} \quad (2.1b)$$

Where

Qinf = influent flow,
QRAS = RAS flow, and
8.34 lb/mil. gal = conversion factor.

Water weighs 8.34 lb/gal; however, the "8.34" used in this equation is a conversion factor used to convert milligrams per liter to pounds per million gallons. The conversion of 1 mg/L into pounds per million gallons is shown below:

$$\frac{1\ \text{mg}}{\text{L}}\left|\frac{1\ \text{g}}{1000\ \text{mg}}\right|\left|\frac{1\ \text{kg}}{1000\ \text{g}}\right|\left|\frac{2.204\ \text{lb}}{1\ \text{kg}}\right|\left|\frac{3.785\ \text{L}}{1\ \text{gal}}\right|\left|\frac{1\,000\,000\ \text{gal}}{1\ \text{mil. gal}}\right| = 8.34\frac{\text{lb}}{\text{mil. gal}}$$

Assuming that the clarifier surface area (number of clarifiers in service) and MLSS concentration remain constant, increasing either the influent flow or the RAS flow increases the SLR to the secondary clarifier. The SLR is a critical operating parameter for secondary clarifiers. If the SLR is too high, the clarifier will not be able to settle the solids and a blanket will build up in the clarifier, eventually washing out over the top of the weirs. Acceptable SLRs for secondary clarifiers are between 122 and 195 kg/m^2·d (25 and 40 lb/sq ft/d). The maximum SLR is affected by how well sludge settles. When MLSS settles poorly, the maximum allowable SLR to the clarifier will be lower than when MLSS is settling well. Chapter 5, "Secondary Clarifiers," explains SLR, settleability, and compaction in greater detail.

Increasing the RAS pushes solids into the clarifier faster, but it also removes solids from the bottom of the clarifier faster and returns them to the activated sludge basin. Within limits, the RAS flow controls how long MLSS remains in the clarifier and controls the blanket depth at the bottom of the clarifier. The RAS flow also controls the solids concentration in the sludge blanket. If the RAS flow is increased, solids spend less time in the clarifier and have less time to compact. This results in a reduction of RAS concentration. If RAS flow is decreased, solids remain in the clarifier longer and have more time to compact. This results in an increase in RAS concentration. The effect of RAS on clarifier solids detention time, blanket depth, and RAS concentration is limited by how well sludge flocculates, settles, and compacts. Chapter 5 includes an in-depth discussion of secondary clarifier process control and how to select an RAS flowrate.

The RAS does not typically affect MLSS concentration in the aeration basin to an appreciable degree. Return activated sludge simply moves solids between the clarifier and the activated sludge basins. Changing the RAS flow can only affect the MLSS concentration in the activated sludge basin if large quantities of solids are stored in the clarifier blanket compared to the quantity of MLSS in the activated sludge basin. As will be discussed in more detail in Chapters 4 and 5, blanket depths should be kept below about 0.6 m (2 ft) whenever possible. Storing solids and maintaining deep blankets in the secondary clarifiers increases the likelihood that solids will be washed out of the clarifier and into the final effluent.

In summary, the RAS flow

- Returns settled solids from the clarifier to the activated sludge basin;
- In conjunction with WAS, makes possible independent control of the HRT and SRT;
- Influences the clarifier SLR in combination with influent flow, MLSS concentration, and clarifier surface area;
- Controls the clarifier solids detention time and blanket depth; and
- Controls the concentration of solids in the RAS within limits fixed by sludge settling and compaction characteristics.

2.4 Waste Activated Sludge

Waste activated sludge is excess sludge that is intentionally removed from the process. For every kilogram (pound) of BOD that is added to the activated sludge basin, bacteria feeding on it will produce approximately 0.6 kg (0.6 lb) of new bacteria (Metcalf & Eddy, Inc./AECOM, 2013). This is equivalent to approximately 0.4 kg (0.4 lb) of new bacteria for every 1 kg (1 lb) of chemical oxygen demand (COD). The yield will vary somewhat depending on the type of organic material available and whether oxygen or nitrate is being used by bacteria. In addition, inert material is being added to the activated sludge basins every day by influent wastewater. If some of those solids are not removed from the treatment process periodically, the MLSS concentration will gradually increase. Eventually, the SLR to the secondary clarifiers will exceed the ability of the clarifiers to effectively settle the solids and they will accumulate in the clarifier. As the blanket approaches the top of the clarifier, solids will be lost to the final effluent, which could result in a discharge permit violation. To maintain a stable treatment process, MLSS must be removed on a regular schedule. Mixed liquor suspended solids can be removed from the bottom of the clarifier or from the aeration basin. In Figure 2.1, the WAS line is shown pulling MLSS from the bottom of the secondary clarifier. The advantage of pulling MLSS from the bottom of the clarifier instead of the aeration basin is that the MLSS will have a higher concentration in milligrams per liter at the bottom of the clarifier than in the aeration basins, which reduces the total volume that must be removed and processed. The advantage of wasting MLSS directly from the aeration basin is that the operator does not need to know the concentration of MLSS. If a 10-day sludge is desired, the operator simply wastes 10% of the total basin volume each day. If a 20-day sludge age is desired, the operator removes one-twentieth, or 5%, of the total basin volume each day. Some early activated sludge process designs were based on hydraulic control, that is, removing solids directly from the aeration basin, but the desire to send thicker sludge to the solids handling side of the water resource recovery facility (WRRF) has limited its implementation.

The flow of WAS, in addition to environmental factors such as water temperature, BOD availability, nutrients, and oxygen, significantly influences process biology and the level of treatment achieved. The amount of WAS removed from the system daily is determined by calculating the sludge age, which is a key parameter for system stability. Sludge age calculations and the relationships among sludge age and other operational parameters are explained in detail in Chapter 4. In summary, the WAS

- Determines sludge age,
- Controls the mass of total solids in the aeration basin,
- Influences sludge quality/settleability,
- Influences the growth rate of microorganisms and the types of microorganisms that can proliferate, and
- Influences the quality of the effluent.

3.0 ANAEROBIC, ANOXIC, AND AEROBIC CONDITIONS

The activated sludge process shown in Figure 2.1 is a conventional activated sludge system. It contains one or more aerated basins followed by one or more secondary clarifiers. A conventional activated sludge process is intended to remove CBOD and TSS but may also remove ammonia by converting it to nitrate. Many activated sludge systems contain other types of basins with different environmental conditions and may also contain additional recycle flows between the basins. Figure 2.2 shows a three-stage activated sludge system that contains an anaerobic zone, anoxic zone, and aerated/oxic zone. This activated sludge process is intended to remove BOD and TSS, convert ammonia to nitrate, convert nitrate to nitrogen gas, and remove phosphorus. The different environments in each basin encourage different groups of bacteria to behave in particular ways. Significant groups of bacteria are discussed in the next section.

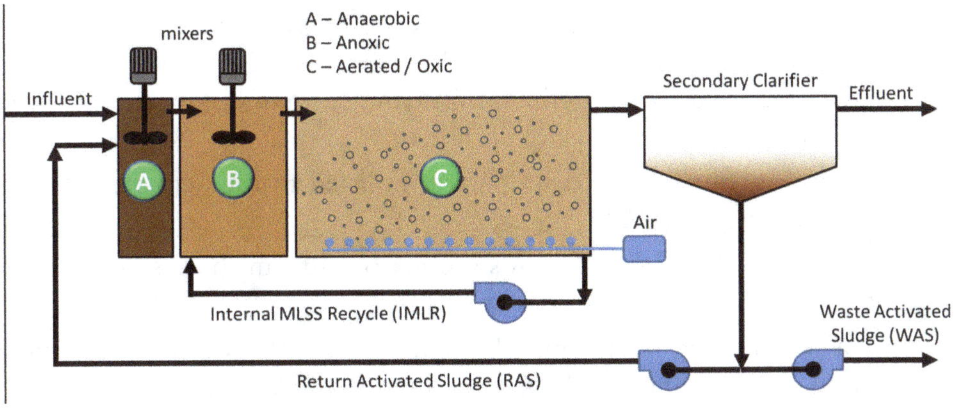

FIGURE 2.2 Three-Stage BNR Activated Sludge System (Reprinted with permission from Indigo Water Group)

The first basin in this example is the anaerobic basin. *Anaerobic* conditions refer to environments without oxygen. In wastewater treatment, this term is used to describe zones or tanks or times during treatment where or when the wastewater does not contain dissolved oxygen or nitrate. The next basin is anoxic. *Anoxic* refers to conditions without dissolved oxygen but with the presence of nitrate. In wastewater treatment, it is important to distinguish between anaerobic and anoxic because different types of treatment are able to take place under each set of conditions. Both the anaerobic and anoxic zones contain mixers to keep MLSS in suspension. The last and largest basin is the aerated basin. *Oxic,* or *aerobic,* means that the wastewater in the tank or zone contains dissolved oxygen and often indicates that oxygen is being added to the wastewater during treatment. An oxic or aerobic zone may also contain nitrite and nitrate. This three-stage activated sludge process contains an additional recycle line that runs from the end of the aerated basin back to the beginning of the anoxic basin. Its purpose is to bring nitrate that is produced in the aerated zone back to the anoxic zone. There are many variations of the activated sludge process that contain more or fewer anaerobic, anoxic, and aerated basins switched around in different ways and with or without internal recycle lines connecting different types of basins. The important thing to remember is that the same fundamental operating principles apply to all of them.

In summary, within the activated sludge process,

- Influent wastewater is combined with a mixture of microorganisms and oxygen.
- Microorganisms consume organic material, reproduce, and form flocs.
- The mixture of microorganisms is called MLSS.

- Once treatment is complete, the mixture of MLSS and treated water pass into the secondary clarifier. The clarifier is not mixed or aerated. Here, the solids settle to the bottom and clean water passes out over the top.
- A portion of the settled solids are returned to the activated sludge basins. This is the RAS.
- A portion of the settled solids are removed from the system. This is the WAS.
- Waste activated sludge may be removed directly from the aeration basin, the RAS line, or through a dedicated WAS line.
- The amount of MLSS removed each day determines the sludge age of the system. Sludge age is an important parameter for activated sludge process control.
- Portions of activated sludge processes may be aerobic, anoxic, or anaerobic. The presence or absence of oxygen and nitrate help determine the types of microbial activity.

4.0 BACTERIAL ENGINES

The activated sludge process depends on a healthy community of bacteria and other microorganisms to produce a final, treated effluent that is clean and clear with little BOD and TSS remaining. Some treatment processes also remove nitrogen and phosphorus. Bacteria in the treatment process can be broken into a few significant groups, depending on their energy source, carbon source, oxygen requirements, and preferred growth patterns. Organisms such as algae that use sunlight as their primary source of energy are called *phototrophs*; all others that use chemicals as a source of energy are called *chemotrophs*. Over the next several sections, the terms *heterotroph*, *autotroph*, *obligate aerobe*, *facultative*, *anaerobic*, *floc former*, and *filament former* will be defined. Many wastewater professionals initially enter the profession as mechanics or have a mechanical background. For this reason, a mechanical approach has been used to describe the needs and functions of the different groups of bacteria. Keep in mind that the following descriptions comparing microorganisms to engines and factories are useful for learning basic concepts, but that there are many factors that contribute to the growth and survival of different microorganisms in the activated sludge process.

Bacteria are somewhat predictable. They can be compared at a basic level to miniature combustion engines. For an engine to function, it requires both a fuel source and an oxygen source. Oxygen is used to burn fuel to produce energy. The byproducts of combustion when burning organic fuel—in this case, sugar ($C_6H_{12}O_6$)—are carbon dioxide (CO_2) and water (H_2O). Burning sugar releases stored chemical energy.

$$C_6H_{12}O_6 + O_2 \rightarrow CO_2 + H_2O \tag{2.2}$$

Bacteria are a little more complicated than simple combustion engines because bacteria have another goal: to reproduce. It is more useful to think of bacteria as simple factories with an engine at their heart. The factory needs energy to both maintain the existing factory and to manufacture more like itself. To build more bacteria, having access to fuel and oxygen is not enough. The factory also needs a ready supply of "spare parts." Bacteria will require carbon, oxygen, nitrogen, phosphorus, and a host of other materials like iron, manganese, and potassium in smaller quantities. The optimal C:N:P for bacterial growth is approximately 100:5:1, critical for balanced nutrient availability. The overall biochemical reaction that occurs for all of the bacteria in the activated sludge process may be expressed as

$$\begin{gathered}\text{Food} + \text{Nutrients} + \text{Organisms} + \\ \text{Oxygen source} \rightarrow \text{New organisms} + \text{Products}\end{gathered} \tag{2.3}$$

4.1 Heterotrophic Bacteria

There are two main types of bacteria in the activated sludge process. The first group is the heterotrophs. *Hetero* means other and *troph* means feeder. *Heterotroph* literally translates to an organism that feeds on others. Bacteria do not consume one another literally, but they do get their energy and carbon from organic compounds produced by other microorganisms and not themselves. This large group of different types of bacteria obtains its carbon and energy from organic molecules like carbohydrates, proteins, and fats in wastewater. The compounds are measured collectively as the CBOD. Bacteria also need a source of oxygen. For heterotrophs using dissolved oxygen, the overall biochemical reaction that occurs may be expressed as

$$\begin{gathered}\text{Organic Matter} + \text{Nutrients} + \text{Organisms} + \\ O_2 \rightarrow \text{New organisms} + CO_2 + H_2O\end{gathered} \tag{2.4}$$

Aerobic bacteria use oxygen to burn fuel. Obligate aerobes are obligated to use dissolved oxygen and cannot use other sources of oxygen like nitrate or sulfate. Humans are obligate aerobes. Humans must have oxygen to breathe and do not survive long without it. Some heterotrophic bacteria are capable of surviving under different environmental conditions because they are adaptable. Facultative heterotrophic bacteria prefer to use dissolved oxygen but can switch to a chemically bound form of oxygen like nitrite or nitrate when oxygen is not available. Most of the bacteria in the activated sludge process are facultative. Nitrite (NO_2) contains one nitrogen atom and two oxygen atoms. Nitrate (NO_3) contains one nitrogen atom and three oxygen atoms. When heterotrophic bacteria use nitrate as an oxygen source instead of dissolved oxygen, the byproducts are nitrogen gas (N_2), carbon dioxide (CO_2), and water. This process is called *denitrification.* Using nitrate instead of oxygen is not as efficient for bacteria. They will always use dissolved oxygen first if it is available. For facultative heterotrophs using nitrate as an oxygen source, the overall biochemical reaction that occurs may be expressed as

$$\begin{aligned}&\text{Organic Matter} + \text{Nutrients} + \text{Organisms} + \\ &\quad NO_3 \rightarrow \text{New organisms} + N_2 + CO_2 + H_2O\end{aligned} \tag{2.5}$$

If facultative heterotrophic bacteria are placed into the aerated zone of an activated sludge process, they will have access to a fuel source (CBOD), an oxygen source (dissolved oxygen), and the "spare parts" (nitrogen, phosphorus, and other micronutrients) they require to build more of themselves. In the aerated or oxic zone, the products will be carbon dioxide and water. When the same facultative heterotrophic bacteria are moved into an anoxic zone, they no longer have access to dissolved oxygen. They will have access to nitrate. Biochemical oxygen demand is also present as are nutrients. In the anoxic zone, facultative heterotrophic bacteria will continue to grow and reproduce, but the products are now nitrogen gas, carbon dioxide, and water. If this same group of organisms is placed into the anaerobic zone, they will not be able to function. There is no nitrate or oxygen for them to use to burn fuel. Heterotrophs will be dormant for the time they are in the anaerobic basin.

4.2 Autotrophic Bacteria

The second main group of bacteria is the autotrophs. *Auto* means self and *troph* means feeder. An autotroph is literally a self-feeder. Autotrophs cannot use organic compounds (CBOD) as an energy or carbon source. Instead, they obtain their energy from inorganic sources. Inorganic carbon available in wastewater treatment processes includes dissolved carbon dioxide gas (CO_2), carbonate ion (CO_3^-), bicarbonate ion (HCO_3^-), and carbonic acid (H_2CO_3). Which form of inorganic carbon is most prevalent depends on pH. Typically, activated sludge processes have a pH between 6.5 and 8. In this pH range, the bicarbonate ion (HCO_3^-) will dominate, although smaller amounts of the other ions will also be present. Below 6.4 pH, carbonic acid (H_2CO_3) is the most common form. Bicarbonate ion, carbonate ion, and hydroxide (OH^-) are the three main components of alkalinity. When bacterial activity produces or consumes carbon dioxide, carbonate, bicarbonate, and hydroxide, it changes the total alkalinity available and may also shift pH.

Although there are many different types of autotrophic bacteria, the most important ones in wastewater treatment are nitrifying bacteria. Nitrifying bacteria use ammonia (NH_3) or nitrite (NO_2^-) as their fuel and dissolved oxygen as their oxygen source. The inputs for nitrifying autotrophic bacteria will be ammonia as fuel plus dissolved oxygen and "spare parts." The outputs from nitrifying autotrophic bacteria will be nitrite (NO_2^-), nitrate (NO_3^-), carbon dioxide (CO_2), and water (H_2O). The "spare parts" list is a little bit different for autotrophic bacteria than it is for the heterotrophic bacteria because their fuel source, ammonia, does not contain any carbon. Carbon is a necessary building block of all life on earth. Autotrophs use inorganic carbon compound bicarbonate (HCO_3^-) as their carbon source. Bicarbonate is a component of wastewater alkalinity. For nitrifying bacteria, the overall biochemical reaction that occurs may be expressed as

$$NH_4^+ + O_2 + HCO_3^- \rightarrow \text{New organisms} + NO_3^- + CO_2 + H_2O \quad (2.6)$$

In the previous paragraph, the conversion of ammonia to nitrate was shown as a single step. There are actually two different groups of nitrifying bacteria: ammonia-oxidizing bacteria (AOB) and nitrite-oxidizing bacteria (NOB). Ammonia-oxidizers convert ammonia to nitrite (NO_2^-). Nitrite-oxidizers then convert the nitrite to nitrate (NO_3^-). The best-known ammonia-oxidizer is a bacterium called *Nitrosomonas* and the best-known nitrite-oxidizer is called *Nitrobacter*. There are, however, many different autotrophic bacteria that can use ammonia or nitrite as fuel. Most nitrifying bacteria are obligate aerobes and must have dissolved oxygen. They function well in the aerated or oxic zone but cannot function in either the anoxic or anaerobic zones because there is not any oxygen. Nitrifying autotrophs will be dormant for the time they are in either anoxic or anaerobic basins. Nitrifying bacteria are discussed in more detail in Chapter 8.

Which group of bacteria grows faster, heterotrophs or autotrophs? Heterotrophs use energy-rich organic compounds as their fuel source, whereas autotrophs use relatively low-energy ammonia as their fuel source. Higher-energy-content fuel enables heterotrophs to grow much faster than autotrophs. Under ideal conditions, heterotrophic bacteria may only need 20 minutes to reproduce, whereas nitrifying bacteria may need between 7 and 13 hours. This means that if a build-up of a stable population of nitrifying bacteria in the activated sludge process is desired, MLSS must remain in the process for a longer period of time. The amount of time that MLSS or solids stay in the activated sludge process is called the *sludge age*. Other terms for sludge age include *mean cell residence time (MCRT)* and SRT. These terms are discussed in detail in Chapter 4.

4.3 Phosphate-Accumulating Organisms

Phosphate-accumulating organisms (PAOs) are a specialized subgroup of facultative heterotrophic bacteria. When these bacteria are cycled between anaerobic and either anoxic or aerated conditions, they will pick up and store excess phosphorus. This process is called *luxury phosphorus uptake* because bacteria are incorporating more phosphorus than they need for growth. As mentioned in Chapter 1, PAOs are able to store up to 15% of their dry weight as phosphorus. The resulting MLSS may be 4% to 12% phosphorus. Without PAOs, MLSS would only be 1% to 2% phosphorus. Phosphorus is removed from wastewater when excess MLSS is wasted.

Understanding why luxury phosphorus uptake occurs begins with understanding what happens to CBOD in the anaerobic zone. In the anaerobic zone, some of the CBOD is broken down through fermentation by anaerobic bacteria into soluble CBOD and simpler organic molecules called volatile fatty acids (VFAs). Volatile fatty acids are chemical compounds that contain five carbons or less, such as acetic acid (two carbons), propionic acid (three carbons), and butyric acid (four carbons). Acetic acid is more commonly known as *vinegar*. Volatile fatty acids are a preferred source of carbon and energy by heterotrophic bacteria, including PAOs, because these compounds are easily absorbed by bacteria. Chapter 3 elaborates on bacteria feeding strategies. The PAOs have a logistical problem—when they are in the anaerobic zone, they have access to VFAs; however, without oxygen or nitrite or nitrate present, they cannot access them. Picking up "fuel" takes energy, which is unavailable to most bacteria in the absence of oxygen, nitrite, and nitrate.

The PAOs have a creative solution to this problem. They build a chemical battery from phosphorus. Most cells use a compound called *adenosine triphosphate (ATP)* to store energy. Adenosine triphosphate consists of three phosphate molecules chemically bound together. The chemical bond between phosphorus atoms stores energy until the cell needs it. Adenosine triphosphate is continuously formed and consumed within cells as energy is stored and recovered. The PAOs take ATP to the next level and form an energy-rich compound called *polyphosphate*, which strings together large numbers of phosphate molecules. Polyphosphate is a chemical "battery." When the PAO needs energy, it breaks the chemical bonds between phosphorus molecules to recover stored energy. This strategy allows PAOs to pick up and store VFAs when they are in anaerobic conditions. Because they have a "battery pack," they are able to outcompete other types of bacteria for available VFAs in the anaerobic zone.

When PAOs are in the anaerobic zone, they use energy stored in polyphosphate to pick up VFAs. The VFAs are then combined into an internal storage product called *poly-B-hydroxybutyrate (PHB)*. The lack of oxygen or nitrite or nitrate prevents PAOs from consuming VFAs immediately. When PAOs move into either an anoxic or aerobic zone, they immediately begin to consume PHB and rebuild their polyphosphate "battery" by uptaking orthophosphate. Cycling PAOs between anaerobic and anoxic or aerated conditions is required for luxury phosphorus uptake to occur. Without exposure to an anaerobic zone, there is no incentive for PAOs to store phosphorus. The PAOs are discussed in more detail in Chapter 8.

Table 2.1 summarizes the characteristics of each basin and lists some of the biological reactions that can take place under each set of conditions.

TABLE 2.1 Anaerobic, Anoxic, and Aerobic Zones (Reprinted with permission from Indigo Water Group)

Environmental condition	Defining characteristics	Possible biological reactions
Anaerobic	No oxygen present No nitrite or nitrate present	CBOD broken down into soluble CBOD and VFAs Heterotrophs are dormant Nitrifiers are dormant PAOs release orthophosphorus PAOs uptake VFAs
Anoxic	No oxygen present Nitrite and/or nitrate present	Heterotrophs convert CBOD to biomass Heterotrophs convert NO_2^- and NO_3^- to nitrogen gas Nitrifiers are dormant PAOs consume stored VFAs PAOs uptake orthophosphorus
Aerobic/Oxic	Oxygen present Nitrite and nitrate may also be present	Heterotrophs convert CBOD to biomass Nitrifiers convert NH_3 to NO_2^- and NO_3^- PAOs consume stored VFAs PAOs uptake orthophosphorus

Within limits, the behavior of different groups of bacteria can be predicted from their environment. Facultative heterotrophs will use oxygen when it is available and nitrate when it is not. Nitrifying autotrophs will convert ammonia into nitrite and nitrate when oxygen is available.

4.4 Growth Patterns: Floc Formers Versus Filament Formers

Bacteria can also be divided into large groups depending on their shape and preferred growth pattern. Bacteria come in one of three basic shapes: spheres (cocci), rods, and spirals. They may grow in clumps or colonies called *flocs* and produce exopolymer to help them stick together or they can grow in long chains called *filaments*. Operation of an activated sludge treatment process depends on balancing these two groups. A certain number of filaments are needed to provide the backbone or scaffolding of the floc particles, but too many can cause settling problems in the secondary clarifier.

Imagine a group of skydivers jumping out of a plane. If a skydiver drew their legs and arms in close to form a tight ball, they would fall faster. If the same skydiver spreads themself out as much as possible, they increase the amount of air resistance and drag and will fall slower. More surface area increases drag. If groups of skydivers join hands and form groups, they increase the total mass of the group. If the group is compact, then the mass of the group will increase more than the surface area of the group and the group will fall faster. Activated sludge particles group together in a similar way through flocculation. If skydivers come together as a group but also spread themselves out, then mass and surface area will both increase, and the entire group will fall even slower. When MLSS goes to the secondary clarifier, floc particles need to be large and dense enough to sink to the bottom. If there are too many filaments extending out of the floc particle or joining one floc particle to another, they may not sink fast enough and could even be washed out of the clarifier and into the final effluent. Read more about filaments and the conditions that promote their growth in Chapter 3.

5.0 MICROBIAL GROWTH KINETICS

Activated sludge process design and operation are based on the growth kinetics of microorganisms. *Kinetics* refers to how fast or slow chemical and biological reactions take place. This section defines several terms that will be used in later chapters—*yield*, *decay*, *maximum specific growth rate*, *maximum substrate utilization rate*, and *saturation coefficient*.

The biomass in an activated sludge process contains many different kinds of microorganisms. Each type of microorganism has a maximum growth rate, which varies depending on temperature. Typically, growth rate increases at higher temperatures, but excessive temperatures can inhibit microbial growth. As a general guide, growth rate doubles when the water temperature is increased by 10 °C (18 °F). Growth rate is measured as the increase in the number of cells per unit of time. Or, alternatively, how long does it take to double the number of cells or the amount of biomass?

5.1 Monod Kinetics

Each kind of bacteria will grow at its maximum rate only when the water surrounding the bacteria has an optimum pH and contains sufficient amounts of food, oxygen, nutrients, and trace elements. Microbial growth rates are often lower than maximum growth rates because one or more of the substances they need to grow and reproduce is not available in sufficient quantities. When a substance is not present in sufficient quantities to support growth, it is said to be limiting. The most common limiting factors include (a) exhaustion of nutrients, (b) accumulation of toxic byproducts, and (c) changes in ion concentrations, especially pH (Monod, 1949).

Monod kinetics relates microbial growth rates to the concentration of a limiting substrate in an aqueous environment. Microbial growth rates are represented by an equation proposed by Jacques Monod (1942, 1949). The Monod equation states that when substrate concentrations are low, the growth rate of the microorganisms will be proportional to the concentration of the limited substrate. Put another way, the growth and reproduction of the microorganism will be restricted by the availability of resources. The Monod equation also states that when resources are plentiful, the growth rate of microorganisms will be maximized. However, the microorganism cannot grow faster than its maximum rate, even when resources are available in excess.

Think back to the factory analogy used earlier in this chapter. Imagine that an assembly line within that factory is producing automobiles. The assembly line needs to have all of the parts available to keep producing cars. What would happen to the speed of the assembly line if a key component was limited? If the automobile factory only has access to 60 radiators on any given day, then the maximum number of automobiles that can be produced is 60. If the automobile factory has access to as many parts as it needs, then the number of automobiles it can produce is limited only by the speed of the assembly line. The speed of the assembly line is the maximum growth rate. It represents the speed at which a particular microorganism is capable of running its internal cellular "assembly lines." The maximum substrate utilization rate is related to the maximum growth rate. When the microorganism is growing at its maximum rate, it will use up resources—BOD, ammonia, oxygen, nitrate, nutrients—at the maximum rate. If the assembly line is only capable of producing 100 new automobiles each day, then the maximum number of radiators it can use each day is also 100.

The Monod equation generates a characteristic growth curve, shown on the graph in Figure 2.3. The microbial growth rate, shown as μ, is determined by the concentration of the limiting substrate until the maximum growth rate (μ_{max}) is reached. The growth rate has units of grams of VSS produced per gram of VSS added to the process per day. Typically, it is shown in mathematical equations simply as "per day" (d^{-1}).

The Monod equation is written as

$$\mu = \mu_{max} \times \left[\frac{S}{K_s + S}\right] \quad (2.7)$$

Where

μ (pronounced "mew") = specific growth rate of microorganisms (pronounced "mew"), per day;
μ_{max} = maximum possible growth rate of the microorganisms, per day;
S = concentration of substrate, mg/L; and
K_s = half-saturation coefficient, mg/L.

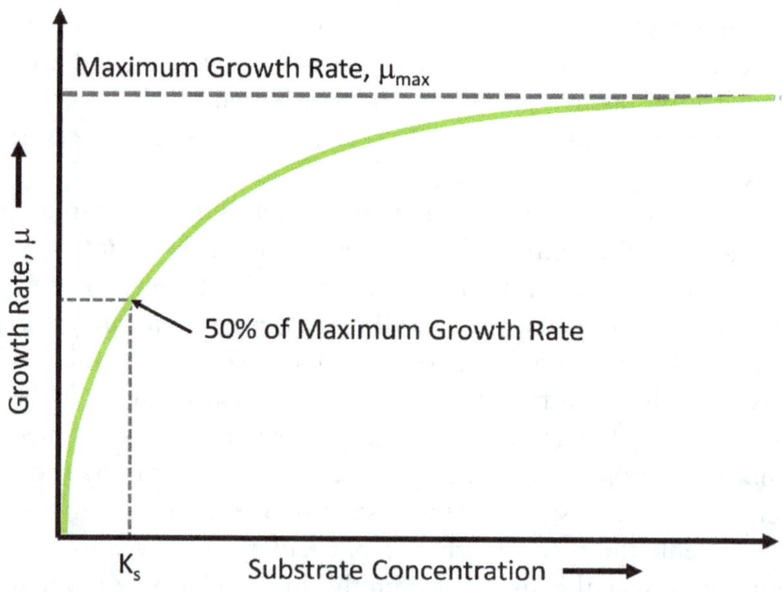

FIGURE 2.3 Graph of the Monod Equation (Reprinted with permission from Indigo Water Group)

5.2 Maximum Specific Growth Rate

The maximum specific growth rate is crucial in determining biological treatment capacity. The maximum specific growth rate is the fastest that a particular microorganism can grow and reproduce. Microorganisms respond to higher concentrations of substrate (food) by increasing their growth rate up to the maximum growth rate. When the substrate is scarce (low food concentration), microorganisms cannot grow as fast as they could if more food were available. In contrast, an abundance of substrate allows each microorganism to consume more food and increase its growth rate. Increasing the amount of substrate reduces competition for food and allows microorganisms to increase their growth rates proportional to the available food source.

In Figure 2.3, the growth rate curve increases rapidly in response to the increasing substrate concentration. It gradually flattens out and approaches the horizontal line, which indicates the maximum growth rate (μ_{max}). When microorganisms are saturated by substrate and attain the maximum growth rate, the growth rate remains constant as substrate concentrations increase further. When surplus substrate is available, growth becomes limited by the number of microorganisms in the system instead of by the amount of substrate available.

The maximum growth rate (μ_{max}) and the half-saturation coefficient (K_s) used in the Monod equation are empirical coefficients. *Empirical* means that they must be measured by experimentation. Growth rates are measured at several different concentrations of the limiting substrate and the results are plotted on a graph. The maximum growth rate and half-saturation coefficient are then estimated by fitting a curved line to the graph.

Each species of bacteria has a different maximum growth rate and half-saturation coefficient for every limiting nutrient and for the surrounding environmental conditions. The maximum growth rate for the biomass in an activated sludge process is simply the average of the maximum growth rates for each of the different types of bacteria present in the biomass. The average maximum growth rate in an activated sludge process can be determined for the biomass in the process using the same methods as for a single type of bacteria and a single limiting substrate. When engineers design secondary treatment processes, they use a μ_{max} between 3 and 13.2 g VSS/g VSS·d for heterotrophic bacteria and a μ_{max} between 0.20 and 0.90 g VSS/g VSS·d for the nitrifying autotrophic bacteria (Metcalf & Eddy, Inc./AECOM, 2013). The big differences in maximum growth rate make it clear that heterotrophic bacteria can reproduce much more rapidly than nitrifying autotrophic bacteria.

The Monod equation represents a stable microbial culture. Changing environmental conditions or changing wastewater characteristics may cause different microorganisms to displace previously dominant species in the treatment process. For example, an influent wastewater with high BOD and low ammonia will favor heterotrophs and increase their numbers rapidly. An influent that is low in BOD and high in ammonia will favor nitrifying autotrophs instead. The changing makeup of the population of microorganisms in the process can change the μ_{max} and K_s values, which thereby affect process performance.

5.3 Saturation Coefficient

The saturation coefficient is the concentration of a resource that a microorganism needs to grow at its maximum rate. Higher concentrations of the resource will not continue to increase growth rates. If an assembly line can produce a maximum of 75 cars a day, then having access to 75 alternators is saturation. The assembly line cannot use more than 75 alternators. If more alternators are available, the assembly line will not produce more cars. This is the saturation coefficient. When resources are available at the saturation coefficient or higher, the microorganism will grow at its maximum rate. Below the saturation coefficient concentration, the "assembly line" will be limited and production of new microorganisms will slow down. Typically, the saturation coefficient is expressed in milligrams per liter.

5.4 Half-Saturation Coefficient

The half-saturation coefficient or half-velocity coefficient, shown as K_s, is the concentration of a resource needed to achieve exactly half of the maximum growth rate. The half-saturation coefficient is typically calculated from measurements of the maximum growth rate and is rarely measured directly. It is typically expressed in milligrams per liter.

The values of the maximum growth rate and half-saturation constant depend on microorganisms and substrates in the biological system. Readily biodegradable substrates have high μ_{max} values and low K_s values. In other words, soluble BOD, which is easy for bacteria to take up, will help bacteria grow quickly. They do not need to waste time and energy breaking down particulate BOD before they can bring it through the cell membrane. In contrast, slowly biodegradable substrates have low μ_{max} values and high K_s values. Therefore, the percentages of readily biodegradable and slowly biodegradable substrates in the wastewater affect μ_{max} and K_s values for each WRRF.

Many wastewater treatment systems are required to meet stringent discharge limits. For this reason, wastewater treatment systems with low effluent limits operate at low substrate concentrations and at growth rates much lower than the maximum growth rate. Looking at Figure 2.3, these facilities are operating toward the left side of the characteristic Monod growth curve. In other words, bacteria are often growing much slower than they would under ideal conditions.

For facultative heterotrophic bacteria, K_s ranges between 25 and 100 mg/L of BOD, with a typical value of 60 mg/L of BOD (Metcalf & Eddy, Inc./AECOM, 2013). This means that when the BOD concentration is in this range, the growth rate of microorganisms is about one-half of maximum. Compare this to an effluent BOD limit of 30 mg/L for most mechanical treatment facilities. For nitrifying bacteria, the K_s for ammonia-nitrogen is 0.50 mg/L (U.S. Environmental Protection Agency, 2010). When the ammonia-nitrogen concentration is higher than 1 mg/L in the activated sludge process, it becomes possible for AOBs to grow at their maximum rate. As the concentration of ammonia-nitrogen gets lower and lower, their growth rate slows down. There is a similar effect with dissolved oxygen concentrations. The K_s for oxygen for nitrifying bacteria ranges between 0.13 and 0.47 mg/L (Manser et al., 2005). Ultimately, the maximum growth rate of microorganisms will be limited by the availability of any or all of the resources they need as well as water temperature.

Although operators will likely never need to calculate growth rates, an example calculation is included here to demonstrate the effects of substrate concentration on growth rate. In this example, bacteria reproduce at a rate of three times per day and the half-saturation coefficient (K_s) is 0.7 mg/L. The substrate concentration will be varied from 0.5 mg/L up to 8 mg/L.

$$\mu = \mu_{max} \times \left[\frac{S}{K_s + S}\right]$$

$$\mu = 3\ \text{d}^{-1} \times \left[\frac{0.5\frac{\text{mg}}{\text{L}}}{0.7\frac{\text{mg}}{\text{L}} + 0.5\frac{\text{mg}}{\text{L}}}\right]$$

$$\mu = 3\ \text{d}^{-1} \times \left[\frac{0.5\frac{\text{mg}}{\text{L}}}{1.2\frac{\text{mg}}{\text{L}}}\right]$$

$$\mu = 1.25\ \text{d}^{-1}$$

The amount of substrate available is low, so the growth rate is also low. If the same calculation is performed with progressively increasing substrate concentrations, the growth rate achieved begins to get close to the maximum growth rate. In Table 2.2, one can see that when the substrate concentration is equal to the half-saturation coefficient of 0.7 mg/L, the resulting growth rate is equal to 1.5 per day, which is exactly one-half of the maximum growth rate. As the substrate continues to increase, the actual growth rate gets closer and closer to the maximum growth rate. However, even when the substrate concentration is higher than twice the half-saturation coefficient, the growth rate can never be higher than the maximum growth rate.

5.5 Effect of Multiple Limiting Substrates

Growth rates in wastewater treatment processes are often limited by two or more substrates at the same time. When growth rates are limited by more than one substrate or growth factor (e.g., organic matter and oxygen are both required by heterotrophic bacteria), multiple Monod terms can be multiplied together to estimate the combined effect of two or more substrates, as shown by the following equation:

$$\mu = \mu_{max} \times \left[\frac{S_1}{K_{s1} + S_1}\right] \times \left[\frac{S_2}{K_{s2} + S_2}\right] \tag{2.8}$$

TABLE 2.2 Effect of Substrate Concentration on Growth Rate (Reprinted with permission from Indigo Water Group)

S, mg/L	Growth rate, μ, d^{-1}
0.7	1.50
1.0	1.76
1.5	2.05
2.0	2.22
2.5	2.34
3.0	2.43
3.5	2.50
4.0	2.55
4.5	2.60
5.0	2.63
5.5	2.66
6.0	2.69
6.5	2.71
7.0	2.73
7.5	2.74
12.5	2.84

Think back to the factory analogy used earlier in this chapter. The "assembly line" for autotrophic nitrifying bacteria requires dissolved oxygen and ammonia. In this example, K_{s1} represents the half-saturation coefficient for dissolved oxygen and S_1 is the dissolved oxygen concentration in the process. The K_{s2} term represents the half-saturation coefficient for ammonia and S_2 is the ammonia concentration in the process.

5.6 Biomass Yield

Yield is the amount of new microorganisms produced per amount of substrate added to the system, as shown in eq 2.9:

$$\text{Yield} = \frac{\text{Mass of new cells formed}}{\text{Mass of substrate removed}} \tag{2.9}$$

Substrates include organic compounds, ammonia, VFAs, oxygen, and nitrate. The availability of each resource affects how efficiently the microorganisms can grow and reproduce. Heterotrophic bacteria produce an average of 0.40 g of VSS, a surrogate measurement for new bacteria, for every 1 g of COD added to the process when oxygen is available (Metcalf & Eddy, Inc./AECOM, 2013). For secondary treatment processes receiving domestic wastewater, this is equivalent to approximately 0.6 g VSS/g BOD. The COD and BOD that do not become new biomass become carbon dioxide and water. Yields are consistent for WRRFs accepting domestic wastewater. It is not possible to produce more new bacteria than what the amount of food in the influent can support. Put another way, if a person consumes a large number of donuts, they can gain weight. The same person cannot gain weight without eating the donuts. Without sufficient substrate, there cannot be an increase in biomass.

Yields are somewhat lower when oxygen is not available and facultative heterotrophs are forced to substitute nitrate or nitrite for oxygen (denitrification). Using nitrate or nitrite is less efficient for bacteria than oxygen.

The AOBs and NOBs produce an average of only 0.15 g and 0.05 g of VSS for every 1 g of ammonia converted to nitrate, respectively (U.S. Environmental Protection Agency, 2010). The yield is lower for nitrifiers than it is for heterotrophs, which means their growth (doubling time) is slower. The amount of new cells produced is essentially constant when energy, carbon, and oxygen sources are known. Yield is unaffected by changes in water temperature between 4 and 20 °C (Sayigh and Malina, Jr., 1978).

5.7 Biomass Decay Rate

Decay accounts for losses from the system caused by microorganisms consuming internal storage products, dying, and being eaten by other microorganisms. The amount of solids lost each day is a fraction of the total solids in the system, typically 0.10 g of VSS lost per gram of VSS in the system per day for heterotrophs and 0.17 g of VSS lost per gram of VSS in the system per day for nitrifying autotrophs (Metcalf & Eddy, Inc./AECOM, 2013, and U.S. Environmental Protection Agency, 2010). In other words, approximately 10% of heterotrophs in the process will die or be reduced in size each day. New microorganisms are produced each day (i.e., yield) and some get smaller and/or die each day (i.e., decay). The difference between the yield and the decay is the net yield or observed yield. This is an important concept for secondary treatment process control. The amount of COD or BOD entering the process each day dictates the amount of new cells that can potentially be produced. When the amount of COD or BOD decreases, fewer new microorganisms are produced. It is not possible to build up a population of live microorganisms without COD or BOD, or, in the case of nitrifying bacteria, ammonia. The decay term used in kinetic equations is calculated by dividing the mass of cells lost per day by the total mass of cells in the system:

$$\text{Decay} = \frac{\text{Mass of cells lost per day}}{\text{Mass of cells in system}} \tag{2.10}$$

Both the maximum specific growth rate and the amount of decay depend on temperature. When the activated sludge process is warmer, bacteria grow faster than when it is colder. A general rule is that the maximum specific growth rate doubles for every 10 °C (18 °F) increase in water temperature (Metcalf & Eddy, Inc./AECOM, 2013). From a practical standpoint, it means that treatment takes less time when the water is warm. Decay also depends on temperature. The amount of bacteria that die or lose mass because they are consuming internal storage products slows down at colder temperatures.

5.8 Modified Monod Kinetics—Substrate Limited Growth

The yield, decay, maximum specific growth rate, maximum substrate utilization rate, and saturation coefficient are related to one another through the following equation (Metcalf & Eddy, Inc./AECOM, 2013):

$$\mu = Y \times \frac{kS}{K_s + S} - K_d \tag{2.11}$$

Alternatively

$$Y = \left[\frac{K_s + S}{kS}\right][\mu + K_d] \tag{2.12}$$

Where

μ = specific growth rate, g VSS/g VSS·d;
Y = yield, g VSS/g BOD or COD;
k = maximum utilization rate, g COD/g VSS·d;
S = substrate concentration, mg/L BOD or COD;
K_s = substrate concentration at one-half the maximum specific substrate utilization rate, mg/L; and
K_d = decay, g VSS/g VSS·d.

This particular example relates the growth rate of heterotrophic bacteria to the availability of BOD or COD. Similar equations exist for nitrifying autotrophic bacteria and are presented in Chapter 8. Operators will not need to calculate growth rates for microorganisms, but it is useful to show the mathematical relationship to understand what happens to the growth rate as available resources diminish.

$$\text{Growth} = (\text{Potential growth})(\text{Resource availability}) - \text{Decay} \quad (2.13)$$

If the amount of resources available to the microorganism is less than 100% of what is needed, then the potential growth will be reduced. It is possible for oxygen sources to be limited as well or for there to be a limitation on the quantity and types of "spare parts" (carbon and nutrients). If the amount of resources available to each bacterium begins to decrease, bacteria will not be able to grow or reproduce as quickly. If available resources continue to decrease, bacteria will begin to use up stored resources. This is called *endogenous respiration* and is similar to what happens when a person goes on a diet. Eventually, bacteria run out of stored resources and will lyse or break open. Their remains then become food for other bacteria in the process.

6.0 PROCESS VARIATIONS

Although the activated sludge process was invented in England in 1914, it was not in widespread use until the 1940s. Small-scale testing for both continuous-flow systems and SBRs was conducted at the Manchester–Davyhulme facility in England in 1914. Both configurations were effective. The first full-scale, continuous-flow activated sludge facility was built 2 years later in the town of Worcester, England, in 1916 (Wanner, 2014). The first full-scale activated sludge facility in the United States was built in 1916 in San Marcos, Texas, followed by facilities in Milwaukee, Wisconsin (1916), and Cleveland, Ohio (1916). By 1927, there were 10 full-scale systems operating in the United States (Alleman & Prakasam, 1983). Variations of the activated sludge process are among the most commonly used treatment methods worldwide.

The design and operation of the activated sludge process have continued to be refined over the last 100 years. Today, there are many different operating modes and process configurations in use for meeting specific treatment goals and effluent limits. Different activated sludge processes may or may not have multiple anaerobic, anoxic, and aerated basins or zones. The same fundamental design and operating principles apply to all activated sludge processes regardless of configuration. The process modification selected depends on treatment objectives, site constraints, operational constraints, operator preference and experience, and design engineer preference and experience, among others. The process can be categorized by loading rates, reactor configuration, feeding and aeration patterns, and other criteria, including numerous biological nutrient removal (BNR) processes.

6.1 Loading Rates

The activated sludge process is often characterized by the volumetric loading rate, MCRT, and food-to-microorganism ratio (F:M). Table 2.3 shows a typical range of these parameters for conventional, high-rate, and low-rate (often called *extended aeration*) systems. Within these parameter ranges, the reactor configuration, number of reactors, and aeration and feed patterns can be selected to achieve target effluent concentrations of water quality variables such as ammonia-nitrogen. Hydraulically, they can be designed as either a continuous-flow process or a batch-flow system.

TABLE 2.3 Typical Process Loading Ranges for the Activated Sludge Process (Water Environment Federation, 2023b).

Loading range	Sludge age, days	Volumetric loading, kg BOD_5/m^3 (lb BOD_5/1000 cu ft)	F:M, kg BOD_5/kg MLVSS·d (lb BOD_5/lb MLVSS·d)
High-rate	1–3	1.60–16.0 (100–1000)	0.5–1.5 (0.5–1.5)
Conventional	5–15	0.32–0.64 (20–40)	0.2–0.5 (0.2–0.5)
Low-rate	20–30	0.16–0.40 (10–25)	0.05–0.15 (0.05–0.15)

Conventional systems provide BOD_5 removal efficiencies of 85% to 95% and typically carry MLSS concentrations varying from 1000 to 3000 mg/L. At conventional loading, some nitrification may occur, especially in warmer climates. This results in higher than estimated oxygen demand and may cause sludge flotation in the final clarifiers. Nitrification can be limited by decreasing the MCRT.

Low-rate systems are typically used for low flows and are characterized by high oxygen requirements and low sludge production rates. These systems are considered to be more stable than conventional systems and thus require less operational attention. Typically, BOD removal efficiencies range from 75% to 95%, and nitrification is complete. However, these systems may suffer from higher effluent suspended solids concentrations because of poor flocculation (pinpoint floc) and clarifier denitrification.

High-rate systems are often used as pretreatment processes in staged biological treatment systems and are also used where only CBOD removal is required. They are characterized by low oxygen requirements and somewhat higher than normal sludge generation compared to conventional facilities. The process may produce BOD removal efficiencies over a wide range, from less than 50% to as high as 95%, depending on loading rates and waste characteristics. Sludge settling can be a problem during instances of high loading when flocculation does not effectively occur.

6.2 Reactor Configuration

Reactor configuration deals primarily with hydraulic characteristics of the process. Continuous-flow systems are often categorized as "ideal plug flow" or "ideal complete-mix" systems, although most operate in a nonideal flow regime somewhere between the two. However, batch systems behave like ideal plug flow systems with respect to biological process performance; the reaction time for batch systems is interchangeable with the space–time for the plug flow reactor.

6.2.1 Ideal Complete-Mix

Ideal completely mixed flow implies that the composition of the mixed liquor is the same throughout the reactor volume. Influent wastewater immediately and completely mixes with the reactor contents so that the concentration of a given component is the same as the effluent concentration. As a result, the mixed liquor oxygen-uptake rate, dissolved oxygen, soluble BOD, VSS, TSS, nitrogen species, phosphorus concentrations, pH, temperature, and other characteristics are identical throughout the reactor. Because the concentration of BOD is at the target effluent concentration for the process, the system has the capability of attenuating wide swings in influent concentration. Completely mixed flow is difficult to achieve, although the use of square or round reactors with intense mixing can approximate the condition. As discussed later in this chapter, it can also be approximated by providing multiple feed points along the reactor periphery. Aeration is provided uniformly throughout the reactor by mechanical aeration equipment or diffused aeration.

The advantages of the complete-mix process include the dilution of shock loads (e.g., high loads or toxic chemicals) and its relative simplicity compared to other configurations. Shock loads can lead to anaerobic conditions or biomass toxicity. There are several disadvantages to completely mixed reactors. First, they are often affected by filamentous bulking problems. This can be overcome, in part, by using selectors, which provide short-term conditioning of the RAS and influent wastewater ahead of the biological reactor (see Section 6.3.5, "Selectors"). Theoretically, the process also suffers because the removal efficiency of the wastewater constituents is lower than in plug flow systems, requiring longer reactor retention times and/or the use of reactors in series to achieve comparable effluent quality. These systems are often used for low-rate applications or for treating industrial or industrial–municipal wastewater, in which large variations in load are anticipated.

6.2.2 Ideal Plug Flow

In an ideal plug flow system, a fluid particle entering the system will move uniformly along the reactor length without dispersing in the fluid (Figure 2.4). The particle will remain in the reactor for a time equal to the theoretical detention time. As a result, the concentration of BOD and oxygen

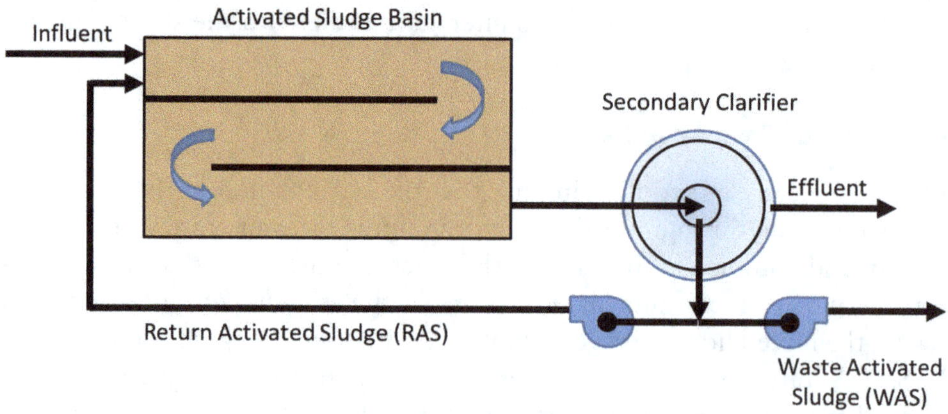

FIGURE 2.4 Plug Flow Activated Sludge (Reprinted with permission from Indigo Water Group)

uptake will decrease along the reactor length. The TSS and VSS will increase along the reactor length as biomass is produced. This type of flow is approximated by long, narrow reactors with length-to-width ratios greater than 10. Plug flow may also be simulated by basins in series or by folding rectangular reactors. Like the complete-mix configuration, it is not practical to produce a true plug flow system.

This configuration results in high organic load and oxygen-uptake rates at the inlet to the reactor because concentrated microorganisms in the RAS are blended with influent wastewater in a small percentage of the total basin. With a complete-mix reactor, influent wastewater is evenly distributed throughout the basin, which dilutes the overall concentration of BOD. With plug flow, there is a concentration gradient. Oxygen-transfer devices must be capable of high mass transfer at the inlet. Distribution of oxygen along the reactor may be tapered to follow oxygen demand. Theoretically, the plug flow reactor delivers the highest removal rate per unit volume. It is also less susceptible to filamentous bulking, provided that sufficient dissolved oxygen is present at the inlet.

6.2.3 Reactors in Series

The reactors-in-series configuration is described as two or more complete-mix reactors operating in a series configuration. It can be approximated with baffles or by folding reactors in a sinusoidal fashion. This configuration, with three or more reactors in series, attempts to simulate a plug flow condition that has a kinetic advantage over mixed-flow systems, as mentioned previously. This configuration also seems to mitigate filamentous bulking.

6.3 Feed and Aeration Patterns

Modifications of the activated sludge process have often been used to enhance process biochemistry or other characteristics to produce a more stable operation at an economical cost. Several of these modifications are described in this section.

6.3.1 Conventional

In the conventional flowsheet (shown in Figure 2.1), influent wastewater is introduced at the influent end of the system. The RAS may be added to the inlet separately or it may be premixed with influent wastewater before introduction to the reactor. Separate addition allows for more flexibility should a step-feed or contact-stabilization configuration be used in the future. In conventional designs, aeration may be provided uniformly along the reactor length or tapered to meet oxygen demand.

6.3.2 Contact-Stabilization and Sludge-Reaeration

Contact stabilization is a modification in which RAS is fed to the head of the biological reactor (reaeration section) and influent wastewater is added downstream (contact section) of the point of sludge addition (Figure 2.5). Wastewater receives a short aeration contact time (typically 30 to 60 minutes) after mixing with reaerated sludge. The idea is to starve bacteria and ensure that they use up any internally stored food so that they will rapidly uptake BOD when exposed to incoming wastewater. Thus, the process relies on rapid uptake (biosorption) of BOD followed by a stabilization step in which the return sludge is aerated to stabilize the sorbed organic matter. During the stabilization step, aeration continues and bacteria continue to consume BOD that has sorbed by excreting enzymes to begin breaking down larger, complex organic molecules. At the same time, bacteria are bringing BOD across their cell membranes. Although the uptake of BOD in the contact section is rapid, it takes time for bacteria to completely absorb and metabolize adsorbed material. These systems are typically designed to operate in the conventional loading rate region (SRTs ranging from approximately 3 to 15 days). Because the RAS flow is only approximately 40% to 70% of the forward flowrate, the total aeration volume is smaller than for a conventional biological reactor with equal HRT. As a result of this modification, the capacity of an existing facility can be increased without adding new process volume. The process has also been found to be applicable in communities with high infiltration and inflow during storms. In conventional facilities, high flows may wash out sludge from the system if clarifiers are not oversized. With contact stabilization, reaerated sludge is isolated from forward flow and, therefore, remains in the system during high-flow events. For additional wet weather operating strategies, please

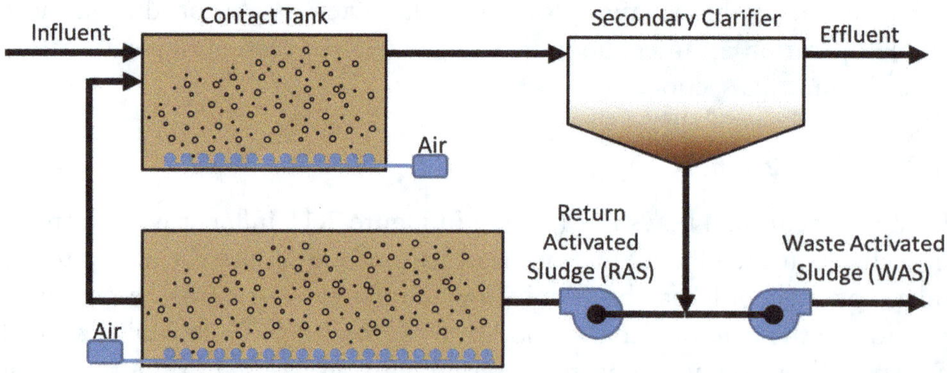

FIGURE 2.5 Contact-Stabilization and Sludge-Reaeration System (Reprinted with permission from Indigo Water Group)

refer to *Wet Weather Design and Operation in Water Resource Recovery Facilities* (Water Environment Federation, 2014).

The contact-stabilization modification, in its original form, does not provide significant nitrification and may not be effective for wastewater high in soluble BOD. However, some facilities have found that the process is effective when the contact time is increased to 4 to 5 hours (often referred to as a *sludge-reaeration process*).

6.3.3 Step-Feed

In the step-feed or step-aeration modification, influent wastewater is added at two or more points along the length of the biological reactor (Figure 2.6). This arrangement evens out the organic load and oxygen-uptake rate in the reactor, thereby simulating a complete-mix reactor. The RAS is typically added at the influent end of the reactor. The advantage of the system is increased facility capacity relative to conventional activated sludge systems without overloaded final clarifiers. For example, for the comparable SRT of a conventional system, the influent MLSS concentration to the clarifier would be lower for the step-feed system. It should be noted that providing additional piping for step feed gives the operator the option of operating the facility in a conventional, contact-stabilization, or step-feed modification.

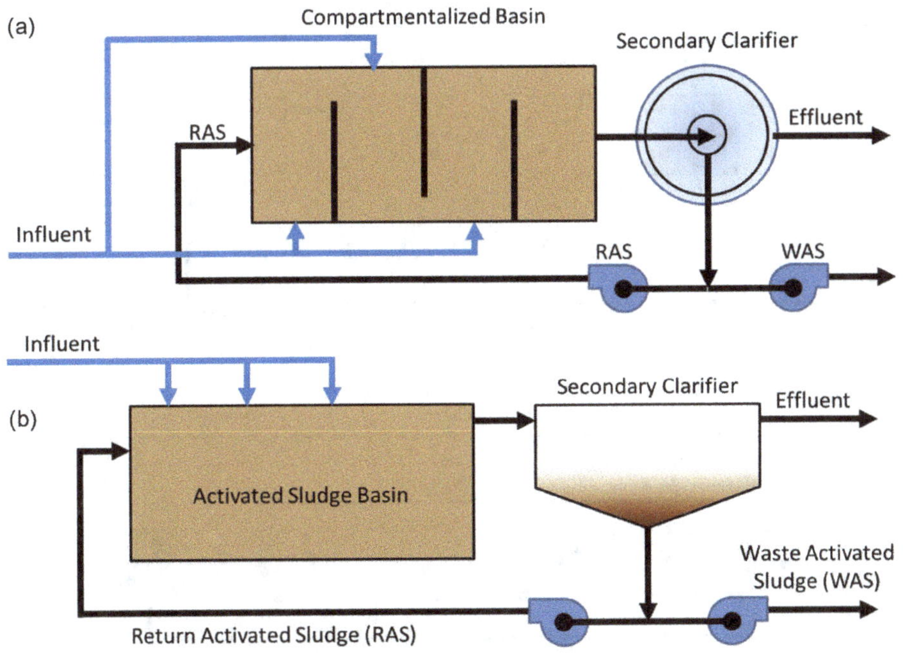

FIGURE 2.6 Step-Feed Process: (a) Plan View (b) Section View (Reprinted with permission from Indigo Water Group)

6.3.4 Tapered Aeration

The design of the aeration system in which oxygen supply approximately parallels oxygen demand is referred to as *tapered aeration*. Such a system would be used in reactors that approximate plug flow systems. Although mostly seen in diffused air systems, the concept can be practiced with mechanical aeration devices.

Airflow in diffused air systems to a given reactor sector can be controlled by airline valves with uniform diffuser placement, or preferably by tapering diffuser density. Typically, mixing requirements will govern the minimum airflow (or power input) rate near the effluent end of the reactor. Tapered aeration makes the most efficient use of power, provides greater operational control, and reduces overaeration at the end of the biological reactor.

6.3.5 Selectors

The use of a selector preceding the biological reactor has effectively controlled filamentous bulking in activated sludge facilities that experience epidemics of poor settling sludge. The selector may be aerobic, anoxic, or anaerobic and, typically, is compartmentalized (Figure 2.7). The basic principle of using a selector is to create environmental conditions that promote the growth of bacteria that settle well over those that do not. In summary,

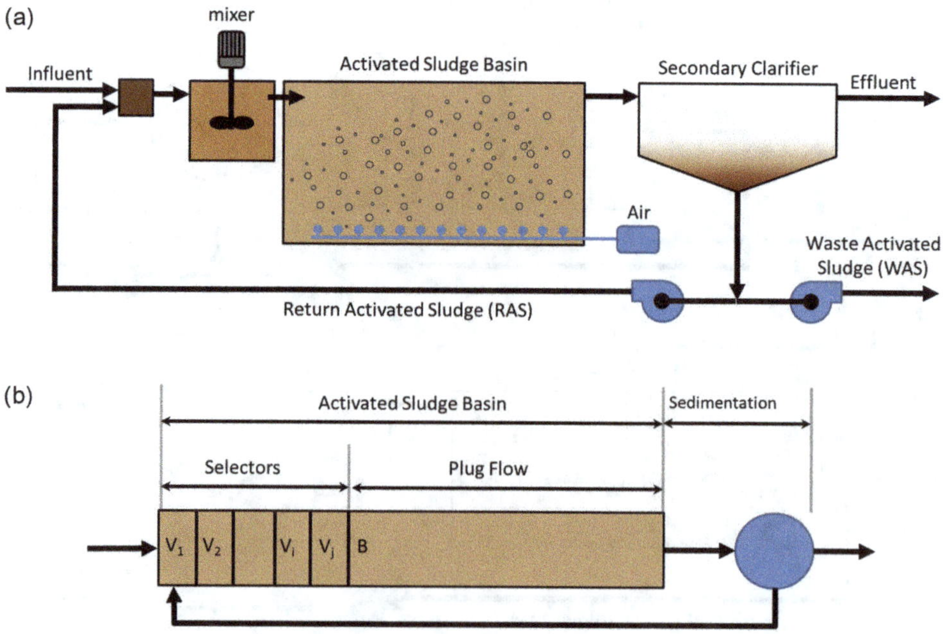

FIGURE 2.7 Selector Configuration: (a) Section View and (b) Plan View (Reprinted with permission from Indigo Water Group)

mixed liquors that are low in dissolved oxygen, or F:M, tend to favor filamentous microorganisms that interfere with effective settling. Many of these organisms can be put at a disadvantage if mixed liquor is subjected to periods of high F:M. Organisms with the greatest ability to rapidly uptake soluble organic matter and store it for later use during low concentration conditions tend to be ones that flocculate and settle better.

Selectors, because of the high organic load imposed on them, are relatively small with short HRTs (10 to 30 minutes for aerobic selectors and somewhat longer for anoxic and anaerobic selectors). If these units are too large, low F:M loads may occur, rendering them ineffective. To date, experience has been excellent with selectors. Details on design and operation of selectors may be found in literature by Jenkins et al. (2003) and Water Environment Federation (2023a).

6.4 Other Modifications

6.4.1 Sequencing Batch Reactors

An SBR is a fill-and-draw activated sludge system in which both steps of aeration and clarification take place in the same reactor. Settling occurs when the air and mixers are turned off and a decanter provides for the withdrawal of treated effluent. Discrete cycles are used during prescribed, programmable time intervals, and MLSS remains in the reactor during all cycles. For conventional systems, there are five steps that are carried out: fill, react, settle, decant, and idle (Figure 2.8). As a result of the batch nature of the process, flow equalization and multiple reactors must be provided to accommodate continuous-flow operation. Mixing and aeration are provided by equipment similar to that found in conventional continuous-flow facilities. Specially designed decanters are provided for effluent withdrawal. An intermittent-cycle extended-aeration system allows influent to be fed continuously to the reactor for all cycles; however, effluent is withdrawn intermittently.

The SBR is typically designed for small-flow applications of 4000 m^3/d (1 mgd) or less, often as extended aeration systems. Some larger installations have been built, however, ranging in size from 150 000 to 700 000 m^3/d (40 to 185 mgd). The process is easily adaptable to BNR by programming different sequences of cycles.

A modified version of the SBR is the *intermittent cycle extended aeration system* (ICEAS). In the ICEAS system, influent wastewater flows into the reactor on a continuous basis. As such, this is not a true batch reactor, as is the conventional SBR. A baffle wall may be used in the ICEAS to buffer this continuous inflow. The design configurations of the ICEAS and the SBR are otherwise similar.

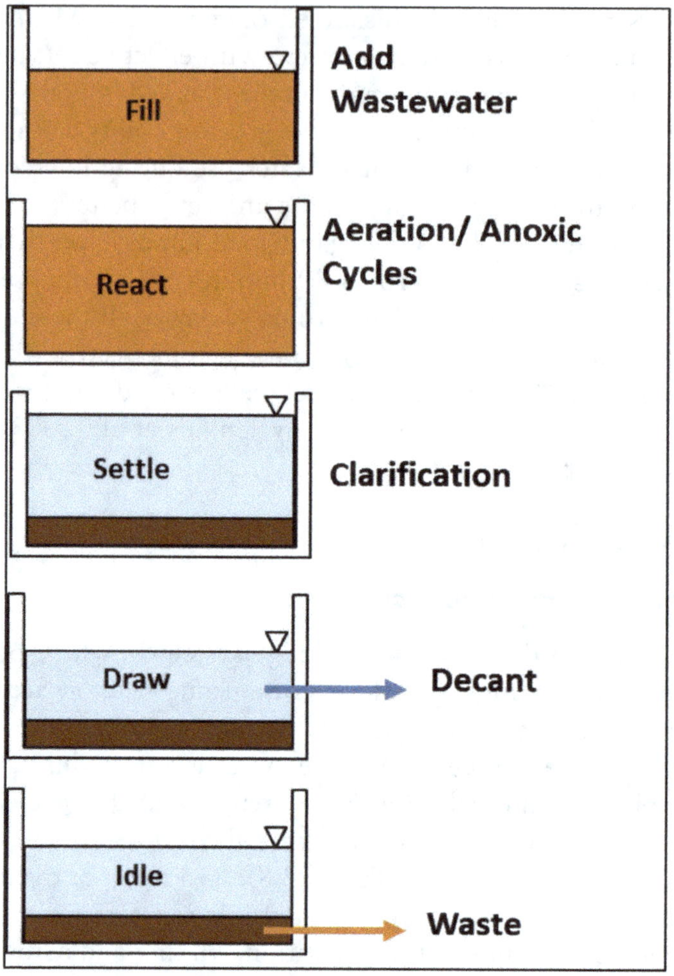

FIGURE 2.8 Sequencing Batch Reactor (Reprinted with permission from Indigo Water Group)

6.4.2 Oxidation Ditch

In the typical oxidation ditch, mixed liquor is pumped around an oval or circular pathway by brushes, rotors, or other mechanical aeration devices and pumping equipment located at one or more points along the flow circuit (Figure 2.9). Mixed liquor is moved at a velocity of 0.24 to 0.37 m/s (0.8 to 1.2 ft/sec) in the channel, resulting in a circuit time of fewer than 15 minutes. Most oxidation ditches are designed as low-rate processes with long HRTs (approximately 24 hours); therefore, the cycle time is short and the hydraulic system may be considered to be completely mixed. Ditches may be designed with a single channel or multiple interconnected concentric channels. Often, they are used for small- to medium-sized communities.

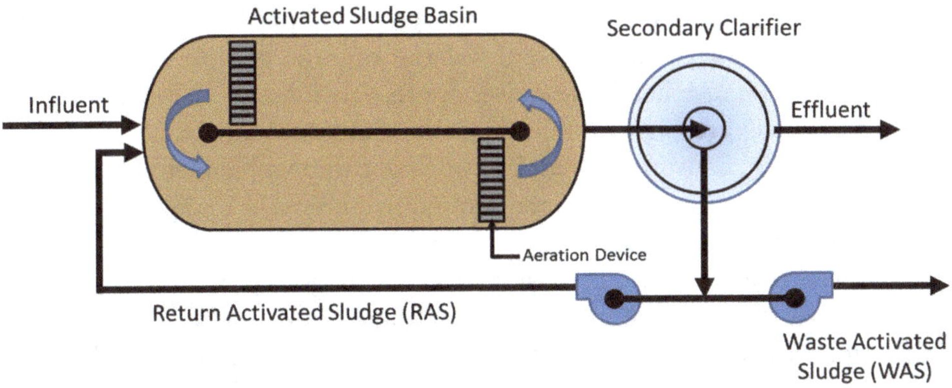

FIGURE 2.9 Oxidation Ditch With Surface Aerators (Reprinted with permission from Indigo Water Group)

6.4.3 High-Purity Oxygen

High-purity oxygen may be used in place of air for activated sludge aeration. Pure oxygen is fed concurrently with wastewater to reactors that are covered typically (Figure 2.10). Oxygen feed is controlled by maintaining constant pressure in the closed reactors. Typically, dissolved oxygen of 4 to 10 mg/L is maintained in the mixed liquor. Less than 10% of the inlet oxygen vents from the last stage of the unit. In the past, these systems were operated at high MLSS concentrations and short HRTs, thereby requiring less area for the facility than conventional activated sludge facilities. Today, these systems are designed with loading comparable to those of conventional activated

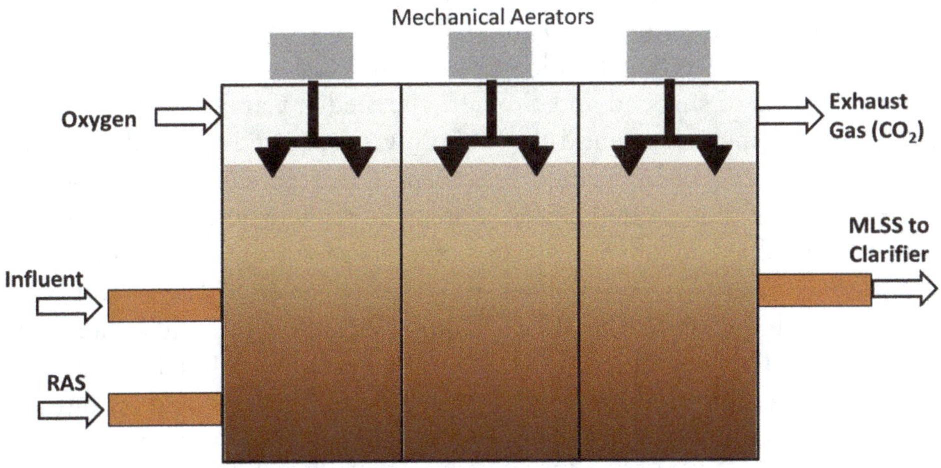

FIGURE 2.10 High-Purity Oxygen Activated Sludge

sludge processes. One significant advantage of the process is that it essentially eliminates stripping potential volatile organic compounds from the mixed liquor. Because headspace gases are recycled, the CO_2 concentrations in the gas phase are high and, as a result, alkalinity can be consumed in the mixed liquor, causing a pH decrease. This can be a problem for systems that are designed for nitrification, often requiring separate stages for CBOD removal and nitrification.

Covered reactors have provisions for warning of potential explosions that could result from the presence of combustible gases from wastewater. Precautions are also taken in the selection of construction materials because of the corrosive and reactive nature of the oxygen–CO_2 gases. The dissolution of oxygen into the mixed liquor is typically accomplished with mechanical surface aerators or submerged sparged turbine systems. The three types of oxygen-separation devices in use today are the cryogenic air-separation process, pressure swing adsorption (PSA), and vacuum swing adsorption (VSA). The PSA system has been used in the past and remains in some facilities but, typically, the VSA system is the more cost-effective system, depending on local energy costs (Water Environment Federation, 2023a).

6.4.4 Coupled Systems

To enhance performance and increase the capacity of biological reactors, the practice of adding inert support media to the biological reactor was developed decades ago. The inert media support a fixed-growth biomass that augments the mixed liquor microbial population. Early applications used panels of asbestos sheeting placed in the biological reactor. More recent applications include the use of plastic trickling filter media, polyurethane foam pads, loops of fiber bundles, or small plastic carriers. One such system is the moving bed biofilm reactor, in which plastic carriers are placed in biological reactors to support biofilm growth. Diffused aeration or mechanical mixers are used to suspend and circulate the media. Carriers are retained in the reactors using screens and no RAS flow is required (Metcalf & Eddy, Inc./AECOM, 2013). Integrated fixed-film activated sludge processes are suspended-growth activated sludge systems that incorporate some form of fixed-film media and have an RAS flow. These systems claim improved performance over conventional aeration basins because of the addition of more active biomass to the system and a reduction in solids loading to the clarifiers. The membrane aerated biofilm reactor is a fixed film process involving a membrane submerged in the biological reactor and air diffused through the membrane. The membrane provides a substrate for the biomass to grow, forming a biofilm, while oxygen is delivered directly to the biofilm to increase the efficiency of the aeration process.

6.4.5 Combined Systems

Most combined systems use a fixed-growth process in series with a suspended-growth system. Typically, the combined system includes a biotower and a biological reactor or aerated channel. Occasionally, a rotating biological contactor process may be used as the fixed-growth system. Combined systems are used when designers attempt to compensate for weaknesses of one of the systems by combining the two. For example, fixed-growth processes are known to resist shock loads and provide low maintenance requirements. By combining this with activated sludge, a process known to produce high-quality effluents with the ability to be operated under a variety of different modes, designers have found that the overall system can achieve a high degree of treatment for certain wastewaters and environmental constraints. Several combined processes are currently used in the United States and may be categorized into two groups: those that have low to moderate organic loading to the fixed-growth reactor and those with high organic loads (roughing filters). Figure 2.11 illustrates alternative methods for returning RAS, or reaeration, that are common practices for combined systems. Terms used to distinguish process modes are also presented. The commonly used combined systems include activated biofilter (ABF), trickling filter (TF)/solids contact (SC), roughing filter/activated sludge (AS), biofilter (BF)/activated sludge, and trickling filter/activated sludge. Details about these systems can be found in *Nutrient Removal* (Water Environment

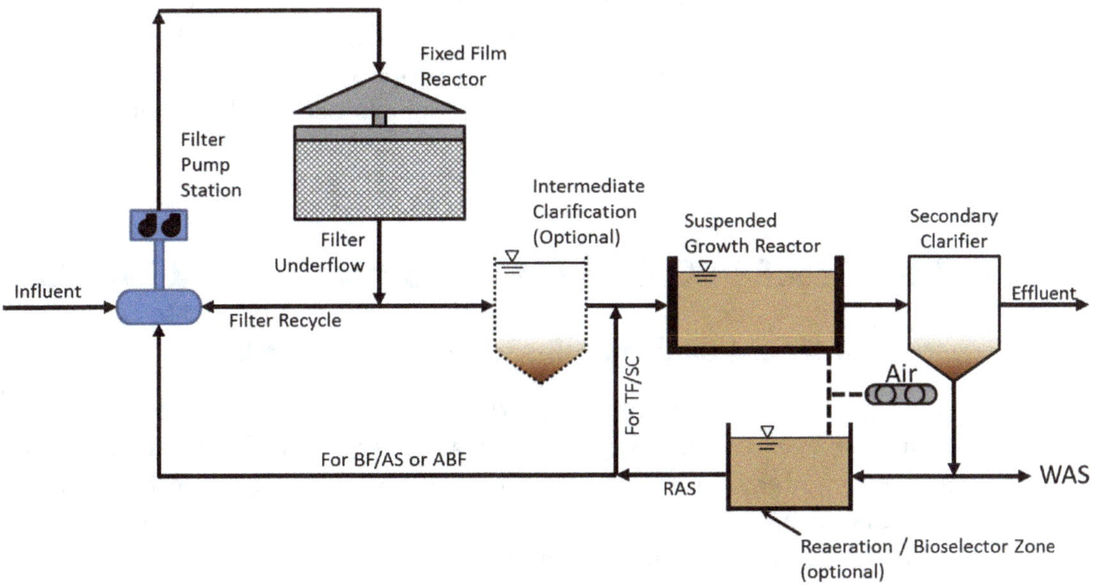

FIGURE 2.11 Combined Fixed-Film and Activated Sludge System (WEF, 2017)

Federation, 2011). Another advantage of the combined process is a lower energy requirement because less aeration is necessary.

6.5 Factors Affecting Process Efficiency

Factors affecting process efficiency may be categorized as environmental, design, operational, and maintenance. Environmental factors that affect performance include wastewater characteristics, system dissolved oxygen, temperature, and pH. Influent wastewater characteristics that affect the process include the nature of the carbonaceous organic matter and nitrogenous compounds, their biodegradability, soluble and particulate fractions, and concentration; the concentration and availability of nutrients; alkalinity; flowrate and its temporal variability; and the presence of toxic or inhibitory compounds.

Design factors that affect process efficiency and operational control include reactor volume (affecting HRT), clarifier sizing, pumping capacities of WAS and RAS systems, recycle pumping capacity, hydraulic design, and aeration system size and configuration. Operational factors include MCRT or F:M loading, dissolved oxygen control, RAS flow, WAS rate, and return stream rates from solids-handling processes. Commonly, various levels of real-time and/or advanced process monitoring technologies, such as sensors for dissolved oxygen, ammonia, and oxidation-reduction potential (ORP) are increasingly used to optimize process control and reduce energy costs. Preliminary and primary treatment operations installed before the activated sludge process also have an effect on performance because they reduce the effect of trash; inert solids; and difficult-to-degrade fats, oils, and grease. Good performance also depends on appropriate equipment maintenance, laboratory quality control, proper sampling protocol, and adequate training of WRRF staff.

7.0 DESCRIPTION OF FACILITIES AND EQUIPMENT USED

7.1 Activated Sludge Basins, or Biological Reactors

Biological reactors are often constructed of reinforced concrete, although some package facilities (e.g., extended air) may use steel. Typically, reactors are rectangular to accommodate common-wall construction for multiple reactors, although some installations may choose to use circular or oval tanks (the oxidation ditch is one example). Reactors may be single-pass, folded, or placed in a series. A minimum of two reactors is desirable, even for small facilities, to accommodate any shutdown for occasional maintenance. Each reactor must be furnished with inlet and outlet gates or valves so they can be removed from service. Proper drains or sumps should be provided for rapid dewatering (approximately 8 to 20 hours). In addition to aeration equipment, reactors should be equipped with a froth-control system to control foaming. Inlet design should ensure that flow to the reactor is as uniform as possible to avoid severe short-circuiting of influent and provide for intermixing of RAS with influent wastewater. Where parallel multiple reactors are used, a method for proper flow splitting should be provided to ensure equal flow to each reactor. In step-feed systems, a positive means of flow control is particularly important.

7.2 Aeration Systems

The supply of oxygen to the biological reactor represents the largest single energy consumer in the activated sludge facility (50% to 90%). Over the years, oxygen-transfer equipment has evolved to a point where engineers have a wide selection of efficient equipment to meet the needs of all types of facilities. Oxygen-transfer devices are used to not only supply oxygen to the process, but also to mix the aerobic compartments of the reactor. Typically, there are two types of aeration devices: diffused-aeration and mechanical-aeration systems. Each of these will be discussed in the following subsections.

7.2.1 Air Delivery

The three basic components of the air-delivery system are air filters or conditioners, blowers, and piping. Air filters remove particulates such as dust from the inlet air to the blowers and protect blowers and diffusers from mechanical damage or clogging. The degree of air cleaning depends on the inlet air quality, type of blower, and diffusers.

Today, many types of dynamic and positive-displacement blowers are used. Dynamic, or centrifugal, blowers are considered constant-pressure machines and are the most efficient and easiest to operate at variable airflow rates, are quieter than positive-displacement blowers, and require less maintenance. Their disadvantages include a limited operating pressure range and reduced delivered air volumes with any increase in backpressure caused by diffuser clogging. The positive-displacement blower is a constant-volume device capable of operating over a wide range of discharge pressures. It has a lower initial cost and requires relatively simple control procedures. Centrifugal blowers have higher efficiency and are favored at higher airflows.

Centrifugal blower options include multistage, gear-driven single-stage, and high-speed turbo blowers. Modern high-speed turbo blowers are single-stage, oil-free, direct-drive centrifugal blowers with advanced bearing design. The reported advantages of high-speed turbo blowers include ease of installation, smaller footprint, lighter weight, higher energy efficiency, higher surge margins, reduced maintenance, and reduced noise (Bell & Abel, 2011). More details on blowers can be found in Water Environment Federation (2023a).

7.2.2 Diffused Aeration System

Diffused aeration is defined as the injection of air or oxygen below the liquid surface. The air or oxygen is supplied by low-head blowers with pressures typically up to 210 kPa absolute (30 psia) or 105 kPa gauge (15 psig). Hybrid devices, including jets and U-tube aerators, which combine gas injection with mechanical pumping or mixing, are also included in this category.

Diffusers include both porous and nonporous devices. Details about these devices can be found in *Standards for Design of Water Resource Recovery Facilities* (Water Environment Federation, 2023a). Porous diffusers, often referred to as *fine-pore* or *fine-bubble diffusers*, are highly efficient and, currently, are the most used of the diffused-air systems. They are typically produced from ceramic materials, porous plastics, or perforated membranes. Nonporous systems include an array of diffusers that have larger orifices ranging from holes in pipes to specially designed valved orifices. Diffusers are placed near the floor of the biological reactor and may be configured in a grid arrangement, along one or both longitudinal sides. Air is delivered in a piping system from the blowers to drop legs, which carry the air down to headers along the bottom of the reactor. Valves in the air system are used to control airflow to individual reactors or sectors of reactors. Diffusers may be placed in uniform densities or arranged in tapered configurations. The diffuser type, pattern of diffuser layout, reactor geometry, submergence, and airflow rate all influence oxygen-transfer performance. Maintenance of porous diffusers is greater than for the large-orifice diffusers. Clogging and fouling of fine pores require occasional cleaning, which can be provided by automatic gas-cleaning systems in situ or by draining reactors and cleaning with high-pressure hosing or acid spritzing (Water Environment Federation, 2023a).

7.2.3 Mechanical Aeration

Mechanical aeration systems include surface aeration devices and submerged turbine aerators. Surface aerators can be grouped into the following four general categories:

1. Radial flow, low-speed aerators;
2. Axial flow, high-speed aerators;

3. Aspirating devices; and
4. Horizontal rotors.

Each system is widely used and has distinct applications. Typically, surface aerators are float-, bridge-, or platform-mounted, and some may be equipped with submerged draft tubes for deep reactor applications. Radial flow, low-speed aerators have seen a recent increase in use and are considered good mixing devices that are more efficient than high-speed aerators. Axial-flow, high-speed aerators are used typically in stabilization lagoons where dispersed growth is found and shearing of biological floc is not an issue. In addition, they may suffer from freezing problems in cold climates. Aspirating devices use a motor-driven propeller aspirator that draws air down a shaft to orifices at the submerged end. Air velocity and propeller action create turbulence, forming small bubbles. These devices can be positioned at various angles to reach different levels of mixing, aeration, and circulation. Aspirating devices are not used often in activated sludge applications. Horizontal rotors, designed in several configurations, are used in oxidation ditches. The impeller agitates the liquid surface, creating a hydraulic jump that is effective in oxygen transfer. The device provides a horizontal velocity component that rotates fluid around the ditch and provides sufficient velocity to maintain MLSS in suspension.

The submerged turbine consists of a motor and gearbox drive mounted over the reactor, one or more submerged impellers, and piped air from a blower to a diffuser ring below impellers. Impellers may be axial or radial flow. Impellers can move mixed liquor down, up, or laterally, depending on impeller design. These devices provide excellent mixing and are capable of high mass-transfer rates. They are used in a number of activated sludge applications.

7.3 Process Control

There are numerous technologies used to improve control in the activated sludge process, including sensors, modulating valves, and airflow meters. There are advanced sensors that can provide real-time data in biological reactors such as dissolved oxygen concentration, ammonia concentration, pH, temperature, and ORP. All of this information can help operators improve the efficiency of the activated sludge process and troubleshoot the process if an issue arises. Furthermore, addition of equipment such as air modulating valves and airflow meters in aeration system piping can help control the duration and volume of air being delivered, which will increase the efficiency of the aeration process.

To reduce the effort of manually recording and responding to changes in process parameters from local readouts, many sensors, motor-operated valves, and flowmeters can send signals to a programmable logic controller (PLC). The PLC can initiate preprogrammed responses to data such as modulating valves or adjusting equipment output and transmit the data into a supervisory control and data acquisition system that allows operators to monitor the process in real time. Operators can also trend data to monitor if any discrepancies arise, which, typically, indicates an issue in the process. For example, if the dissolved oxygen concentration drops drastically and suddenly in the biological reactor, it may indicate of an issue with the aeration system or a sudden increase in BOD loading. There are also Internet of Things sensors and probes that can be installed in a reactor and connected to the Internet to allow process monitoring from a cell phone, tablet, or computer. Typically, these types of sensors are used in process troubleshooting.

7.4 Mixing

In the activated sludge process, mixing is important for maintaining the MLSS in suspension. In most applications, aerators serve as both oxygen-transfer devices and mixers. With the exception of the effluent end of plug flow reactors, typically, oxygen requirements control aerator design and operation. However, at the effluent end of plug flow reactors, in anoxic/anaerobic zones and in some low-loaded, complete-mix systems, mixing may be the controlling factor, especially during low-flow conditions. In most cases, the aeration system is used for mixing in these situations; however, use of low-speed, submerged-propeller, horizontal mixing devices (such as banana blade mixers) to move mixed liquor horizontally along the reactor is another option. A combination of this mechanical mixing with aeration may result in significant power savings when mixing controls the process.

Both submerged propeller or turbine mixers have been used for mixing anoxic and anaerobic compartments of biological reactors. These devices mix without breaking the water surface and are capable of maintaining biological solids in suspension at minimal energy inputs. The number and placement of these units is critical to effective suspension of MLSS.

7.5 Clarification

Separation of MLSS from the liquid stream is vital to the operation and performance of activated sludge systems. Typically, this is achieved by gravity separators, although membranes can also be used. Clarification not only separates the MLSS, but also, in some situations, may be designed to thicken settled sludge before returning it to the aeration process or to wasting (WAS).

Clarifier shapes include rectangular, square, circular, and others such as hexagonal or octagonal. There seems to be no observable difference in performance at average or peak flow based on shape alone (WEF, 2023a). Typically, sludge is removed by chain-and-flight collectors in rectangular clarifiers. Hydraulic suction using floating or bridge-mounted mechanisms has been used. Circular clarifier mechanisms with hydraulic pickups use mechanical seals to allow continuous solids withdrawal. In addition, mechanical plows are used on these rotary collectors. Interchannel clarifiers have been used in some small-flow oxidation-ditch applications. A clarifier is installed directly in the ditch, thereby eliminating the need for a separate clarifier. Scum and sludge are returned directly to the ditch by hydraulic means.

Clarifiers should also be equipped with a froth-control system to control foaming, typically achieved using water sprays, foam skimmers, or addition of antifoaming chemicals. This will prevent a buildup of clarifier foam, which can enter the effluent or be blown into walkways that can be hazardous as some foam contains human pathogens. More details on different types of foam and ways to combat it are provided in Chapter 7. Most clarifiers are equipped with a scum-removal system that consists of a skimmer arm with blades that slowly travels along the water surface of the clarifier. The skimmer arm will collect floating debris and direct it towards a scum box to be removed from the clarifier.

Clarifier depth continues to be a debated issue relative to its importance in performance. There are many factors that will affect the selection of depth, including clarifier diameter, collector characteristics, sludge-withdrawal point, and inlet design. Typically, minimum clarifier depths of 3.7 to 4.6 m (12 to 15 ft) are recommended. Deeper clarifiers may be desired for large-diameter clarifiers as long as they are no deeper than 4.6 to 5.0 m (15 to 16 ft). Clarifier performance also depends on SLR, surface overflow rate, inlet design, effluent weir arrangement, and settling characteristics of the MLSS.

Clarifier performance and characteristics are reviewed in literature by Water Environment Federation (2005, 2023a) and Ekama et al. (1997).

When membranes are used for solids separation, the process is known as membrane bioreactor process. Membranes are submerged in biological reactors or separate membrane tanks and achieve almost complete removal of suspended solids.

7.6 Return and Waste Activated Sludge Systems

The RAS system pumps settled sludge, thickened as much as practical in the clarifier, from the clarifier back to the biological reactors. Most RAS stations use centrifugal pumps for this application. For intermediate-sized facilities, pumps are often connected directly to sludge-withdrawal pipes. In larger facilities, a wet well is provided for these pumps to draw suction. In addition, screw pumps are used in many facilities in the United States and Europe. The RAS system must also allow for accurate measurement and control of sludge flow. Typically, RAS pumping systems are controlled by a positive variable-flow control device capable of changing flow within the minimum to maximum range for proper process control. The variable-flow control device can be controlled by PLCs, which can receive and send signals proportional to flow or preprogrammed patterns. In small facilities, throttling of the force main from RAS pumps has been used.

All activated sludge processes must have a WAS system to remove excess biomass from the system. Sludge may be wasted from the clarifier under-flow or directly from biological reactors. The WAS system needs flow-metering and pumping equipment independent of other activated sludge control devices. The most positive and flexible system will include an independent pumping system with flow adjustability and a flowmeter that provides feedback to a flow-control device.

7.7 Recirculation Pumping

Typically, mixed liquor recycling in several of the BNR processes is accomplished by low-head, submersible, nonclog pumps; propeller pumps; or nonclog, vertical, turbine pumps mounted on the biological reactors. Because water level in the zones is virtually the same, the only pumping head is attributable to pipe friction and fitting losses. However, offsetting the low-head requirement is the high pumping volume required (recycle ratios may range from 1:1 to as high as 4:1 based on forward flow). Typically, flow is conveyed in a pipe rather than a channel to avoid dissolved oxygen entrainment when discharged to an anoxic or anaerobic zone. Also, intakes should be located away from aeration diffusers, if possible, for the same reason. Constant-speed pumps may be used because it is not essential that recycle flows parallel forward flow exactly. In most cases, multiple pumps can provide sufficient flexibility.

8.0 PRACTICE QUESTIONS

1. The purpose of the secondary clarifier is to
    a. Separate solids from treated wastewater
    b. Promote denitrification
    c. Manipulate MLSS concentration
    d. Reduce soluble BOD

2. Which of the following parameters is the best estimate of live microorganisms in the activated sludge process?
    a. Influent BOD concentration
    b. Mixed liquor suspended solids
    c. Return activated sludge
    d. Mixed liquor volatile suspended solids

3. One reason air is added to some activated sludge basins is to
    a. Prevent freezing
    b. Keep the MLSS in suspension
    c. Circulate waste products
    d. Maintain water temperature

4. Settled solids at the bottom of the clarifier are called the
    a. Blanket
    b. Septic layer
    c. Schmutzdecke
    d. Scum

5. What is the most likely outcome if excess solids are not removed from the activated sludge basin on a regular basis?
    a. Treatment efficiency will increase
    b. Sludge settleability will improve
    c. Solids may be lost to the final effluent
    d. Oxygen use will decrease

6. The purpose of the RAS line is to
    a. Transfer settled solids to solids-handling processes
    b. Return settled solids to the activated sludge basin
    c. Manipulate MLSS concentrations
    d. Control sludge age

7. The RAS is a fixed percentage of influent flow. The clarifier blanket depth is less than 0.6 m (2 ft). Influent flow to the WRRF is increasing. What is happening to the clarifier SLR?
    a. Held constant by constant RAS percentage
    b. Decreasing with increasing influent flow
    c. Held constant by blanket depth
    d. Increasing with increasing influent flow

8. The MLSS is settling poorly. The operator must reduce the SLR to the secondary clarifiers to compensate. Which action should they take?
   a. Increase MLSS concentration by decreasing wasting
   b. Increase surface area by placing more clarifiers in service
   c. Increase RAS flow
   d. Increase blower output to activated sludge basins

9. Increasing the RAS flowrate will
   a. Decrease the RAS concentration
   b. Decrease the SLR to the secondary clarifier
   c. Increase the WAS concentration
   d. Increase sludge age

10. Waste activated sludge influences all of the following EXCEPT
    a. Sludge age
    b. Sludge settleability
    c. Sludge concentration
    d. Sludge color

11. Aerobic processes always contain
    a. Sulfate
    b. Oxygen
    c. Carbon dioxide
    d. Sodium chloride

12. By definition, anoxic zones contain ____________ but not ____________.
    a. Oxygen, nitrate
    b. Nitrate, sulfate
    c. Nitrate, oxygen
    d. Nitrate, nitrite

13. Heterotrophic bacteria obtain their carbon from
    a. Organic compounds
    b. Dissolved carbon dioxide
    c. Graphene
    d. Inorganic compounds

14. Nitrifying autotrophic bacteria obtain their energy from
    a. Organic compounds

b. Ammonia and nitrite
c. Nitrite and nitrate
d. Oxygen

15. Facultative heterotrophic bacteria are moved from the aerobic basin to the anoxic basin. Which two resources will bacteria use under anoxic conditions?

a. Carbonate and oxygen
b. BOD and oxygen
c. Carbonate and nitrate
d. BOD and nitrate

16. Which biochemical reaction may take place in an aerobic basin?

a. Conversion of nitrate to nitrogen gas
b. Reduction of BOD to carbonate
c. Conversion of ammonia to nitrate
d. Production of volatile fatty acids

17. Phosphate-accumulating organisms must be cycled between anaerobic and ____________ conditions for luxury uptake of phosphorus to take place.

a. Aerobic
b. Septic
c. Anaerobic
d. Reduced

18. Polyphosphate, a compound made by PAOs, is used for

a. Carbon storage
b. ATP reduction
c. Energy storage
d. VFA reduction

19. The water temperature in the activated sludge basins dropped from 20 °C (68 °F) to 10 °C (50 °F). The growth rate of microorganisms will

a. Be unaffected
b. Decrease by 50%
c. Increase by 25%
d. Lag by two weeks

20. Match the process variation to its defining characteristic:

a. Complete-mix	1. Long, narrow basins
b. Plug flow	2. RAS returned to a separate tank before mixing with the influent
c. Sequencing batch reactor	3. Influent added at multiple locations
d. Oxidation ditch	4. Uses high-purity oxygen instead of air
e. Contact-stabilization	5. Combines a fixed-film process with activated sludge
f. Step-feed	6. Composition of MLSS is the same everywhere
g. High-purity oxygen	7. Treatment and settling within the same basin
h. Combined system	8. Looped racetrack design

Answer key:

1. a, 2. d, 3. b, 4. a, 5. c, 6. b, 7. d, 8. b, 9. a, 10. d, 11. b, 12. c, 13. a, 14. b, 15. d, 16. c, 17. a, 18. c, 19. b, 20. a. 6, b. 1, c. 7, d. 8, e. 2, f. 3, g. 4, h. 5

9.0 REFERENCES

Alleman, J. E., & Prakasam, T. B. S. (1983). Reflections on seven decades of activated sludge history. *Journal of the Water Pollution Control Federation*, *55*, 436–443.

Bell, K. Y., & Abel, S. (2011). Optimization of WWTP aeration process upgrades for energy efficiency. *Water Practice Technology*, *6*(2).

Ekama, G. A., Barnard, J. L., Günthert, F. W., Krebs, P., McCorquodale, J. A., Parker, D. S., & Wahlberg, E. J. (1997). Secondary settling tanks: theory, modelling, design and operation. *International Association on Water Quality*. Science and Technology Report Series.

Jenkins, D., Richard, M. G., & Daigger, G. T. (2003). *Manual on the causes and control of activated sludge bulking and foaming* (3rd ed.). Lewis Publishers.

Manser, R., Gujer, W., & Siegrist, H. (2005). Consequences of mass transfer effects on the kinetics of nitrifiers. *Water Research*, *39*(19), 4633–4642.

Metcalf & Eddy, Inc./AECOM. (2013). *Wastewater engineering: Treatment and resource recovery* (5th ed.). McGraw-Hill.

Monod, J. (1942). *Recherches sur la croissance des cultures bacteriennes*, Herman et Cie.

Monod, J. (1949). The growth of bacterial cultures. *Annual Review of Microbiology, 3,* 371–394.

Sayigh, B. A., & Malina, J. F., Jr. (1978). Temperature effects on the activated sludge process. *Journal of the Water Pollution Control Federation, 50*(4), 678–687.

U.S. Environmental Protection Agency. (1974). *Process design manual for upgrading existing wastewater treatment plants*. EPA-625/171004A.

U.S. Environmental Protection Agency. (1978). *Design of wastewater treatment facilities major systems.* EPA-430/9-79-008.

U.S. Environmental Protection Agency. (2010). *Nutrient control design manual.* EPA-600/R-10-100.

Wanner, J. (2014). IWA conference: activated sludge—100 years and counting.

Water Environment Federation. (2005). *Clarifier design* (2nd ed., Manual of Practice FD-8).

Water Environment Federation. (2011). *Nutrient removal* (Manual of Practice No. 34).

Water Environment Federation. (2014). *Wet weather design and operation in water resource recovery facilities.*

Water Environment Federation. (2023a). *Standards for design of water resource recovery facilities* (WEF 8).

Water Environment Federation. (2023b). *Standards for operation of water resource recovery facilities* (WEF 11).

3

Microbiology

1.0 SUMMARY OF BACTERIAL ENGINE

Chapter 2 provided an overview of the activated sludge process and discussed its basic control using sludge wasting, sludge return, and dissolved oxygen concentration to govern the system by controlling the growth of the microorganisms. Because activated sludge microorganism mass, typically, is approximately 95% to 98% bacteria by weight, the operator is primarily managing bacterial growth, which, in turn, controls the growth of protozoa and *Metazoa*. Types of bacteria can be differentiated by carbon source (heterotroph vs. autotroph), use of oxygen (aerobic, facultative, or anaerobic), or physical characteristics (floc formers or filamentous microorganisms). This chapter will delve deeper into these organisms, describing the basic cell structure, the mechanisms driving their functions, why this information is important to you, and how you can use it to improve activated sludge facility performance. However, this is just an overview; there is much more to know about the bacteria that treat wastewater. We cannot even grow most activated sludge organisms in the laboratory, so they are difficult to identify. Genetic testing approaches, however, will shed more light on that in the future. For now, just understand that only approximately 30% of bacteria in activated sludge can be identified by observation and biochemical means. Although we cannot grow most of them in pure culture in the laboratory and identify them, understanding the ecology of activated sludge is important to controlling the process. Ecologically, it is important to note that few activated sludge organisms are of fecal origin; most are soil-type organisms that can withstand varying environmental conditions found in activated sludge. Further, a specific set of environmental conditions will promote growth of certain groups of bacteria. When those environmental conditions change, other bacterial groups begin to grow, possibly resulting in changes in process conditions (e.g., different sludge settleability). Various bacterial groups are all competing for a limited food supply, and the ones that thrive in present environmental conditions will outcompete the others. If the resulting growth of one of those groups is unfavorable, the environment must be modified to minimize the growth of that group and stimulate the growth of a different, preferable group. For instance, if a septicity-type filament that can cause poor sludge settleability is seen growing more abundant, the cause of the septicity must be identified and eliminated. Environmental conditions should be adjusted so that floc formers can outcompete and overcome septicity filament growth.

2.0 THE BACTERIAL CELL AND HOW IT WORKS

2.1 Description

Describing the cell will provide the groundwork for understanding how and why it works the way it does. There have been many basic descriptions of a cell, but one of the most beneficial may be Dr. Seth Terry's portrait of them as a "spiteful bag of enzymes," meaning that the minute we think we have figured them out, bacteria let us know that there is much left to be learned. A typical activated sludge bacterial cell is approximately 1 to 2 μm in length and diameter; if it is longer and narrow, it has a rod or bacillus shape, and if it is round, it has a coccus shape. The small size provides a large surface-area-to-volume ratio, which means that cells rapidly absorb nutrients and grow under optimal environmental conditions. For instance, it has been estimated that a 378-m^3 (100 000-gal) reactor with 2500 mg/L of mixed liquor suspended solids (MLSS) has an area approaching 8 825 789 m^2 (95 000 000 sq ft), so most soluble biochemical oxygen demand (BOD) will be consumed within minutes of bacteria contacting the substrate.

2.2 Parts and Function

Figure 3.1 shows the important parts of the cell. The bacterial cell is something like a balloon or bag in a box, the balloon representing the periplasmic or cell membrane that surrounds the contents of the cell and the box representing the cell wall that gives shape to the cell and supports it structurally.

The single strand of DNA serves as the nucleoid, which is just simple genetic material and is not nearly as complicated as a nucleus of a human

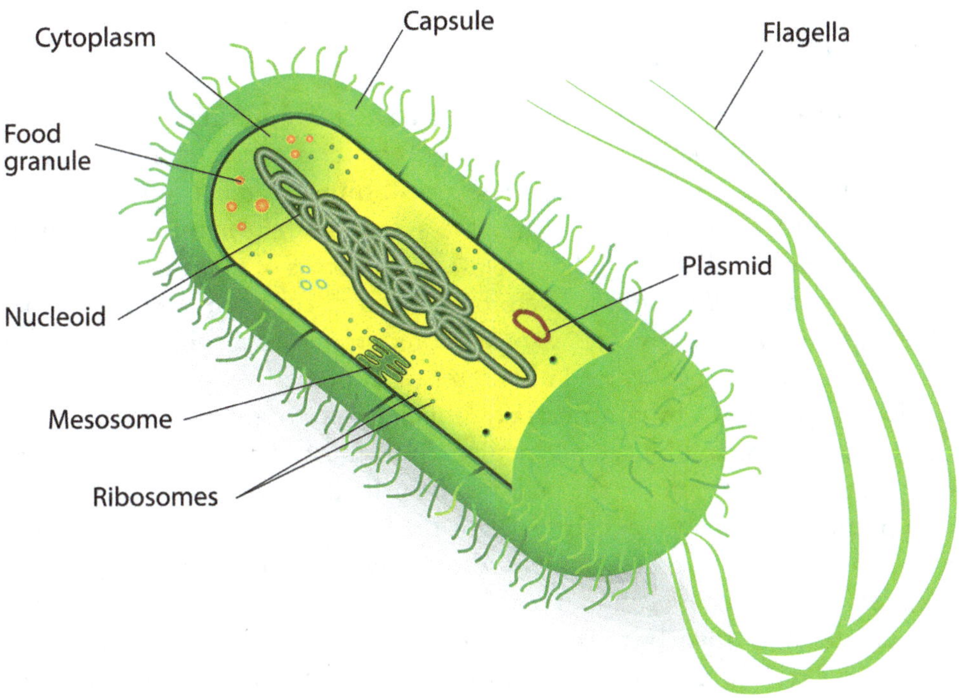

FIGURE 3.1 Bacterial Cell Components

cell. Much of the cellular contents are large protein molecules known as *enzymes* that are manufactured in ribosomes. Enzymes control biochemical reactions inside and around the cell. Other cellular contents may include inclusion bodies such as polyphosphate, sulfur, poly-β-hydroxybutyric acid, or glycogen. The "balloon" is important in that it controls intake of food and other nutrients and discharge of waste products through tiny holes. It keeps "in" what should be inside (e.g., enzymes and food) and keeps "out" what should be outside (e.g., excess water and metabolic byproducts). The cell wall provides structural support and maintains the shape of the cell. As discussed in Chapter 2, outside of the cell wall is a layer of exopolymer (slime layer), typically made up of accumulated waste products and storage products that help the cells stick together and form flocs. Bacteria must stick together to form larger particles that are heavy enough to settle in the clarifier. If sludge does not settle, it is wasted into the facility effluent. Thus, activated sludge selects for floc formers by allowing nonsettleable organisms to pass through the process and into the facility effluent. Also, in the slime layer and next to the cell wall are exocellular enzymes that will help break down particulate BOD adsorbed to the bacterial cell. Lastly, some bacteria have a flagellum or many flagella that allow them to be motile, or move through the water. Hopefully, there are not many bacteria with flagella in the system because swimming cells would not settle well in the clarifier.

2.3 How They Feed

Bacteria can feed on both dissolved and particulate matter. A bacterium, however, can only take in soluble (dissolved) materials through its cell membrane. The holes in the cell membrane are so tiny that most molecules will not cross the membrane unless they are small enough or ionized (broken down into electrically charged particles). The process of taking up dissolved particles is called *absorption*. Therefore, bacterial cells can take up soluble BOD extremely fast, which is why, as noted previously, soluble BOD can be removed from wastewater in a matter of minutes.

Particulate BOD, by contrast takes much longer to take up because particles must first be solubilized. Particulate BOD must first contact and adhere to the outer membrane of the cell and then be dissolved or broken down into particles that can be absorbed into the cell. The process of contacting and adhering to the cells is called *adsorption*, and the dissolving process is called *exocellular digestion*, meaning digestion outside of the cell. When particulate BOD adsorbs to the outside of a cell, exocellular enzymes (proteins) are excreted on the outside of the cell, which break down particles that cannot be absorbed into dissolved, ionized particles that can then enter the cell and be metabolized.

2.4 How They Grow

Living bacteria are present in a medium because environmental conditions favor their growth. If conditions are unfavorable, such as when resources become limited or there is a build-up of toxic products, bacteria will not grow and may even die or be wasted from the system. When their growth environment meets their demands, bacteria grow by ingesting and metabolizing soluble BOD. Individual cells reproduce by binary fission in which the cell feeds until it has accumulated sufficient mass and then splits into two smaller cells that can then grow through the same process. When bacterial cells feed, the first energy produced is used to keep the cell alive and function as intended. Excess energy produced beyond that can be used for reproduction.

Actively feeding cells replicate when they have processed enough food, whereas nonfeeding cells can use stored materials to stay alive. This can be shown by a heterotrophic bacterial growth curve such as that given in Figure 3.2. Bacterial cells can be grown in the laboratory, counted, and the number of living bacteria plotted against culture time. At the start of a culture's growth, the number of cells does not increase. This is called the lag phase, when the cells are taking up food and increasing in size. They are removing BOD at a high rate, but not yet replicating. Once cells are as large as they can get, they begin to divide during the accelerated growth phase. Soon they are growing just as fast as they can in the log phase. However, when the bacterial cells run out of food or accumulate toxic byproducts, growth slows in the declining growth phase. In the stationary phase, they

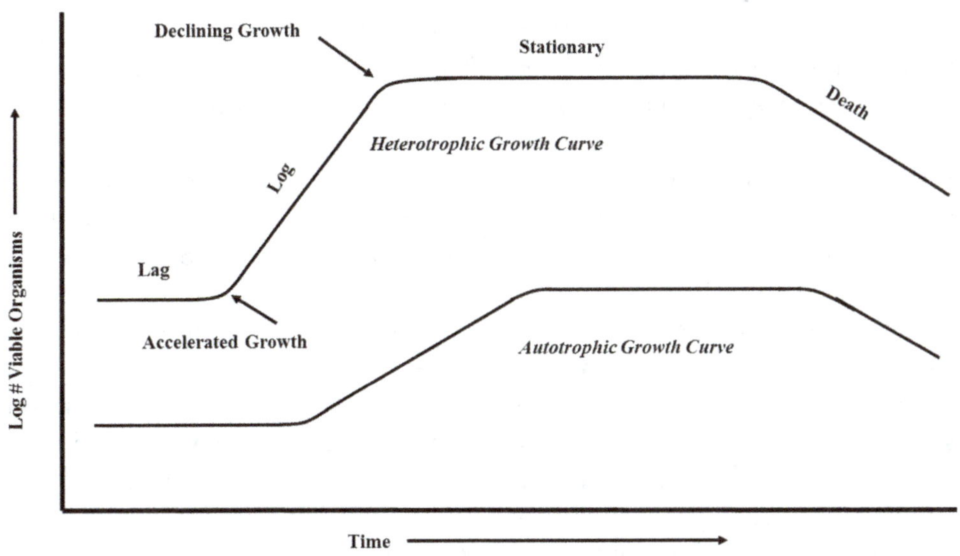

FIGURE 3.2 Bacterial Growth Curves

are not increasing in number because growth is equaled by death and the number of living cells stays relatively constant. When cells die, they lyse (i.e., break apart), releasing contents that are then available as food for other organisms. After some period of time in the stationary phase, they enter the death phase in which more cells die than grow (think of aerobic digestion). The culture time on the x-axis of the growth curve can be related to mean cell residence time (MCRT). Activated sludge does not work well until the declining growth or stationary phase is reached. Until then, cells are growing too fast and do not flocculate well. If they are fed too much per cell, cells replicate quickly, do not form a slime layer, and do not stick together; thus, activated sludge functions best when cells are fed just the right amount to cause the slime layer and exocellular polymeric substances (EPS) to form. However, activated sludge also does not work well when growth is too slow or in the death phase, where flocs break apart and effluent quality deteriorates. Because most activated sludge is in the stationary phase, one can see that activated sludge organisms grow in a relatively starved condition (i.e., slow growth).

Nitrifying organisms are somewhat different in that nitrifiers cannot store food. There appear to be no known chemical polymers that can be used by these organisms to store ammonia, so nitrifiers have to immediately oxidize any ammonia molecules within which they come in contact. This is one reason that nitrifier growth rate is slower than their carbonaceous BOD (CBOD)-eating cousins. Another reason nitrifiers grow slower is that oxidation of ammonia is not as energetically favorable as using organic carbon or even hydrogen gas as a substrate. Therefore, ammonia is not consumed and processed in the aeration tank as quickly as CBOD is processed.

The growth curve shows separate curves for heterotrophs, the CBOD feeders, and autotrophs such as nitrifiers. The autotrophic curve shows a longer lag phase, a flatter log phase, and a lower number once the stationary phase is reached. As stated previously, autotrophs typically grow slower, and they never attain large numbers as heterotrophs do; thus, an autotrophic process such as nitrification requires use of a higher solids retention time (SRT) to provide enough time for slow-growing nitrifiers. Slower growth also suggests that they cannot address shock loads as well as heterotrophs can, and, because nitrifiers do not store food, it is difficult for the biomass to address quick changes in ammonia load.

How fast can bacteria grow? Under optimal growth conditions (never seen in activated sludge), some *Bacillus* spp. can divide every 12 to 15 minutes. Under this feeding condition, cells will not flocculate; therefore, activated sludge settles at a much slower rate. Provided their growth environment is to their liking, Figure 3.3 suggests that typical activated sludge organisms can maintain approximately a 90% growth rate at a soluble

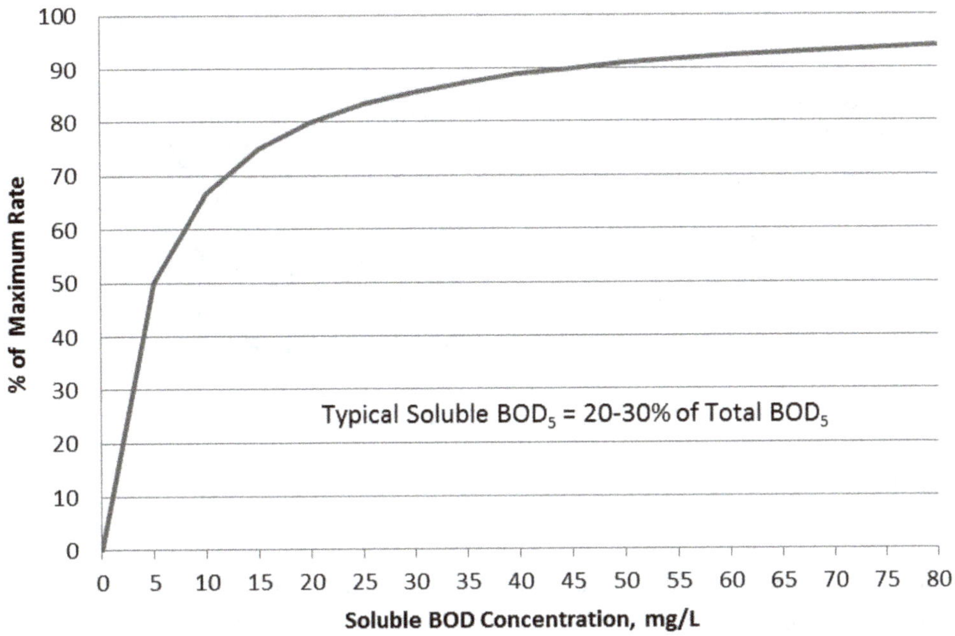

FIGURE 3.3 Bacterial Growth Rate Versus Soluble BOD_5 Concentration

5-day BOD (BOD_5) of about 45 mg/L, which is equivalent to approximately 150 to 225 mg/L BOD_5. If the concentration seen by organisms decreases below these values, the growth rate slows down significantly. The BOD_5 concentration actually seen by the organisms is not the influent BOD_5 concentration, but, instead, the influent BOD_5 concentration diluted by return sludge flow and any recycle flow such as mixed liquor recycle in a denitrifying system. For example, an influent BOD_5 of 300 mg/L, return rate of $0.4Q$ (Q = influent flowrate), and a recycle of $4Q$ would result in a total BOD_5 concentration of approximately 55 mg/L and 15 mg/L soluble BOD_5, resulting in a growth rate of approximately 75% of maximum. Dilution of ammonia causes an even more drastic reduction in the growth rate of nitrifying bacteria. Biochemical oxygen demand dilution becomes very important when considering filament growth, which will be discussed in Section 3.0.

3.0 BACTERIA DEFINED BY GROWTH PATTERN: FLOC FORMERS VERSUS FILAMENTOUS MICROORGANISMS

Discussion up to this point has focused on floc-forming bacteria, which grow in clumps; those clumps are held together by EPS to form flocs (Figure 3.4). Typically, such flocs settle relatively fast because they are compact, rounded,

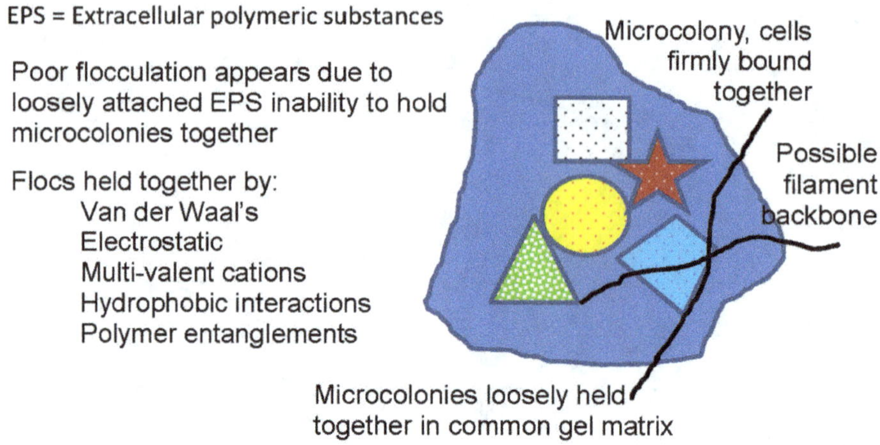

FIGURE 3.4 Floc Components

and dense. However, there are some bacteria that grow as filaments, that is, single lines of cells growing end to end in straight, bent, curved, or coiled fashion. A small number of filaments may actually help hold flocs together, but excessive growths of filaments produce diffuse flocs of low density, flocs with filaments on the edge that slow settling, or flocs held apart by bridging filaments that do not allow them to settle well and/or produce dense sludge to return from the clarifier.

It is now known that some filamentous microorganisms can grow either as single cells or in small clusters within the floc, growing only as filaments when the environment favors growth in its filamentous form. This helps explain why blooms of filaments can occur quickly, often causing settling problems.

As discussed later in Section 5.0, certain filaments favor certain growth environments; therefore, if a filament is observed, environmental conditions are present in which that filament can grow. Thus, to control the filament's growth, environmental conditions favoring growth of that filament must be changed to those favoring growth of floc formers; then, floc formers can outcompete the filamentous organism and the sludge can settle faster. For instance, if one wants to control *Sphaerotilus natans*, a low dissolved oxygen organism, one would increase the dissolved oxygen or take it down to 0.0 mg/L, controlling its growth because *S. natans* is an aerobic microorganism.

4.0 BACTERIA DEFINED BY DISSOLVED OXYGEN LEVEL

Bacteria can be grouped according to five different oxygen levels/conditions shown in Table 3.1. As the table indicates, different oxygen environments

TABLE 3.1 Oxygen Classifications of Microorganisms

Group	Oxygen condition	Metabolism	Example
Aerobic	Required	Aerobic respiration	Nitrifiers, most filaments
Microaerophilic	Low level	Aerobic respiration	Spirillum (stored effluent samples)
Facultative	Not required, but more efficient	Aerobic, anaerobic respiration, or fermentation	Most activated sludge bacteria
Aerotolerant	Not required, oxygen does not help	Fermentation	Streptococcus pyogenes (skin pimples)
Obligate anaerobic	Harmful, toxic	Fermentation or anaerobic respiration	Methanobacterium

can be used to foster the growth of different microorganism populations. Obligate anaerobes are rarely seen in activated sludge because it is primarily an aerobic process. The few that might grow, say in an anaerobic selector, would die or be inhibited upon entering the aeration tank. Similarly, aerotolerant anaerobes can stay alive but cannot function well in an aerobic environment. Microaerophilic organisms might grow at the head of an anaerobic or facultative selector, where just a tiny amount of oxygen is present, but they could not compete elsewhere in the system. Aerobic microorganisms can proliferate in the aeration tank, or wherever dissolved oxygen is present. Also growing in the aeration tank are facultative bacteria, which can grow under anaerobic or anoxic (anaerobic, but with nitrate available) conditions. This means that aerobic and facultative bacteria will be the primary organisms in an activated sludge system.

Typically, oxygen is supplied so that aerobic and facultative bacteria can use it to collect electrons released during metabolism. In anoxic selectors, instead of using oxygen as the electron acceptor, facultative bacteria use nitrate for electron collection, whereas, in anaerobic selectors, fermenting facultative bacteria produce their own electron acceptor, an organic compound (ethanol or low-molecular-weight organic acids).

Previously, *selectors* were defined as small tanks ahead of the aeration tank to provide a high food-to-microorganism ratio (F:M) environment. That small tank can be designed to be anaerobic, in which facultative bacteria grow, but in which most filaments, which are aerobic, cannot grow. Thus, facultative bacteria get the food while in the selector, whereas aerobic filaments are starved. One can also use an aerobic selector, but aerobic is not as selective as anaerobic because aerobic filaments can still grow if the BOD concentration is too low. Last is the anoxic selector, which is an anaerobic tank but with a source of nitrate for the electron acceptor. This selector allows denitrification, discussed in Chapters 7 and 8, where facultative bacteria feed on CBOD and use nitrate to accept electrons released during BOD_5 metabolism. For instance, an anoxic selector can be used in a nitrifying activated sludge system to denitrify. Anoxic selectors are not nearly as good at filament control as the anaerobic selector.

Over the last 20 to 30 years, activated sludge process design was developed to take advantage of three different oxygen-growth environments (aerobic, anoxic and anaerobic), spurred on primarily by the growing need to remove nutrients from the wastewater stream. Each environment has traditionally been maintained to promote the growth of microorganism populations that perform specific biochemical reactions within those environments to obtain desired treatment results. Until recently, activated sludge bioreactors were understood to affect those specific biochemical reactions within a strictly prescribed oxygen environment meant to sustain each reaction. In more recent years, the strict definition of what environment affects what biochemical reaction has had to be modified to reflect a growing understanding of the dynamics of microorganism populations and their ability to shift under finer gradations of the oxygen environment. Much of the advance in our understanding has come as a result of activated sludge operation venturing into lower and lower dissolved oxygen ranges and achieving unexpected results. Investigation of observed overlaps of the traditional biochemical reactions, previously thought to be restricted to well-defined dissolved oxygen ranges, uncovered factors such as the role of microenvironments within activated sludge floc and previously unrecognized characteristics of certain bacteria under low dissolved oxygen conditions. Two examples of bacteria with previously unrecognized characteristics are *comammox* bacteria and *anammox* bacteria, both of which have significant implications for activated sludge design and operation.

Recently discovered, *comammox* bacteria can perform complete ammonia oxidation (NH_4 to NO_3) as a single organism. Commamox bacteria compress the well known two-step nitrification process by oxidizing intermediate nitrite directly to nitrate, with the obvious implication of aeration savings. Comammox have been found to be part of the general bacterial population in activated sludge systems, coexisting alongside traditional nitrifiers. Comammox, which have a high affinity for ammonia and oxygen but a slower growth rate than traditional nitrifiers, have been found by some researchers to be the dominant nitrifiers in systems with long SRTs, limited dissolved oxygen, and limited nitrite conditions (Mehrani, 2021). The predominant comammox representative found in wastewater treatment systems is *Nitrospira*.

Anammox bacteria, first identified in the 1990s, can perform ammonia oxidation under anaerobic conditions without the need for oxygen or organic carbon, which implies savings in both aeration and external carbon addition. Anammox bacteria belong to the phylum Planctomycetes. Anammox bacteria oxidize ammonia using nitrite as the electron acceptor (Ma, 2016). The nitrite used by anammox can be obtained from partial *nitritation*, which is the oxidation of ammonia to nitrite, or partial *denitrification*, denitrification of nitrate to nitrite. In partial nitritation, anammox bacteria essentially "shunt" the traditional nitrification/denitrification process by converting ammonium and nitrite directly to nitrogen gas without the need to complete the traditional nitrification process to nitrate. Unlike partial nitritation, however, partial denitrification does require an external carbon source. Partial denitrification anammox is an active and promising avenue for development of novel activated sludge processes.

For further details regarding growth kinetics and suboxic operation, see Chapter 8.

5.0 FILAMENTOUS MICROORGANISMS

Filaments either grow inside the floc to produce lightweight, diffuse flocs, on the surface of the flocs, or bridge between the flocs, holding them apart. In any case, settleability is slowed and control of the process may be lost until filament growth is controlled. Filaments make use of their long, narrow form to be able to contact low concentrations of nutrients, whereas floc formers prefer higher concentrations of nutrients. Think of filaments as snacking on potato chips and soft drinks for many hours, while floc formers tend to prefer a big meal of "steak and potatoes" and then want to be left alone for hours while they digest the last feeding. To control nuisance filaments, it is important to identify the primary filaments involved. Once identified, a control strategy can be developed based on the cause(s) of the growth.

Approximately 27 basic filament groups are common (Jenkins et al., 2004), and they grow under specific environmental conditions (Table 3.2), including low dissolved oxygen, low F:M (high MCRT), septicity (low-molecular-weight organic acids and/or hydrogen sulfide), deficiency of available nitrogen (ammonia) or phosphorus (orthophosphate), or low pH (fungus starts to grow between 6.0 and 6.5 pH but really grows quickly below 6.0 pH).

TABLE 3.2 Conditions Favorable to Filamentous Microorganisms in Activated Sludge and Observed Effects (Adapted from Jenkins et al., 2004)

Filament type	Low dissolved oxygen	Low F:M	High SRT	Fats, oils, and greases	Reduced sulfur	Deficiencies		Observed effects	
						Nitrogen	Phosphorus	Bulking	Foaming
Beggiatoa	X				X			X	
Halisomenobacter Hydrossis	X						X	X	
Microthrix parvicella	X	X	X	X	X			X	X
Nostocoida limicola				X			X	X	X
Nocardia		X	X	X					X
S. natans	X						X	X	
Thiothrix	X			X	X	X		X	
Type 0041		X						X	
Type 0092				X				X	
Type 021N				X	X	X		X	
Type 0914				X	X			X	
Type 1701	X							X	
Type 1863		X		X					X

5.1 Low Dissolved Oxygen

Low dissolved oxygen in the feeding zone allows low dissolved oxygen filaments to grow. Historically, it was thought that any dissolved oxygen below approximately 1.0 to 1.5 mg/L would cause low dissolved oxygen filament growth; however, it has been found that, often, dissolved oxygen values below approximately 0.5 mg/L can reduce low dissolved oxygen filament growth. So, the term *low dissolved oxygen* should possibly be redefined.

5.2 Low Food-to-Microorganism Ratio

Low F:M (high MCRT) means that the filament can probably find low amounts of food available in such environments; thus, they grow faster at low BOD concentrations. Figure 3.5 shows that filaments grow faster than floc formers up to some value, but, above that value, floc formers grow faster. Thus, many filaments outcompete floc formers at low F:M. Low F:M filaments may also be able to feed on slowly metabolized materials only broken down at higher MCRTs.

5.3 Low Nutrients—Phosphorus or Nitrogen

Phosphorus is involved in energy reactions within the cell, so low phosphorus concentration may limit BOD metabolism, causing slime development

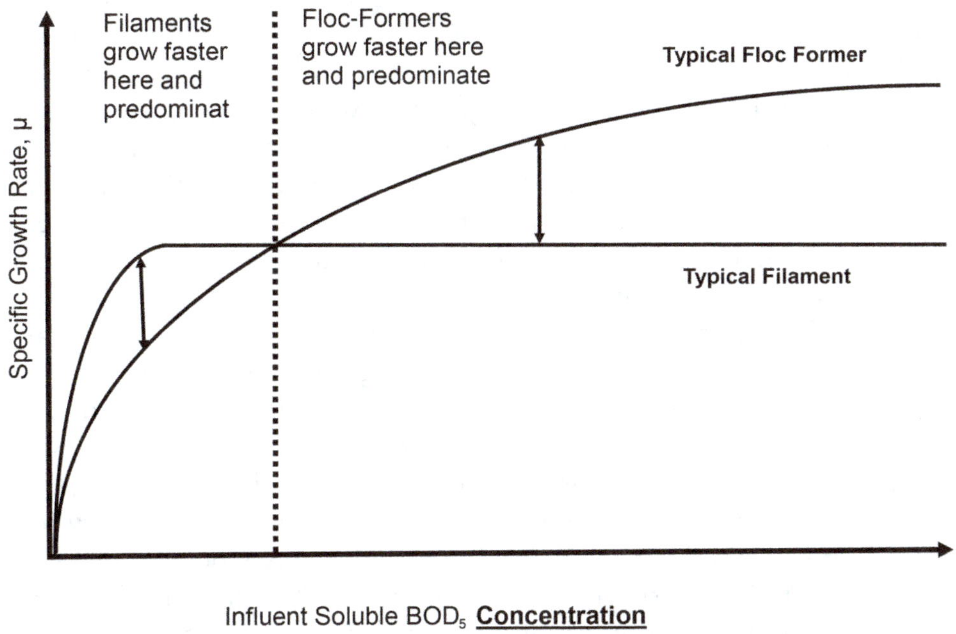

FIGURE 3.5 Kinetic Selection Theory

or certain filaments. Phosphorus must be in the orthophosphate form to be used by bacteria. Nitrogen is a building block for all proteins and enzymes; therefore, low available nitrogen (NH_3) limits enzyme production, resulting in filament growth or slime development.

Organic acids and/or sulfide produced from septic environments can be used by some aerobic filaments in an aerobic environment. These filaments do not produce the septicity; rather, they feed on the materials resulting from septicity. For instance, septicity could come from long sewer lines, pumping stations, primary clarifiers, and even secondary clarifiers with too-low sludge return rates.

5.4 Filament Control

Do not make filament control more difficult than it really is. Remember that a filament is there because it likes the environment; therefore, to control it, change the environment to one that the filament does not like (e.g., adjust dissolved oxygen up or down, remove environment-producing organic acids and sulfide, reduce MCRT [increase F:M, selector], increase pH, and so on).

There are also some modifiers to the list presented in Table 3.2. The first is completely mixed, uniformly aerated, and continuously fed aeration basins. Many filamentous blooms can be stopped if feeding can be started and stopped (sequencing batch reactor); complete-mix converted to long, narrow tanks (baffles); higher dissolved oxygen provided at the head of the tank and low dissolved oxygen at the tail of the aeration tank; or a high F:M feed zone (selector) provided ahead of the aeration tank. Selectors are small tanks ahead of the aeration tank that provide a high F:M that floc formers prefer. The second is pH. Fungi can grow well below approximately 6.5 pH, whereas other organisms may be inhibited. The third is temperature. For example, *Microthrix parvicella* tend to prefer lower aeration tank temperatures, whereas nocardioforms prefer higher temperatures; therefore, it is common to find more *M. parvicella* growth in the winter and spring and nocardioforms in the summer and fall.

Anaerobic selectors, which are small, unaerated tanks ahead of the aeration tanks, tend to work well at controlling non-low-F:M filaments. It appears that facultative bacteria quickly degrade soluble BOD to organic acids, which are fed upon by other organisms (e.g., phosphorus-accumulating organisms) so that little food is available for filaments in the aeration tank.

Some filaments, such as Type 1863, nocardioforms, and *M. parvicella* can produce foam. Of these, only *M. parvicella* causes poor settleability. While Type 1863 is a short, high F:M (low MCRT) filament that floats with aeration, nocardioforms and *M. parvicella* incorporate oleic acid in their cell walls and are hydrophobic, causing them to float.

Identification of filaments is important to determine the required control approach and is based on a number of different cell, filament, and differential stain characteristics. Typical identification keys can be found in *Wastewater Biology: The Microlife* (Water Environment Federation, 2017) and *Manual on the Causes and Control of Activated Sludge Bulking, Foaming, and Other Solids Separation Problems* (Jenkins et al., 2004). These keys depend heavily on staining; however, another key (Figures 3.6a, 3.6b, 3.6c, and 3.6d) based on the Stover Group's *Fil-Indentpro* software is less dependent on staining. In fact, all of the organisms shown in Figures 3.6a, 3.6b, and 3.6c can be identified without staining, but the use of Figure 3.6d requires staining.

Physical observations using phase contrast at 10× give protozoa and *Metazoa* diversity and overall floc and filament characteristics, including filament length and shape, rosettes, and branching. To see the following details, 400× or 1000× oil immersion is required: false/true branching, motility, sheath, cross-walls present, filament and cell widths, cell shape and size, deposits, and gonidia.

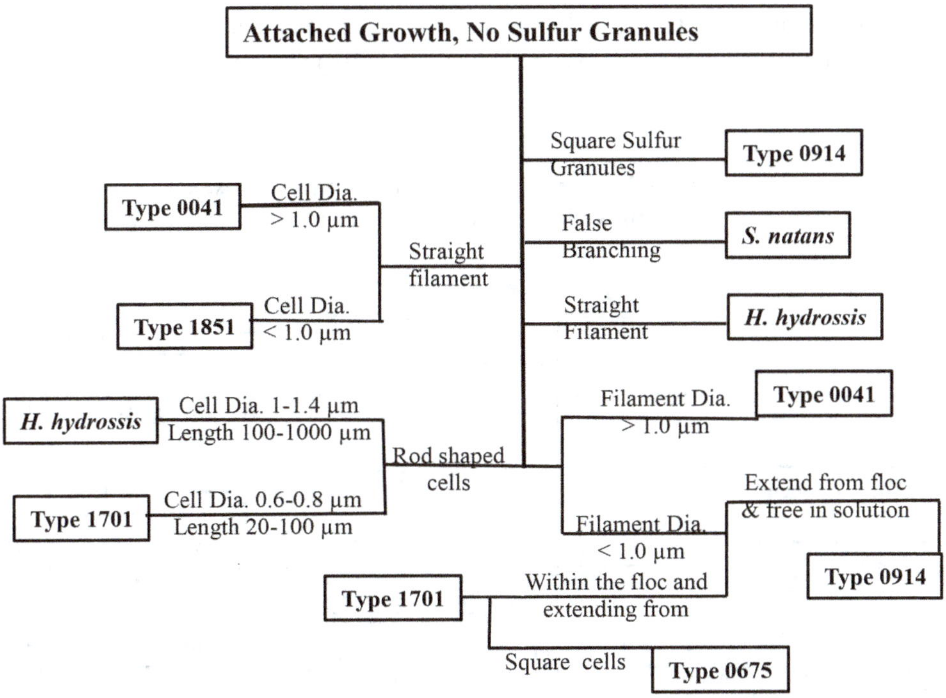

FIGURE 3.6a Filament Identification—Attached Growth, No Sulfur Granules (Developed from The Stover Group, 1999–2002, FIL-IDENTPRO (R) version 1.5.70.)

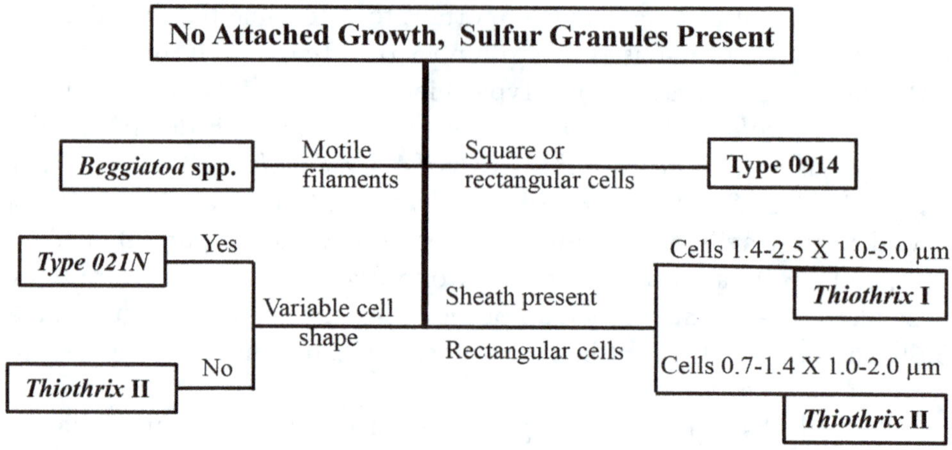

FIGURE 3.6b Filament Identification—No Attached Growth, Sulfur Granules Present (Developed from The Stover Group, 1999–2002, FIL-IDENTPRO (R) version 1.5.70.)

Differential stains include Gram and Neisser stains. The latter is fairly easy, but Gram staining is difficult and takes significant practice. A bad Gram stain can suggest the wrong organism, which, in turn, can suggest the wrong solution. To make better Gram stains, it is suggested that ethanol be used for decolorization instead of the decolorizer supplied with a Gram stain kit.

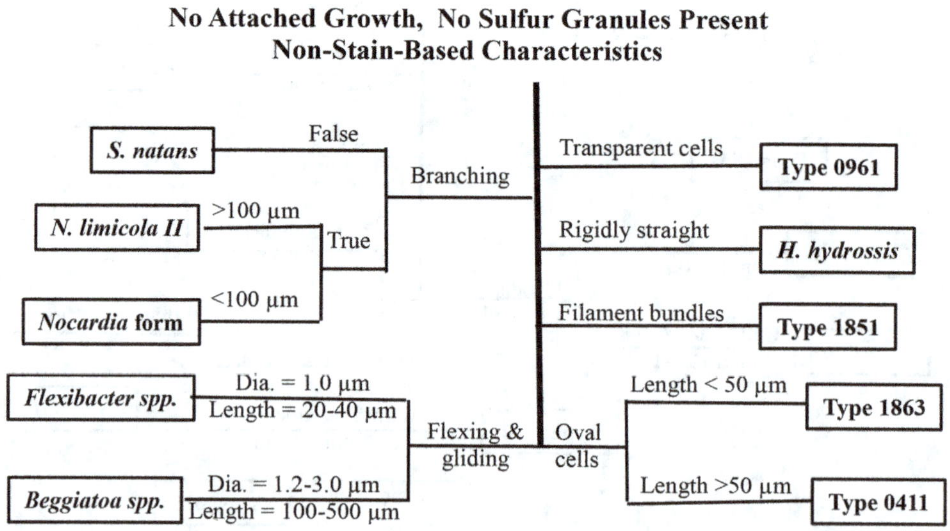

FIGURE 3.6c Filament Identification—No Attached Growth, No Sulfur Granules, Non-Stain-Based Characteristics (Developed from The Stover Group, 1999–2002, FIL-IDENTPRO (R) version 1.5.70.)

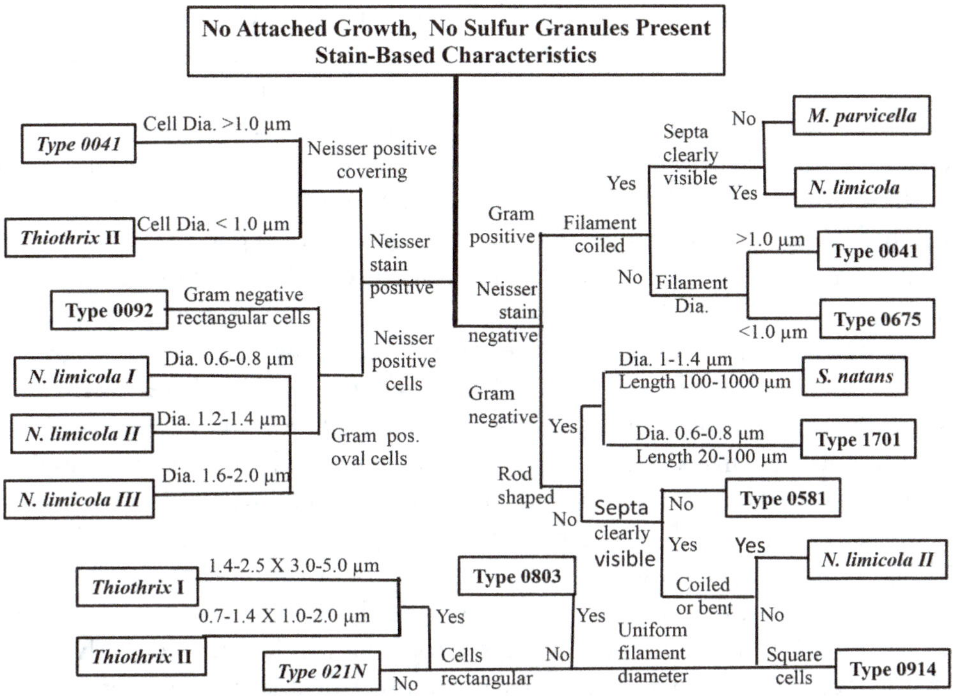

FIGURE 3.6d Filament Identification—No Attached Growth, No Sulfur Granules, Stain-Based Characteristics (Developed from The Stover Group, 1999–2002, FIL-IDENTPRO (R) version 1.5.70.)

With physical observation and differential stain data, Figures 3.6a to 3.6d can be used to identify the filament. Once the filament is correctly identified, Table 3.2 can be used to determine the potential cause, and appropriate wasting, dissolved oxygen, or sludge return rate changes can be initiated.

6.0 PROTOZOA AND *METAZOA*

6.1 Protozoa

Protozoa are a group of primarily single-celled organisms, which may or may not have a test (shell), and can be either sedentary or motile. Locomotion is provided by pseudopodia (in *Amoeba*), flagella (in flagellates), or cilia (in ciliates). They were first identified as "little animals" by Antonie van Leeuwenhoek in the 17th century. Their primary function in wastewater is the removal of nonflocculated bacteria and small floc particles. There are some that may use some of the organics in the water as well as other higher lifeforms. There have been numerous hypotheses that protozoa could devour biomass and reduce overall MLSS; however, based on the growth rates of most protozoa and bacteria, this could never happen. For the purpose of identifying indicator organisms in wastewater, protozoa are broken down into the following groups: flagellates, *Amoeba* (shelled and naked), free-swimming ciliates, stalk ciliates, crawler ciliates, carnivore ciliates, and suctoria. Most of these organisms can be viewed easily with a phase-contrast microscope at 100× total magnification. Viewing using a brightfield microscope may show the organisms, but the level of detail and organism-identifying keys may be difficult to see. This section will provide a basic description of the organisms and their importance. Because there are so many different protozoa in the environment, it may be helpful to gather more resources, such as the listed references at the end of this section. There are numerous publications, posters, and Internet resources that may help with identification as well (Figure 3.7). When the identification of higher lifeforms is combined with floc interpretation and bulk water characteristics, it is easier to understand the health of the biomass. Floc size and density, along with filament presence, will have a direct effect on settling, and bulk water directly influences the clarity of the discharge water. All of these characteristics are important when performing a microscopic analysis on the wastewater biomass. The following subsections describe the organisms that are most recognized and their meaning to the analysis.

6.1.1 Flagellates

Flagellates are some of the most active organisms found in wastewater. They can vary in size from 1.5 to 25 μm. They all possess at least one flagellum, and many have more than one. The flagellum is used for both movement and feeding (Figure 3.8a). Some species may reside within a small test (shell) and only expose the flagellum to capture bacteria (Figure 3.8b). They use a lot of energy to feed and require a considerable amount of bacteria, which is primarily found in bulk water. This is why we see more in a young sludge or after a recent upset. If viewed in high numbers compared to previous exams, chances are a recent shock has affected the biomass. The smallest flagellates (Figure 3.8c) are the most active and most indicative of recent upsets to a system.

6.1.2 Amoeba (Testate and Naked)

Amoeba are unique organisms, especially in terms of movement. They use pseudopodia to help propel them through the water. There are two primary types of locomotion: two-way pulling, thin pseudopodia and tubular, contractile pseudopodia (Figure 3.9a). There are variations within these types. *Amoeba* typically feed on organics, free bacteria, and other protozoa and *Metazoa* (Figure 3.9b shows a rotifer engulfed by a large *Amoeba*). They

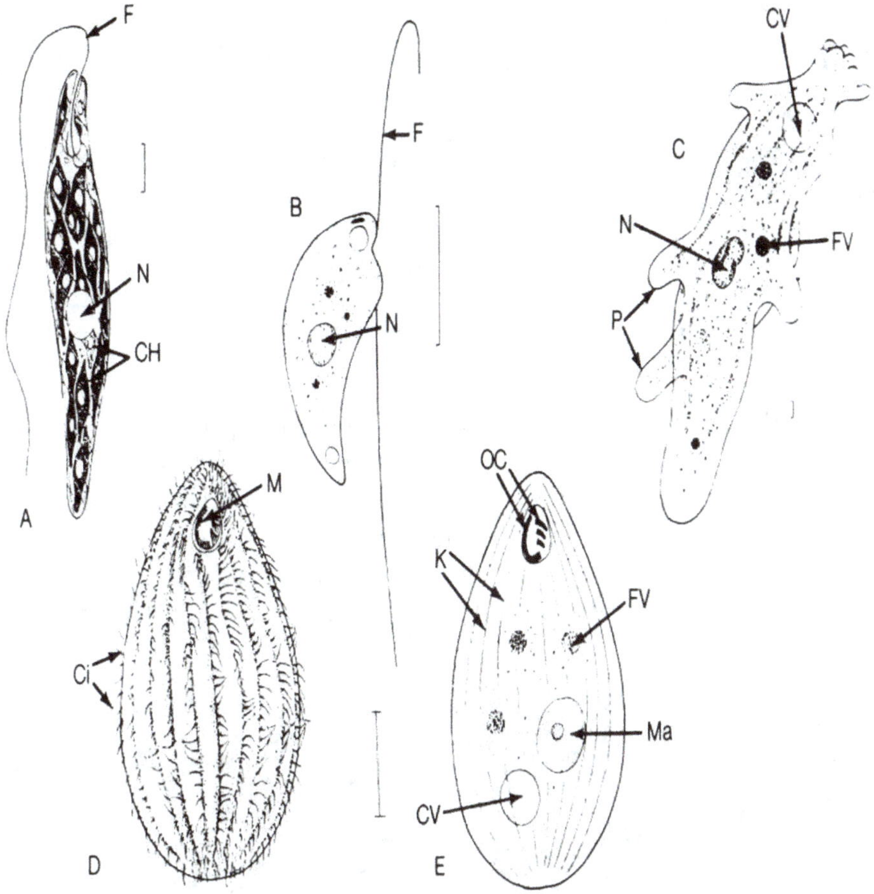

FIGURE 3.7 Representatives of the Primary Protozoan Groups Found in Wastewater Treatment Processes (scale bars = 10 µm): (a) *Euglena gracilis*, (b) *Bodo caudatus*, (c) *Amoeba proteus*, (d) *Tetrahymena pyriformis ciliation*, and (e) *Tetrahymena pyriformis*. Ci: cilia, CH: chloroplasts, CV: contractile vacuole, F: flagellum, FV: food vacuole, K: kineties, M: mouth, Ma: macronucleus, P: pseudopodium, and OC: oral cilia (Water Environment Federation, 2017)

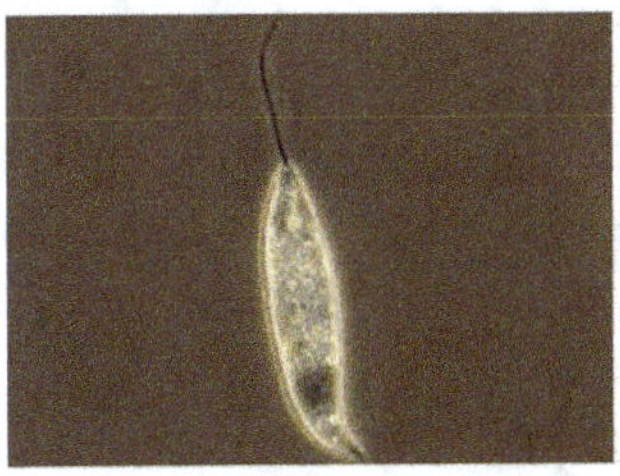

(a) Flagellate

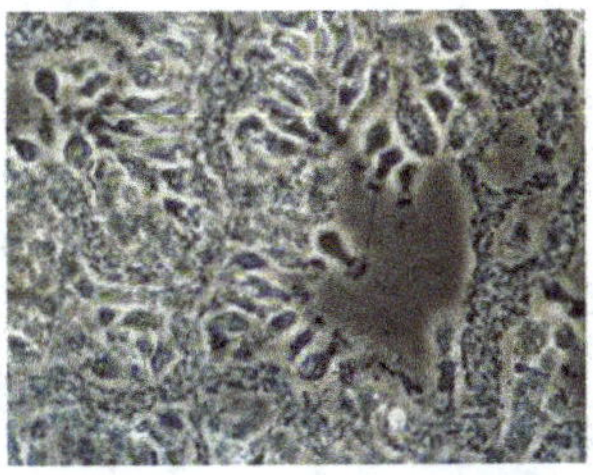

(b) Flagellates w/in tests

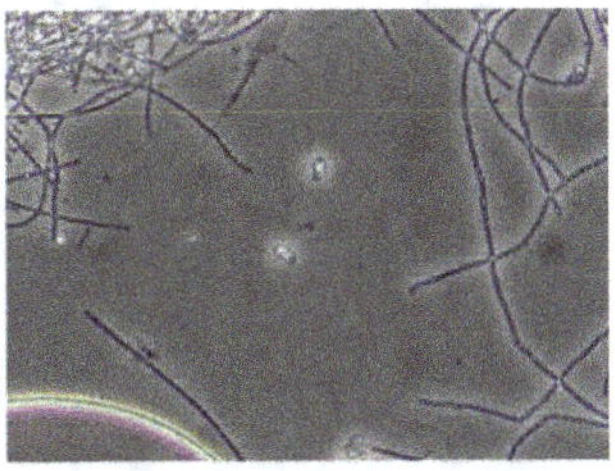

(c) 2 small flagellates

FIGURE 3.8 Flagellated Protozoa (Reprinted with permission from Steven C. Leach)

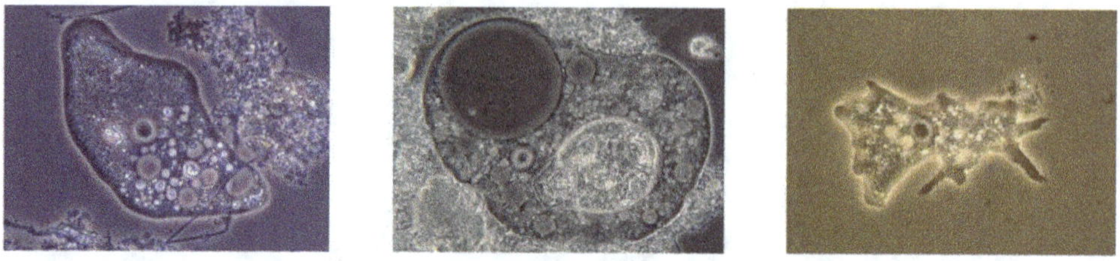

(a) Naked amoeba (b) Naked amoeba with rotifer inside (c) Naked amoeba

FIGURE 3.9 Naked *Amoeba* (Reprinted with permission from Steven C. Leach)

are not deliberate feeders, such as some of the other protozoa, but may be more opportunistic. Typically, more *Amoeba* are seen in younger sludge or as a reaction to a recent shock, when bulk water is loaded with free motile bacteria (Figure 3.9c). Sizes will range from 2 to 3 to 5000 μm. Testate (shelled) *Amoeba* (Figure 3.10a) have been observed in various systems operating in a steady-state and non-steady-state condition (Figure 3.10b).

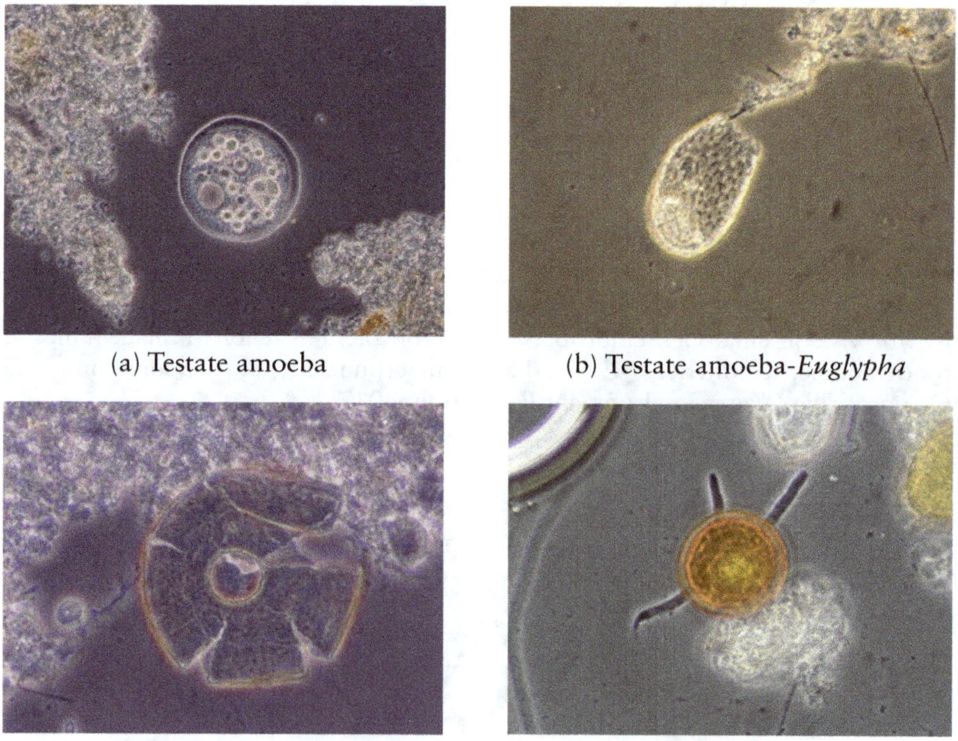

(a) Testate amoeba (b) Testate amoeba-*Euglypha*

(c) Crushed test (d) Testate amoeba with indeterminate pseudopodia

FIGURE 3.10 Testate *Amoeba* (Reprinted with permission from Steven C. Leach)

The presence of a crushed shell (Figure 3.10c) may be the result of the cover slip and not a toxic shock. Typically, toxins will not break the test but will harm the *Amoeba* itself. They are not necessarily the best indicator organisms; however, the color of the test will suggest how long they have been in the system. If the test is clear, they are relatively new to the system; however, if the test is copper-colored (Figure 3.10d), they have been in the system much longer. If they have been in the system longer, chances are there has been no change to growth conditions of the biomass.

6.1.3 Free-Swimming Ciliates

Ciliates are classified based on how their rows of cilia (tiny hairs) (Figure 3.11a) on the body are arranged. They use these cilia for movement and for gathering food (bacteria). Cilia cover the entire body and, in some organisms, surround the oral region. All ciliates will contain a water expulsion vesicle (WEV) (Figure 3.11b) that releases the water either absorbed through the cell wall or ingested with food. This is recognized easily by observing the opening

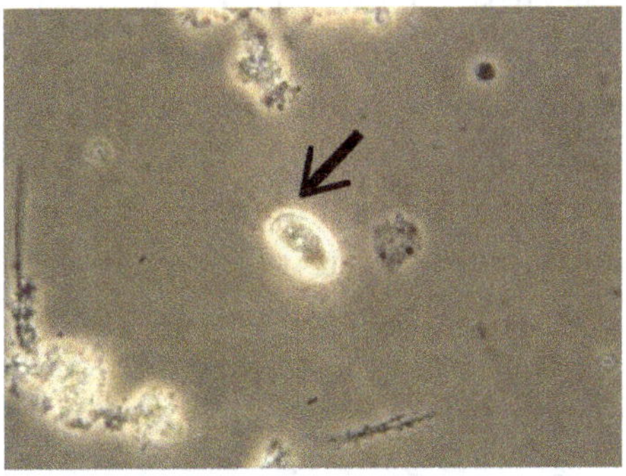

(a) Arrow pointing to cilia state amoeba

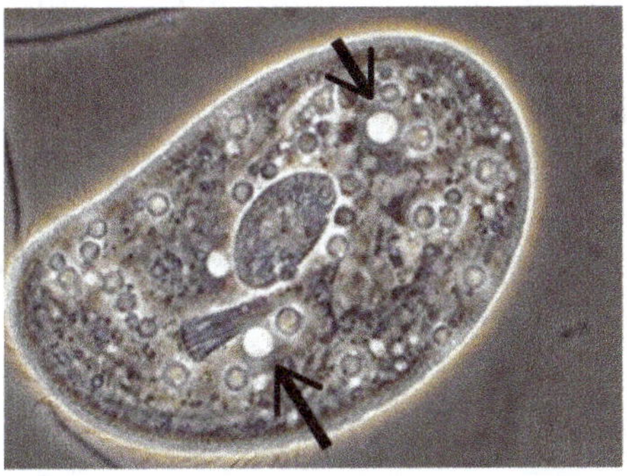

(b) Arrows pointing to WEVs

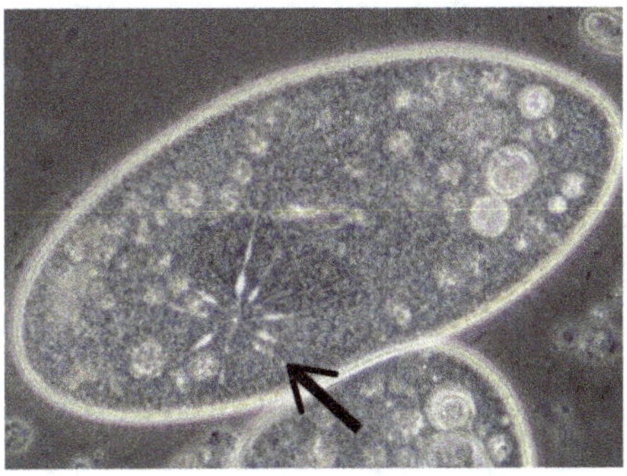

(c) Paramecium with WEV

FIGURE 3.11 Free-Swimming Ciliates (Reprinted with permission from Steven C. Leach)

and closing of the expulsion vacuole. Certain *Paramecium* sp. will have a radiated canal feeding the WEV (Figure 3.11c). This makes identifying this particular organism a little easier. The majority of ciliates will divide from 1 to 8 times per day, depending on a number of environmental conditions. Free-swimming ciliates are active and constantly feeding. The majority will feed on free bacteria found in the bulk water, indicating the presence of nonflocculated bacteria. Typically, this would indicate a nonstable system; however, free-swimming ciliates have been found in stable systems as well. Observing these organisms regularly in biomass may also be a result of infiltration. Therefore, documenting their presence during optimum performance and poor operations will help the operator recognize changes to the biomass.

6.1.4 Attached (Stalked) Ciliates

Stalked ciliates are the largest order of ciliates, and *Vorticella* is the most familiar. They include solitary (*Vorticella* sp., Figure 3.12a) and colonial forms (*Opercularia* sp., Figure 3.12b). Some stalked ciliates may contract along their stalk and others may not. There are a few that even reside within lorica (shell) (*Cothurnia* sp., Figure 3.12c). These ciliates are primarily bacteria feeders. Cilia that circle the oral region create a current that helps bring bacteria into the buccal (mouth) cavity. These organisms will attach to floc, inert material, or anything that may provide a base. There are no body cilia unless the organism is experiencing division (Figure 3.12d) or a change in location is needed as a reaction to a change in the environment. They will then develop a trochal band, which is a ring of cilia toward the base of the ciliate that is used to move to an area with a more satisfactory growth environment. This then becomes a telotroch (Figure 3.12e).

6.1.5 Crawler Ciliates

Crawler ciliates are easier to identify because of movement and cirri (i.e., fused-together cilia) (Figure 3.13a). Cirri resemble legs as they move along the floc and are easier to see than cilia. These protists will crawl around the floc (Figure 3.13b) and feed similar to cattle. They possess no cilia on their body, except near their oral region, and do not swim well (Figure 3.13c), as compared to free-swimming ciliates. Because these organisms primarily feed crawling on floc, it makes sense that an ideal floc is present. Crawler ciliates are typically found in a stable biomass.

6.1.6 Suctoria

Suctoria are unique protozoa in that they may have cilia in their younger stage or as free-swimming larvae. Once they have matured, they have

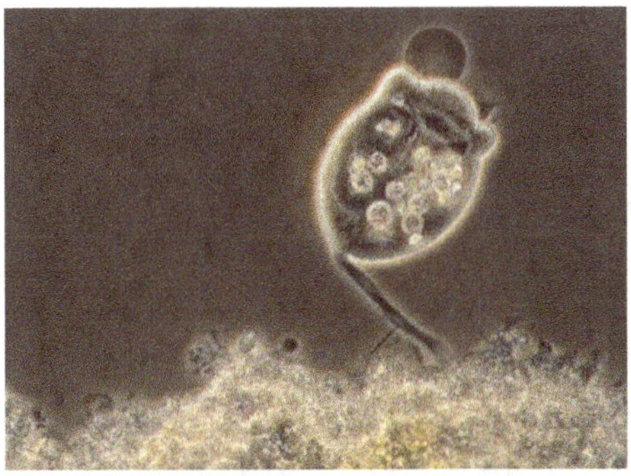

(a) *Vorticella*

(b) *Opercularia*

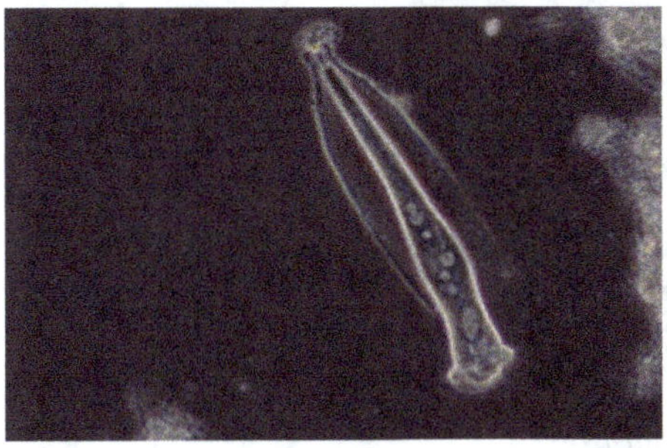

(c) *Cothurnia* sp.

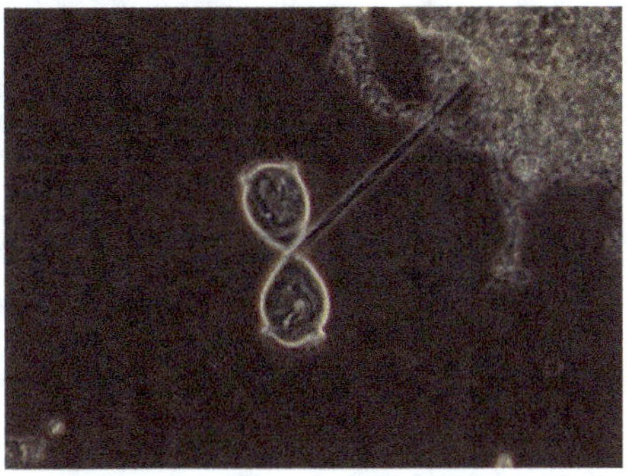

(d) Daughter cell on top

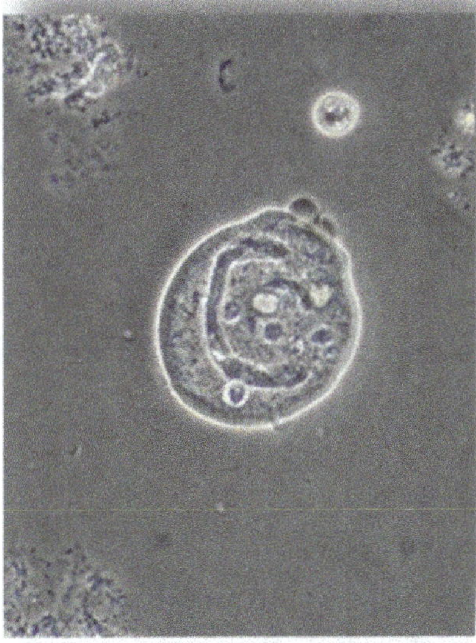

(e) Telotroch (detached ciliate from stalk)

FIGURE 3.12 Stalk Ciliates (Reprinted with permission from Steven C. Leach)

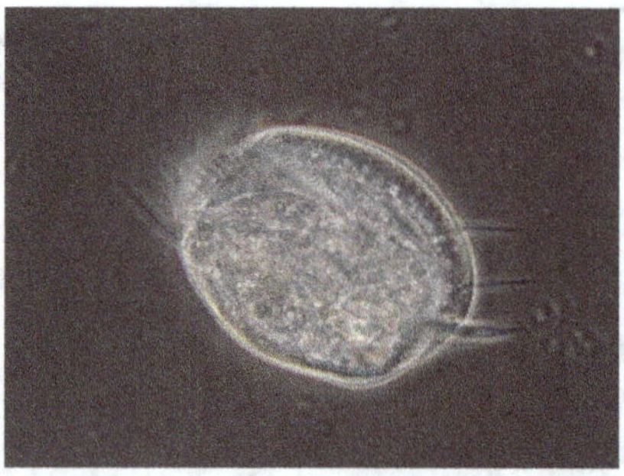

(a) *Euplotes sp.*

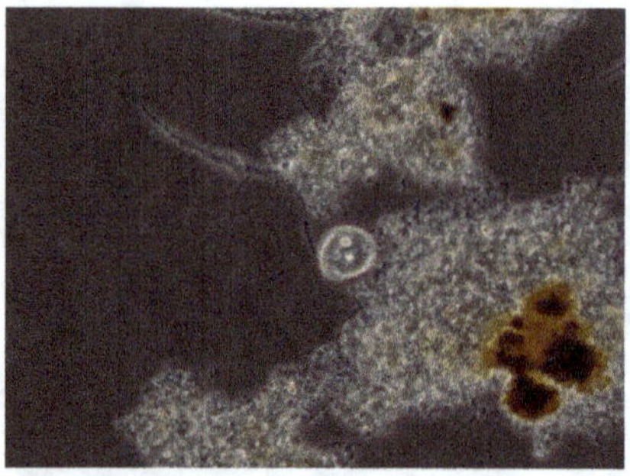

(b) Crawler ciliate on floc

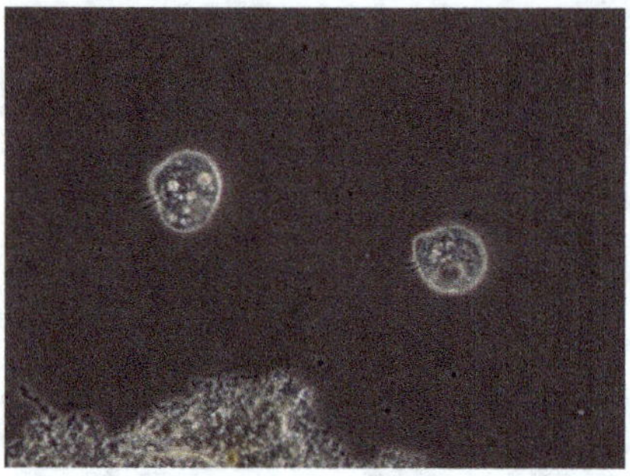

(c) *Crawler ciliates*

FIGURE 3.13 Crawler Ciliates (Reprinted with permission from Steven C. Leach)

tentacles and lose the cilia. The tentacles (Figure 3.14a) may move in and out from the main body as the organism waits for its prey. These tentacles can both pierce the cell wall of its prey and keep it held while another tentacle will penetrate the captured organism and extract the cytoplasm. Some species will attach through a stalk (Figure 3.14b), while others may free-float (Figure 3.14c) through the water. Typically, these organisms have been viewed in steady operating systems that nitrify and, primarily, do not like higher ammonia concentrations. Suctoria are commonly found in a stable biomass.

6.1.7 Carnivorous Ciliates

There are some ciliates that would be considered predaceous and use toxicysts (small venom-containing capsules) to anesthetize their prey to make it easier to swallow. Some species will even produce pexicysts, which are like tiny harpoons, to attach themselves to their prey. Many of these organisms may be more difficult to identify without further experience, but a few are easily recognizable (Figures 3.15a, 3.15b, and 3.15c). The presence of these types of ciliates is indicative of a stable system. Based on growth rates, these organisms will never be able to feed on all the prey in the biomass.

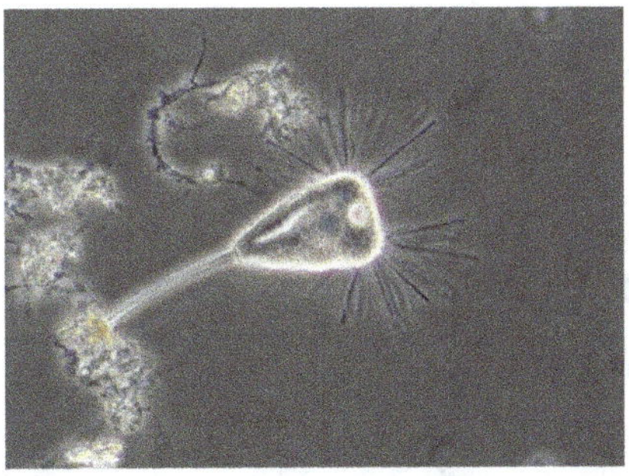

(a) *Tokophyra sp.*

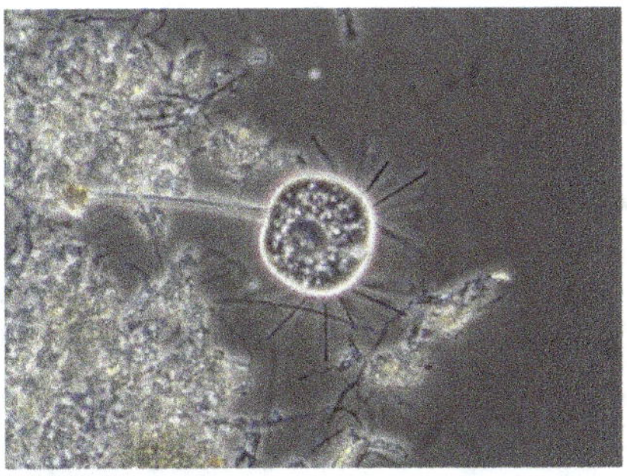

(b) *Podophyra sp.*

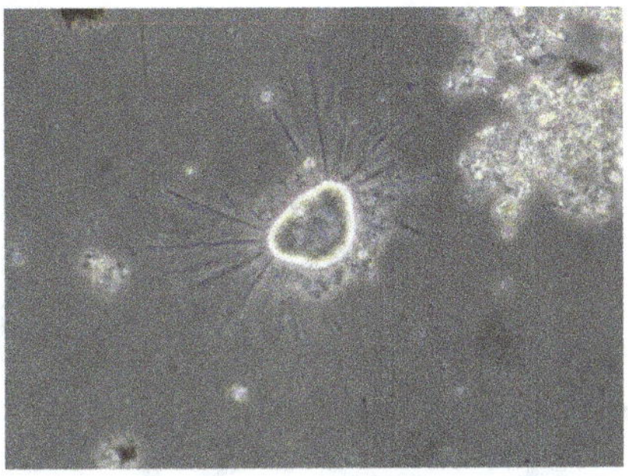

(c) Unidentified suctoria

FIGURE 3.14 Suctoria (Reprinted with permission from Steven C. Leach)

6.1.8 Protozoa Reproduction

Reproduction in protozoa can occur in a number of ways. The most common is binary fission, during which the parent cell divides equally into two new daughter cells. A second type of reproduction is multiple fission, during which the nucleus and cytoplasm divide into numerous smaller cells. A third method is external budding, which occurs when the parent cell produces smaller cells that contain a micronucleus and small branch of macronucleus. The parent cell remains intact while the smaller cells move off to mature and develop. The final method of reproduction in protozoa is internal budding. This occurs when the bud is formed within the cytoplasm and is released through the birth pore or via the parent cell rupturing. The frequency of reproduction is controlled by environmental conditions such as the presence of oxygen, temperature, and pH.

6.1.9 Cell Size

Protozoa found in activated sludge systems exist in a rather wide size range, from as small as a few microns to as large as several millimeters. The following examples (from Jahn et al., 1979) are illustrative of the protozoa

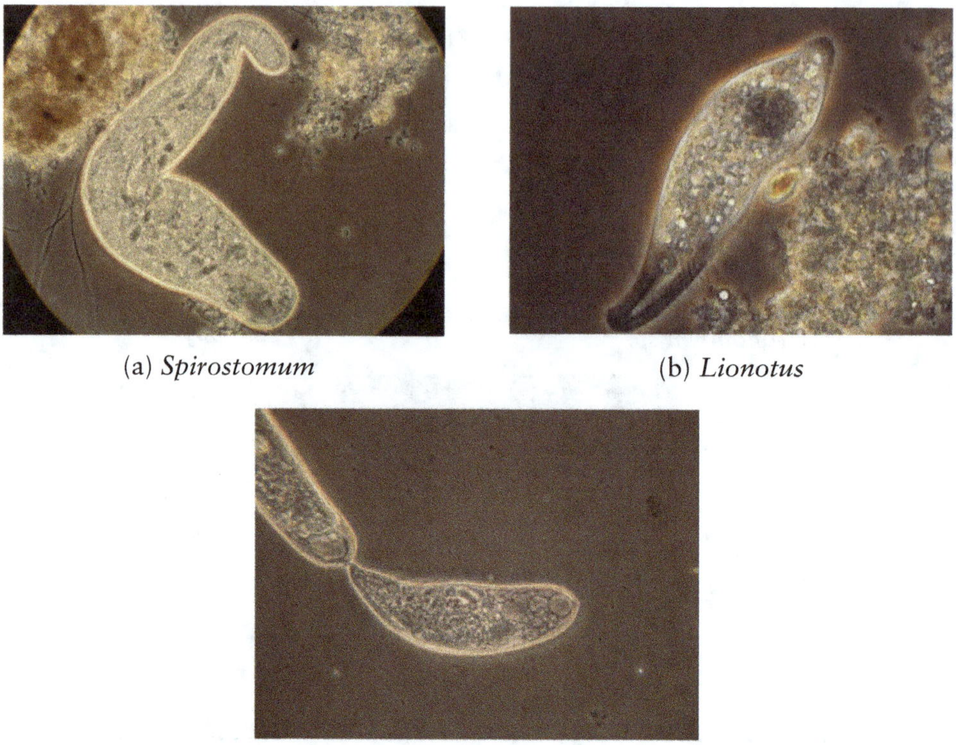

(a) *Spirostomum* (b) *Lionotus*

(c) *Blepharisma*

FIGURE 3.15 Carnivore Ciliates (Reprinted with permission from Steven C. Leach)

size ranges observed in activated sludge systems. The smallest size ranges include the flagellates, such as *Bodo,* which range in size from 5 to 20 μm. Ciliates, such as *Paramecium*, can range from 100 to 350 μmin size. Toward the higher end of the scale, long ciliates such as *Spirostomum* can range in size from 1000 to 3000 μm. The widest range in size for a specific group is that of the *Amoeba*, with includes small *Amoeba* such as *Vahlkampfia* ranging in size from 30 to 50 μm, and larger *Amoeba* such as *Chaos* which can range in size from 1000 to 5000 μm, large enough to be seen with the unaided eye.

6.2 Metazoa

Metazoa are a group of multicellular animals that may or may not have a digestive tract and central nervous system. Typical *Metazoa* found in wastewater treatment systems include but are not limited to rotifers, gastrotrichs, bristle worms, and tardigrades. Typically, these organisms are larger than protozoa and feed heterotrophically, meaning that they use organic material as food. Because these organisms reproduce slowly, they have been associated with longer sludge ages.

6.2.1 Rotifers

Rotifers are commonly found in wastewater treatment systems with longer sludge ages but have been viewed in industrial facilities with SRTs less than 5 days (Figure 3.16). There are approximately 2500 different species currently identified, and more are found every year. Rotifers are divided into two classes based on the number of gonads. Monogononta have one gonad and digononta possess two gonads. These animals move either by crawling or swimming and feed primarily on bacteria, detritus, and some protozoa. Their mastax (jaws), which are used to grind or crush their food, have often been confused for a beating heart. Past research has determined that these organisms help promote oxygen transfer within the floc by "snipping off the top layer." Rotifers can produce young sexually or asexually and will have live young or eggs.

6.2.2 Water Bears, Bristle Worms, Nematodes, and Gastrotrichs

These organisms are rarely seen in conventional activated sludge systems and are more commonly found in fixed-film or attached-growth systems. If these organisms are found in activated sludge, they are indicative of an older sludge and may be cause for further investigation if system performance is an issue.

6.2.3 Water Bear (Tardigrade)

Water bears are found in many wastewater systems (although not commonly) and have been associated with stable biomass. These animals possess eight legs (Figure 3.17a) with small claws on each. They primarily feed on plant material; however, there are a few species that are carnivorous and feed on rotifers, nematodes, and protozoa. The majority of water bears are

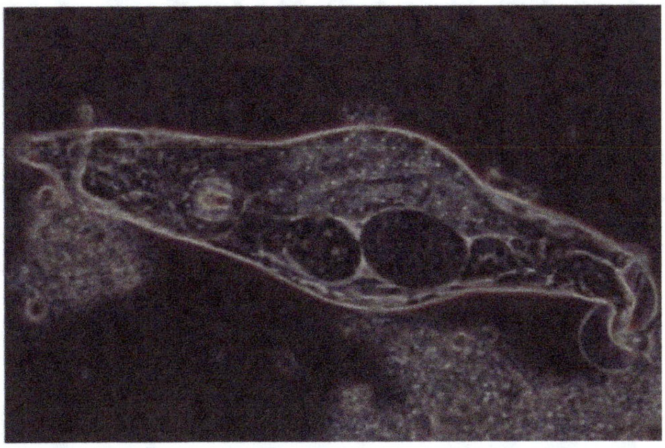

FIGURE 3.16 Rotifer

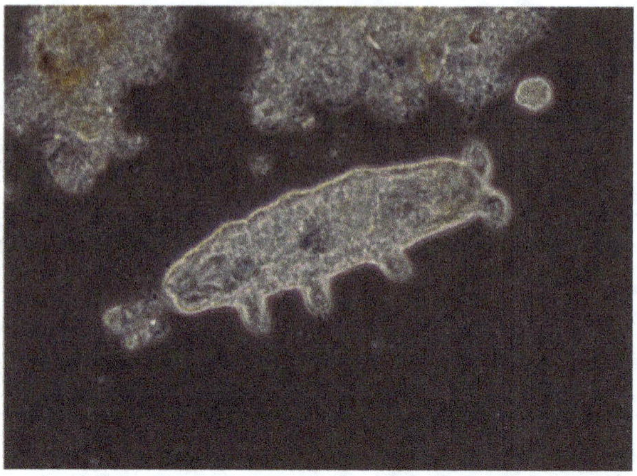

FIGURE 3.17a Water Bear (Reprinted with permission from Steven C. Leach)

FIGURE 3.17b Large Pregnant Water Bear (Reprinted with permission from Steven C. Leach)

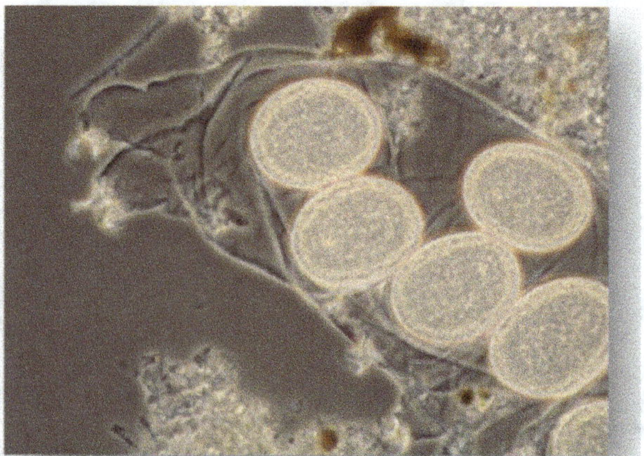

FIGURE 3.17c Eggs in Cuticle (Reprinted with permission from Steven C. Leach)

female and can produce between 2 and 30 eggs (Figure 3.17b). They deposit the eggs into the old cuticle (shell) (Figure 3.17c) and may emerge in 3 to 14 days. The young emerge much smaller and will grow in size as they mature. There have been recent papers and research showing this organism's ability to handle desiccation and emerge after extended periods of time. The National Aeronautics and Space Administration even used water bears for testing in space.

Other typical *Metazoa* include bristle worms (*Aeoleosoma*) (Figure 3.18a), gastrotrichs (Figure 3.18b), and nematodes (Figure 3.18c).

7.0 NUMBER AND TYPE OF PROTOZOA AND *METAZOA* VERSUS VARIOUS PARAMETERS

There are numerous correlated reports (Figure 3.19) relating protozoa and *Metazoa* to facility performance. Many of these are based on laboratory-scale pilot facilities and not actual systems. Therefore, inserting an individual system's operation with higher lifeform presence and abundance will provide a more detailed representation of what to expect from that system's biomass. There are certain organisms that become more apparent during higher loading rates, such as flagellates and *Amoeba*. This may also occur with a younger sludge age because, typically, these organisms reproduce faster than many other protozoa and *Metazoa* and require a larger amount of free motile bacteria to survive. As organics are reduced and more floc becomes present, ciliates such as free-swimmers and crawler ciliates become abundant. This will also progress to stalked ciliates; suctoria; and, as the sludge age increases, the presence of *Metazoa* such as rotifers. The primary key is how this relates to each system, and the only way to use that is to collect the system's individual data and microscopic analysis.

8.0 USE OF PROTOZOA AND *METAZOA* AS INDICATOR ORGANISMS

In most properly functioning biological wastewater treatment systems, a hierarchy of microscopic organisms exists. These organisms include bacteria, fungi, algae, protozoa, and *Metazoa*. Together, they are responsible for effective wastewater treatment. Bacteria do the majority of the work by using soluble and particulate organic substrates as sources of carbon and energy. The growth and performance of bacteria depend on the growth pressures that constitute their presence, such as substrate, temperature, and dissolved oxygen. Changes in biomass, particularly the bacterial population, can be

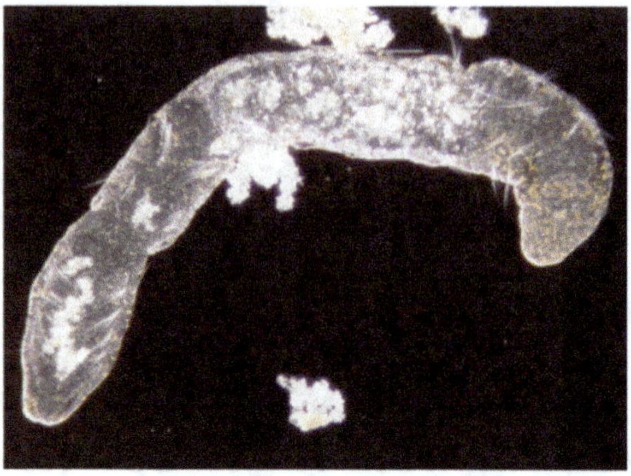

FIGURE 3.18a Bristle Worm (Reprinted with permission from Steven C. Leach)

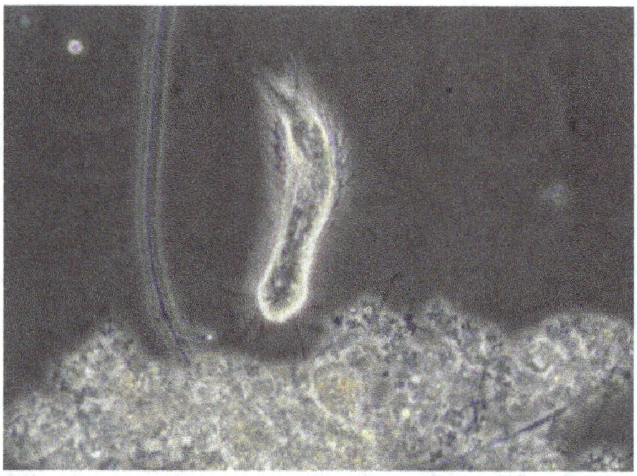

FIGURE 3.18b Gastrotrich (Reprinted with permission from Steven C. Leach)

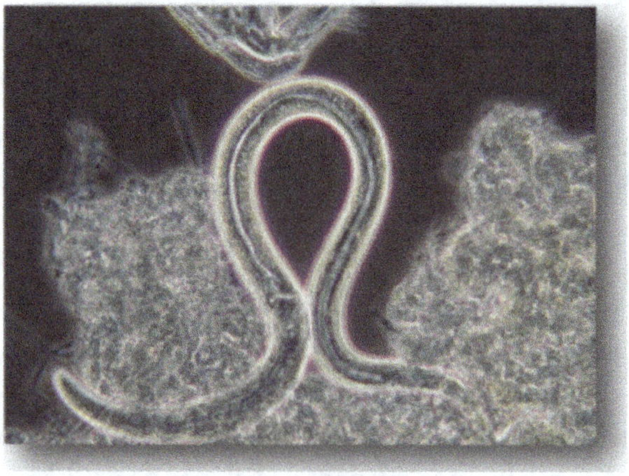

FIGURE 3.18c Nematode (Reprinted with permission from Steven C. Leach)

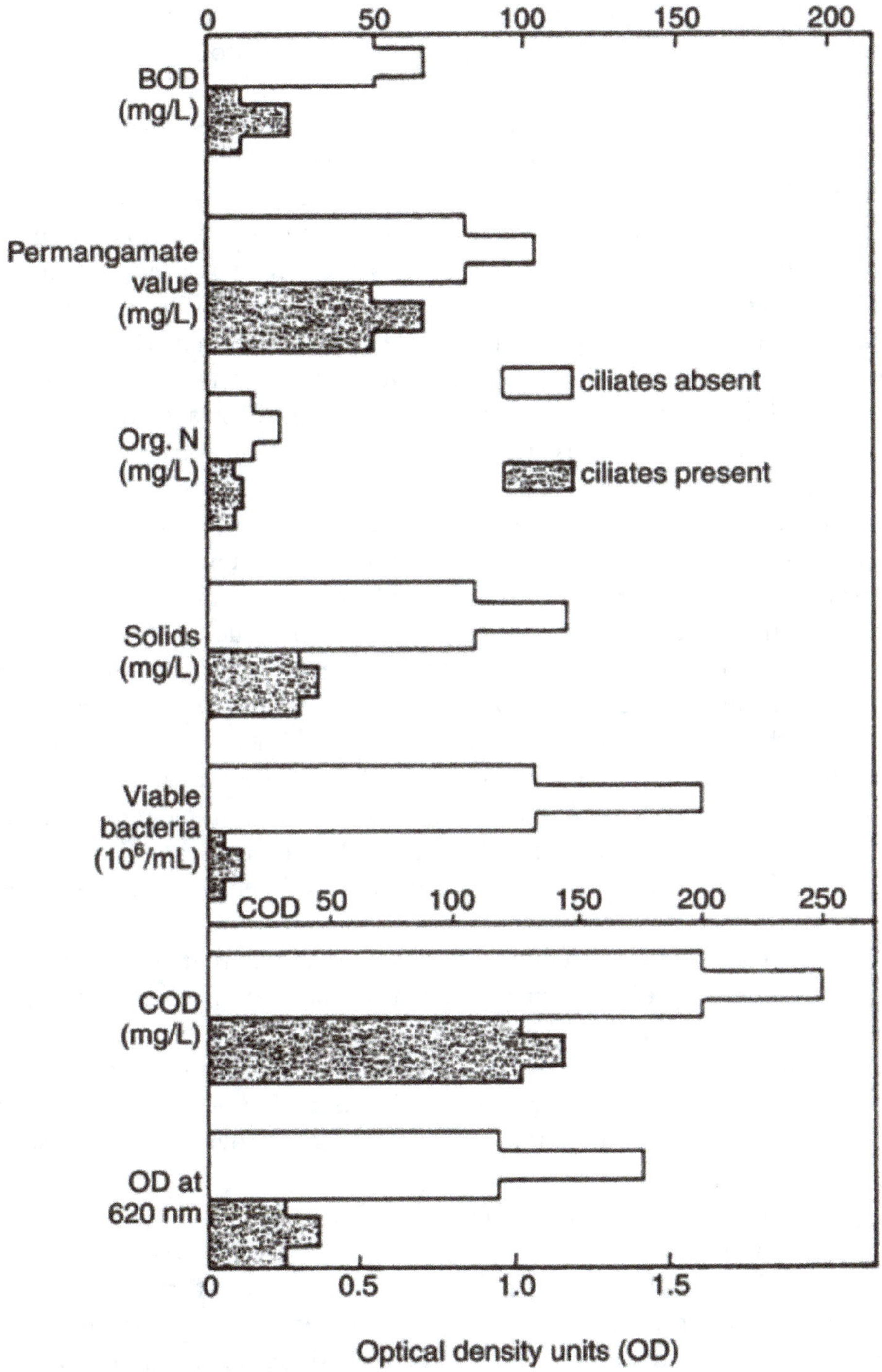

FIGURE 3.19 Bar Chart of Effluent Quality Issuing from Laboratory-Scale Pilot Facilities Operating in Presence and Absence of Ciliated Protozoa (the shoulders on the bars indicate ranges of variation between replicate facilities) (Reproduced with permission from Curds [1992])

influenced quickly by pH, temperature, substrate, and so on. These changes may be observed in floc formation, presence of filamentous bacteria, or even in the sudden presence of dispersed or free, motile bacteria. It is easy for someone with an understanding of basic microscopic observation to recognize the difference between a dense or weak floc. However, bacteria can be found as individual cells in bulk water (free bacteria); as filaments (growing in long chains); or, ideally, as flocculated particles (floc). The dominant form often has a direct correlation to how well the wastewater treatment system is performing. Determining why they may be present or identifying to a genus level requires more sophisticated analysis such as DNA analysis. Most facilities, whether they are a municipal or industrial site, do not have these capabilities yet. Bacteria also act as primary substrates in the ecosystem's food chain and represent an excellent source of nutrition for higher lifeforms.

Because identification of most bacteria requires more advanced techniques, using higher lifeforms commonly found in wastewater biomass such as protozoa (the majority are single-celled organisms) and *Metazoa* (multicellular) makes a simple microscopic observation much easier. Both these groups of organisms may be used as indicators of steady-state or non-steady-state operation. An operator who can determine what the organism present means to the overall biomass will have a better chance of understanding or preventing significant upsets. The difficulty becomes the understanding and transfer of this knowledge from facility to facility. Not all facilities will have a wide variety of higher lifeforms but may still operate ideally. Just because crawler ciliates or stalked ciliates are not seen and performance is optimal, there is no reason to panic.

In the past, many manuals and books would state that the higher presence of crawler ciliates, combined with a few other types, indicated a healthy biomass. During the past 15 years of microscopic observations, more variations in these indicators are being seen. The best organisms are often those seen when the facility is performing at its best. Therefore, observing the biomass when the system is running well will help provide a good baseline of what is common.

Another key in using higher lifeforms as indicators is diversity. The more diverse the population, the healthier and more stable the biomass will be. Protozoa and *Metazoa* are orders of magnitude larger than bacteria and can be easily observed microscopically. The majority of these organisms are affected by BOD and ammonia concentrations, sludge age, and toxic shocks. Many higher lifeforms flourish and decline in certain environmental conditions. In addition, they can be found in different forms, depending on their surrounding conditions. As such, the appearance or lack of these organisms provides a basis for understanding how well the microlife is performing. For example, the majority of protozoa reproduce through binary fission, resulting in two daughter cells. Therefore, the presence and abundance of daughter cells (Figure 3.20) may be a function of ideal environmental conditions.

Another example is the presence of protozoan or metazoan cysts. Many higher lifeforms have the ability to produce a cyst (Figure 3.21), which is a protective coating that enables the organism to withstand unfavorable conditions. An operator who recognizes the presence of cysts may realize that the biomass is experiencing a change, possibly because of an organic shock load. If cysts are identified quickly enough, operational changes might be made to reduce the effect of the shock. In water resource recovery facilities with no influence of industrial waste, it would not be common to see cysts unless it was a function of reproduction. However, the industrial system with variations in substrate or operations could see cysts on a regular basis. An important point to make is that cysts take time to form, so a quick toxin dump or shock to the system will not produce cysts. The change has to be a slower response for an organism to encyst.

A final example is the presence of detached stalked ciliates. If it is a reaction to reproduction, one may be able to view elongated nuclei. If not, chances are that growth pressures have changed slowly and that the protist is detaching to find a more favorable environment. If an operator has taken a fresh sample of the biomass and placed it under the microscope with no

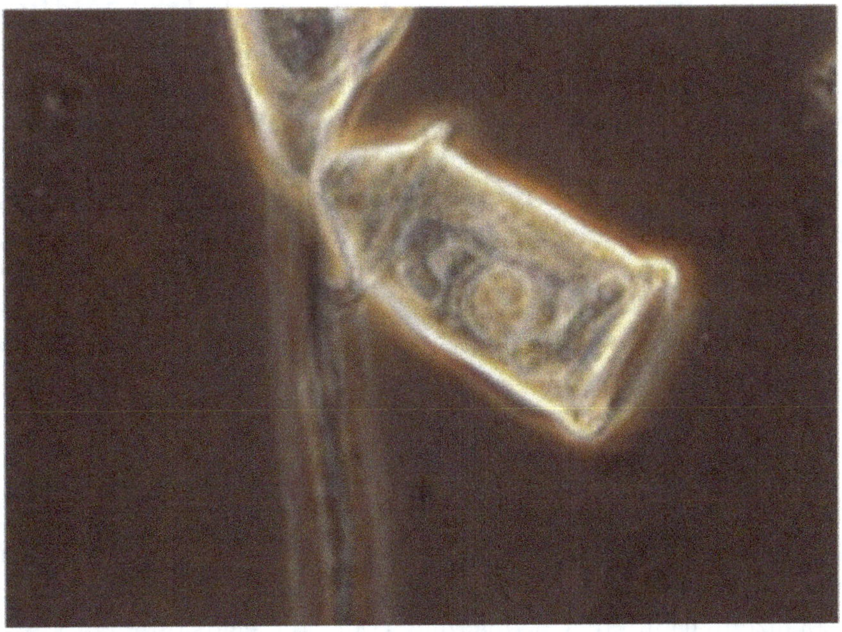

FIGURE 3.20 Daughter Cell with Trochal Band (Reprinted with permission from Steven C. Leach)

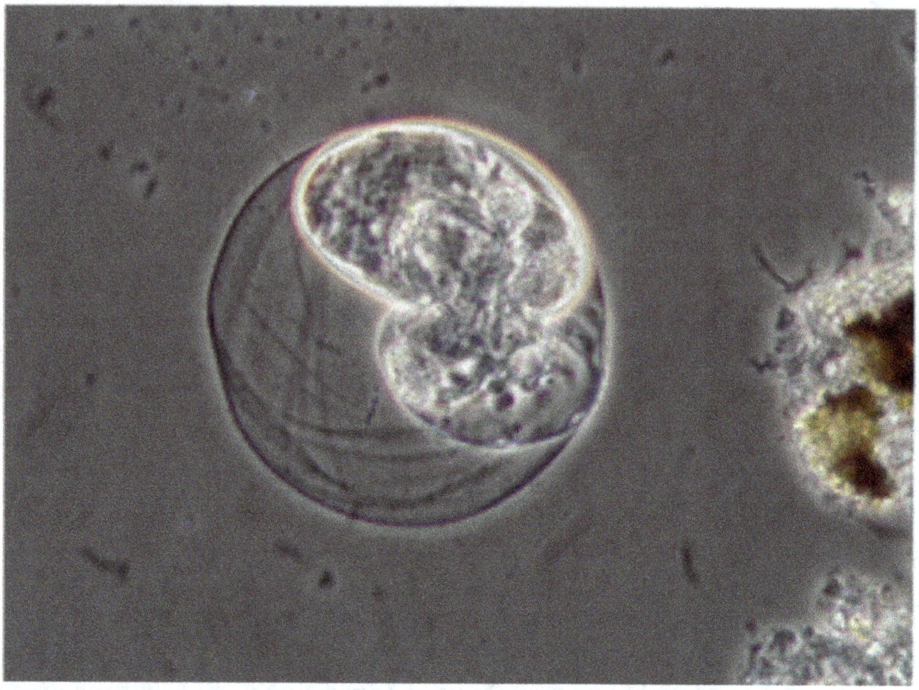

FIGURE 3.21 Cyst (Reprinted with permission from Steven C. Leach)

delay, it is a real-time example of what is happening in the basin at that time. If one observes an entire slide showing detached stalked ciliates (teletrochs, Figure 3.20) or empty stalks (Figure 3.22), it is likely that the biomass is experiencing a shock to the system and further testing or analysis is needed.

An important point to recognize is that some facilities may never have a large diversity or, in some instances, protozoa or *Metazoa*, yet their treatment systems run very well. Every facility is its own environmental system, with various growth pressures and operating conditions. The presence or absence of higher lifeforms will not indicate exactly what is wrong with the system but does suggest that something is happening; therefore, a further in-depth look at the operation may be needed. With this in mind, there are a couple of methods for tracking higher lifeforms.

8.1 Higher Lifeform Counting Form

Forms for counting protozoa and *Metazoa* have been around for years, and many people still use them on a regular basis. Figure 3.23 presents a sample form that operators have used for counting higher lifeforms. It is important to recognize that different people may count the fields differently or even certain organisms incorrectly, which may have an effect on the observation. If possible, using the same person for counting would be a best practice.

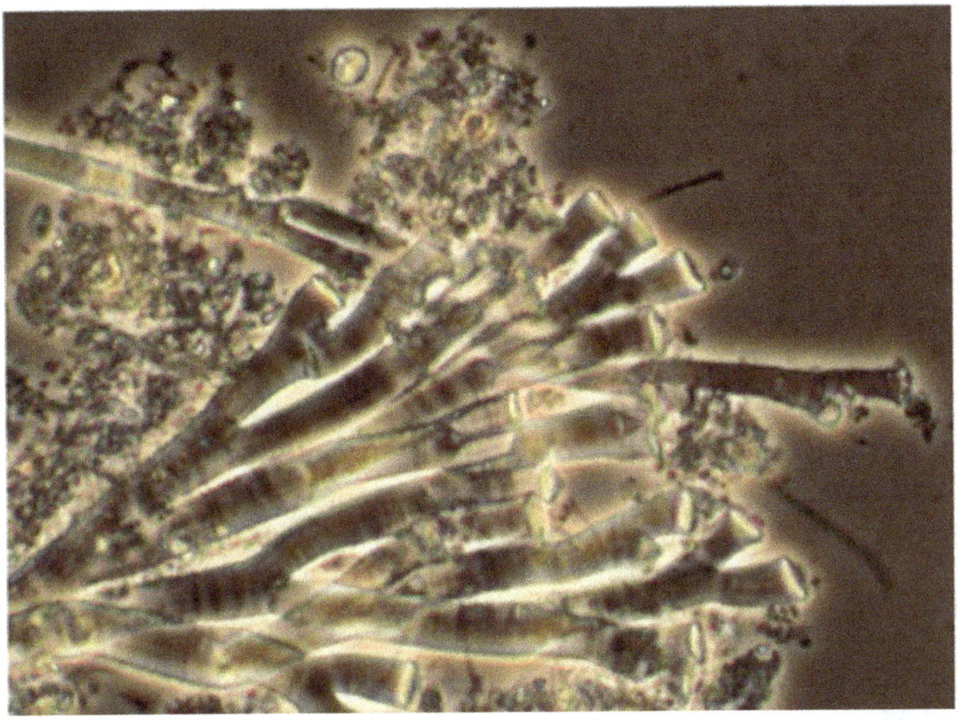

FIGURE 3.22 Empty Stalks (Reprinted with permission from Steven C. Leach)

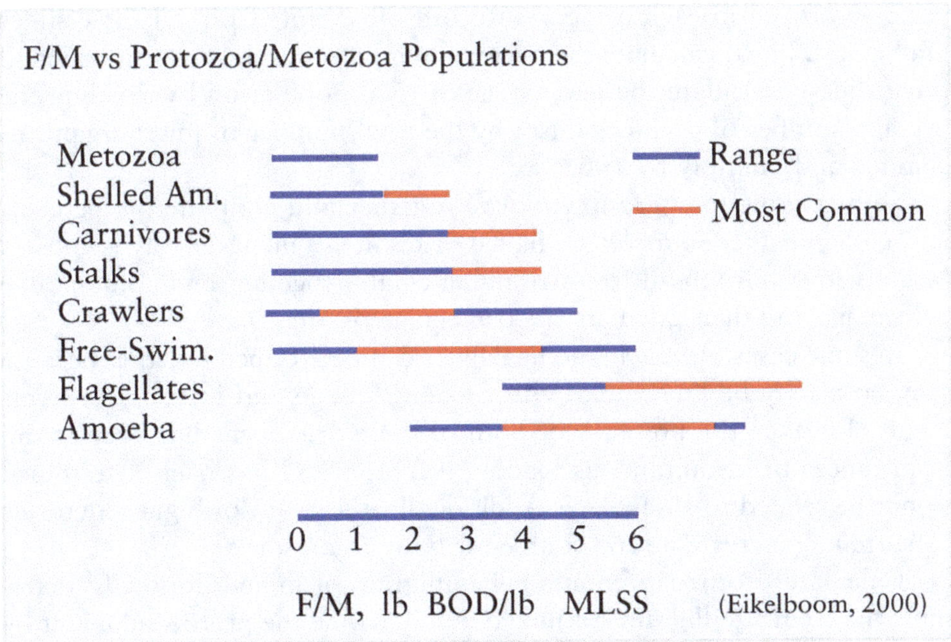

FIGURE 3.23 Higher Lifeform Counting Example (Eikelboom, 2000)

8.2 The Counting Procedure

Collect a fresh, well-mixed, representative sample of mixed liquor from the discharge end of the aeration basin. If the concentration of the mixed liquor is high, the sample should be diluted. If you choose to dilute the sample, always use the same dilution if you plan to do routine counts. Ideally, the sample should be observed under the 20× objective using the phase-contrast condenser. Using a smaller objective (10× or below) will make it difficult to see smaller flagellates and using a larger objective (40× or higher) will narrow the field of view.

Place one drop of mixed liquor on a clean, grease-free slide and cover with a cover slip. Next, scan the slide using five passes. Focus the objective at the top left corner of the cover slip. Moving down the cover slip, count and record the number of each type of protozoa that you see. This completes one pass. Move over to the right slightly and, this time, move up the cover slip, keeping a running total of the number of each type of protozoa. This will complete the second pass. Move up and down the slide until five passes have been completed.

Record the running total number of each protozoa type. Remember, if colonies of stalked ciliates are encountered, count (or estimate) the total number of heads. For the best results, scan three or more slides and average the number of each protozoan type.

To calculate relative dominance, count and record the total number of protozoa in each category. Record the counts from all three slides (Table 3.3). Next, calculate the average number of each organism from all three slides. Calculate the percentage of total population by dividing the average number of each organism by the total number of microorganisms counted and multiply by 100.

Some people may group *Amoeba* together and combine the percentage. Note whether *Amoeba* are naked or testate. Whichever way is chosen, be consistent. Group all free-swimming ciliates together and combine the percentage and then calculate the percentage of *Metazoa*.

In some cases, *Amoeba* and flagellates can be grouped together because they both may be associated with a young sludge and incomplete treatment. However, the presence of both of these organisms has been found in instances of instantaneous loading increases and has been determined to not be related to sludge age. Additionally, the death of higher lifeforms (*Metazoa*) can create a food source for both organisms as well.

Calculate protozoan counts per milligram of mixed liquor. To determine the counts/milligrams of mixed liquor, divide the protozoan count by the mixed liquor volatile suspended solids. Protozoan counts are not exact

TABLE 3.3 Example of Higher Lifeforms Counting Worksheet (Reprinted with permission from Steven C. Leach)

Organism	Slide 1	Slide 2	Slide 3	Total	Average
Protozoa					
Amoeba (naked)					
Amoeba (testate)					
Flagellates					
Free-swimming ciliates					
Crawler ciliates					
Carnivore ciliates					
Stalk ciliates					
Suctoria					
Metazoa					
Rotifers					
Nematodes					
Gastrotrichs					
Bristle worms					

science and should not be relied on solely for determining treatment system conditions. They are useful, however, in helping the operator to assess conditions within the aeration basin.

9.0 MICROSCOPIC EXAM FORM

The microscopic exam form has had many variations, most of which are based on the example in the manual by Jenkins et al. (2004). The key point in any exam is understanding and qualifying the four most important points in any wastewater biomass analysis; floc, filaments, higher lifeforms, and bulk water. As long as the form has these four areas covered, the operator will be able to provide a detailed exam of biomass quality. Modifications can be made to focus on particular areas but this form is a good description of an overall analysis.

10.0 PRACTICE QUESTIONS

1. Should a microscopic exam be performed on the aeration basin, clarifiers, or return activated sludge (RAS)?
2. What is the difference between protozoa and *Metazoa*?
3. Are suctoria good indicators of nitrification?
4. Are filaments beneficial?
5. What are the four focus points in performing a microscopic exam?
6. What is the difference between a free-swimming ciliate and a crawler ciliate?
7. Are rotifers protozoa or metazoan?
8. Can protozoa eat all of the bacteria in an aeration basin?
9. Will a phase-contrast microscope provide more detail than a typical brightfield microscope?
10. Under proper environmental conditions, what physical characteristic allows bacteria to grow so quickly?

Answer key:

1. Aeration basin; this is where the biomass is most active.
2. Protozoa are single celled and *Metazoa* are multicellular.
3. Yes, they do not like ammonia.
4. Yes, as backbone of floc and for BOD and turbidity removal but can foam or create settling issues.
5. Floc, filaments, higher lifeforms, and bulk water.
6. Free-swimming ciliates have cilia covering the entire organism, whereas crawler ciliates have fused-together cilia called cirri.
7. Metazoan.
8. No, growth rates for bacteria are much faster.
9. Yes.
10. Very large surface-to-volume ratio.

11.0 REFERENCES

Curds, C. R. (1992). *Protozoa in the water industry*. Cambridge University Press.

Eikelboom, D. H. (2000). *Process control of activated sludge plants by microscopic investigation.* IWA Publishing.

Jahn, T. L., Bovee, E. C., & Jahn, F. F. (1979). *How to know the protozoa* (2nd ed.). The Pictured Key Nature Series. Wm. C. Brown Publishers.

Jenkins, D., Richard, M., & Daigger, G. (2004). *Manual on the causes and control of activated sludge bulking, foaming, and other solids separation problems* (3rd ed.). Lewis Publishers.

Ma, B. (2016). Biological nitrogen removal from sewage via anammox: Recent advances. *Bioresource Technology, 200,* 981-990.

Mehrani, M. J. (2021). Incorporation of the complete ammonia oxidation (comammox) process for modeling nitrification in suspended growth wastewater treatment systems. *Journal of Environmental Management, 297,* 113223.

Water Environment Federation. (2017). *Wastewater biology: The microlife* (3rd ed.).

12.0 SUGGESTED READINGS

Brock, T. D. (1966). *Principles of microbial ecology.* Prentice-Hall.

Gaudy, A. F., & Gaudy, E. T. (1980). *Microbiology for environmental scientists and engineers.* McGraw-Hill.

Hausman, K., & Hulsmann, N. (1996). *Protozoology.* Thieme Medical Publishers.

Minchin, E. A. (2006). *Protozoa microbiology and guide to microscopic identification.* Merchant Books.

Patterson, D. J. (1992). *Free-living freshwater protozoa: A color guide.* John Wiley & Sons.

Sleigh, M. A. (1973). *The biology of protozoa.* William Clowes & Sons.

Society of Protozoologists. (2000). *The illustrated guide to the protozoa* (2nd ed.) Allen Press.

Willey, J. M., Sandman, K., & Wood, D. (2022). *Prescott's microbiology* (12th ed.). McGraw-Hill.

4

Process Control for the Activated Sludge Basin

1.0 INTRODUCTION

Activated sludge is extremely flexible and can be manipulated to meet a variety of process goals. Operators can influence mixed liquor suspended solids (MLSS) settleability and final effluent quality by adjusting five process variables: the number of basins and clarifiers in service; the quantity of waste activated sludge (WAS) removed from the system each day; the return activated sludge (RAS) flowrate; the availability of oxygen; and, in some systems, internal mixed liquor recycle flows. These five process variables give operators an immense amount of control over sludge settleability, treatment level, and effluent quality. Other factors that can affect activated sludge performance include pH, alkalinity, and characteristics of the influent wastewater. Emerging research also highlights the role of trace elements and micropollutants influencing microbial diversity and treatment performance in activated sludge systems. In this chapter, three different methods of process control are discussed, including maintaining constant sludge age, maintaining constant food-to-microorganism ratio (F:M), and maintaining constant MLSS concentration. All three of these methods are currently widely practiced; however, the method of process control recommended by numerous authorities since the mid-1960s is constant sludge age (Ekama, 2010; Jenkins and Garrison, 1968; Pretorius et al., 2016; U.S. Environmental Protection Agency, 1977; West, 1973a, 1973b, 1978). Sludge age directly controls both F:M and the MLSS concentration. It is not possible to control these variables independently of one another. This chapter goes on to discuss how to select a target sludge age, target dissolved oxygen concentrations, and target oxidation–potential reduction values.

Table 4.1 gives operating ranges for various process control variables for the most common operating modes in domestic water resource recovery facilities (WRRFs). These are not the only operating modes available. Keep in mind that numerical ranges given in Table 4.1 are functional definitions and that it is entirely possible for a single, well-functioning activated sludge process to have some variables fall within more than one definition. These functional definitions are from the 1960s and 1970s and reflect operating practices and research conducted during that time. The solids retention time (SRT) recommended for extended aeration activated sludge is quite long at 20 to 30 days. Today, we know that sludge ages as long as 20 days can encourage the growth of filamentous bacteria, which can affect sludge settleability, and we know that much younger sludge ages are capable of producing an effluent low in ammonia and nitrate. Guidance on how to select the right sludge age for a facility is given in Section 5.0.

2.0 SLUDGE AGE

The fundamental difference between a lagoon treatment system and an activated sludge process is the presence of an RAS line, which takes settled solids from the clarifier and returns them to the activated sludge basins. Return activated sludge makes it possible to independently control the hydraulic

TABLE 4.1 Design Parameters for Activated Sludge Processes (U.S. Environmental Protection Agency, 1974, 1978)

Process modification	SRT, days	F:M, kg BOD applied/kg MLVSS·d (lb BOD_5/lb MLVSS·d)	Volumetric loading, kg BOD_5/m^3·d (lb BOD_5/cu ft·d)	MLSS, mg/L	HRT, h
Conventional (plug flow)	5–15	0.2–0.4 (0.2–0.4)	20–40 (1.25–2.5)	1500–3000	4–8
Extended aeration	20–30*	0.05–0.15 (0.05–0.15)	10–25 (0.6–1.6)	3000–6000	18–36
Step feed	3–7	0.2–0.4 (0.2–0.4)	40–60 (2.5–3.75)	2000–3500	3–5
Pure oxygen	3–20	0.25–1.0 (0.25–1.0)	150–200 (9.3–12.5)	4000–7000	2–4

*Excessively long SRTs are no longer recommended because of the growth of filamentous bacteria and the deterioration of final effluent quality that can occur at long SRTs. Refer to Section 5.0, Selecting a Target Mean Cell Residence Time, Solids Retention Time, or Solids Retention Time$_{Aerobic}$ for guidance.

residence time (HRT), which is the amount of time that the water spends in the process, and SRT, which is the amount of time that solids (bacteria) spend in the process. This allows more microorganisms to build up in the process and decreases the total time needed for treatment.

The definition of sludge age has evolved over the years as our understanding of the process has increased. Sludge age is the amount of time that solids or MLSS spend moving between the activated sludge basin(s) and secondary clarifier(s) before they are removed from the system either through the WAS line or by escaping into the final effluent. In the case of a sequencing batch reactor (SBR), treatment and clarification take place in the same tank. Sludge age affects every aspect of the activated sludge process operation, including MLSS concentration and quality; sludge settling characteristics; types of microorganisms present in the process; F:M; treatment capacity; ability to remove ammonia, nitrate, and phosphorus; oxygen demand; oxygen-transfer efficiency; sludge production; and final effluent quality. *Sludge age should be the primary process control variable for any activated sludge system.* It should be adjusted, as needed, to optimize sludge settleability, minimize energy use, and meet final effluent discharge limits. In locations that experience seasonal changes in influent water temperature, sludge age should be adjusted seasonally to compensate for increased or decreased microbial activity. The sludge age selected will dictate whether an activated sludge process will be able to meet treatment goals for effluent total suspended solids (TSS), biochemical oxygen demand (BOD), and ammonia.

Sludge age is a broad term that may refer to any of the following more specific terms: Gould sludge age (GSA), mean cell residence time (MCRT), SRT, or aerobic SRT ($SRT_{aerobic}$). Of these, GSA is no longer in common use, although it is still referenced in many publications. The GSA was one of the earliest developed parameters for activated sludge process control (Gould, 1953). It is based on the idea that activated sludge is primarily a physical–chemical removal process. It assumes that TSS in influent wastewater are merely flocculated and are not otherwise changed during treatment. Gould's equation does not account for biological activity. It assumes that all of the solids in the activated sludge basin came from influent wastewater, that no new solids are produced during treatment, and that solids are not consumed or broken down. This physical theory was disproven in 1946 and shifted the focus to activated sludge microbiology (National Research Council Subcommittee, 1946). The National Research Council Subcommittee report recognized that organic TSS in the influent are a food source for microorganisms in the process and that the makeup of those solids changed during treatment. Treatment time, temperature, availability of nutrients, and dissolved oxygen concentration also became recognized as important process control variables. The report also examined the concept of F:M, which is discussed later in this chapter. A modification to the GSA calculation was proposed by the Water Environment Federation that used influent BOD instead of suspended solids but was never in common use. The inverse of the modified GSA equation is the F:M.

The $SRT_{aerobic}$ is used by operators and design engineers to ensure that the system has adequate sludge age for nitrification. Depending on which sludge age definition is in use, the mass of MLSS in the aerated zones, unaerated zones, and clarifier blankets may or may not be included in process control calculations, as shown in Table 4.2.

TABLE 4.2 Solids Included in Sludge Age Calculations

Sludge age definition	Influent solids	Anaerobic and anoxic zones	Aerated zones	Clarifier blanket	Waste solids
GSA	Yes	Yes	Yes	Maybe	No
MCRT	No	Yes	Yes	Yes	Yes
SRT	No	Yes	Yes	No	Yes
$SRT_{aerobic}$	No	No	Yes	No	Yes

2.1 Mean Cell Residence Time

By 1955, many of the tools of modern-day process control were already in common use. Settleometer tests were being conducted in 1-L graduated cylinders and GSA calculations were performed daily at many facilities. Wasting rates were adjusted daily to achieve a target GSA (Torpey and Chasick, 1955). It was widely recognized that sludge age needed to be controlled within a narrow range to maintain sludge settleability and effluent quality. If sludge age was too low, sludge did not flocculate and compact well. If sludge age was too high, the growth of filamentous bacteria caused sludge bulking and eventual process failure (Torpey and Chasick, 1955). Despite these advances, many activated sludge processes had difficulty controlling effluent TSS and BOD. This was partly because of the reactive nature of the GSA and its sensitivity to small changes in the WAS rate.

The first example of modern MCRT process control was published in 1958 by M. T. Garrett, Jr., who called it *hydraulic control*. Hydraulic control, as proposed by Garrett, involved managing the flow and residence times of mixed liquor in the treatment basin, targeting optimization of sludge age and microbial growth for better process control. Garrett noted that one of the significant drawbacks of the GSA calculation was that it forced the operator to keep up with and predict changes in the incoming load as well as solids in the facility. Operators were always reacting to past conditions. Garrett's previous work showed that there is a relatively uniform relationship between the amount of BOD removed by activated sludge and the amount of mixed liquor volatile suspended solids (MLVSS) produced (Garrett and Sawyer, 1952). Recall the discussion on yield in Chapter 2, where approximately 0.42 kg of MLVSS are produced for every 1 kg of chemical oxygen demand (COD) in the influent or approximately 0.67 kg of MLVSS for every 1 kg of BOD (Metcalf & Eddy, Inc./AECOM, 2013). For domestic wastewater, the influent COD-to-BOD ratio is typically between 1.8 and 2.2 (EnviroSim Associates, Ltd., 2005). Because the amount of MLVSS produced each day is directly related to the mass of BOD in the influent, Garrett and others proposed that the sludge age and growth rate could be controlled simply by controlling the amount of waste sludge removed from the system each day (Garrett, 1958). If the influent BOD load increases, more MLVSS will be produced and the MLVSS concentration will increase. Removing the same percentage of sludge from the system each day results in a constant sludge age. Long-term operation of a full-scale facility using this idea demonstrated that controlling the wasting rate also controlled the balance between influent wastewater BOD load and MLVSS concentration and directly controlled effluent BOD concentrations (Garrett, 1958). In other words, controlling MCRT also controls MLVSS concentration, F:M, and effluent quality.

2.1.1 Definition

The term *MCRT* was coined by E. A. Pearson (1966) and was later applied to activated sludge process control by Jenkins and Garrison (1968). *Mean* is another word for *average*. Mean cell residence time is the average amount of time a bacterial cell spends in the activated sludge process. By using this specific terminology, a distinction is made between the TSS used in the GSA calculation and biological solids in MLVSS. *Mean cell residence time* was defined by Jenkins and Garrison (1968) as the total mass of volatile suspended solids (VSS) in the activated sludge process divided by the mass of the volatile suspended solids leaving the system. Mass is calculated in kilograms (pounds). The calculation includes solids from all basins, regardless of whether they are aerated as well as the solids in the clarifier blanket. Originally, the MCRT calculation included only MLVSS, but modern use of MCRT includes MLSS rather than MLVSS. For most domestic WRRFs, the MLVSS-to-MLSS ratio is fairly constant over long periods of time as long as (1) sludge age is well controlled, (2) the WRRF does not receive water treatment facility residuals, and (3) substantial quantities of inert material are not added through chemical addition. The near-constant, slowly changing ratio observed in many activated sludge processes allows for the use of MLSS, a simpler laboratory test, in place of MLVSS.

2.1.2 Calculation Method

Mean cell residence time is calculated using the following equation:

$$\text{Mean cell residence time} = \frac{\text{Activated sludge basin mass} + \text{Clarifier mass}}{\text{Waste activated sludge mass} + \text{Effluent solids mass}} \tag{4.1}$$

Mass is calculated in kilograms (pounds). In a well-functioning activated sludge process, the mass of effluent solids should be small compared to the mass of solids present in the activated sludge basin, clarifier blanket, and waste activated sludge. Often, effluent solids can be ignored and excluded from the calculation. Section 2.6 discusses when effluent TSS should be included in the calculation.

In International Standard units:

$$\text{Mass, kg} = \frac{\left[(\text{Concentration, } \frac{\text{mg}}{\text{L}})(\text{Volume, m}^3)\right]}{1000} \tag{4.2}$$

Use dimensional analysis to convert the units in Equation 4.2 and arrive at kg:

$$\frac{X\text{ mg}}{\text{L}}\left|\frac{1\text{ g}}{1000\text{ mg}}\right|\left|\frac{1\text{ kg}}{1000\text{ g}}\right|\left|\frac{X\text{m}^3}{}\right| = \text{kg} \tag{4.3}$$

In U.S. customary units:

To find pounds per day, use the flowrate in million gallons per day in place of volume in million gallons.

$$\text{Mass, lb} = (\text{Concentration, }\frac{\text{mg}}{\text{L}})(\text{Volume, mil. gal})(8.34\frac{\text{lb}}{\text{mil. gal}}) \tag{4.4}$$

Use dimensional analysis to result in lb:

$$\frac{X\text{ mg}}{\text{L}}\left|\frac{1\text{ g}}{1000\text{ mg}}\right|\left|\frac{1\text{ kg}}{1000\text{ g}}\right|\left|\frac{2.204\text{ lb}}{1\text{ kg}}\right|\left|\frac{3.785\text{ L}}{\text{gal}}\right|\left|\frac{1\,000\,000\text{ gal}}{\text{mil. gal}}\right|\left|\frac{X\text{ mil. gal}}{}\right| = \text{lb} \tag{4.5}$$

To maintain an MCRT of 10 days, for example, the operator must remove one-tenth (10%) of the total mass of solids from the process each day. To maintain an MCRT of 20 days, the operator would remove one-twentieth (5%) of the total mass each day. Note that influent flow to the WRRF does not affect the percentage of WAS removed each day. Mass or weight of solids (kg or lb) removed is not the same as the volume removed (m^3 or mil. gal). Because WAS concentration can change frequently, it is critically important that mass calculations be performed regularly.

2.1.3 Calculation Example

Find the MCRT for a WRRF with the following characteristics:

Activated sludge basin volume = 7570 m^3 (2 mil. gal)
MLSS concentration = 2800 mg/L
Clarifier volume = 1514 m^3 (0.4 mil. gal)
Clarifier solids concentration = 600 mg/L
WAS flowrate = 303 m^3/d (0.08 mgd)
WAS concentration = 7400 mg/L
Effluent TSS concentration = negligible

Solution in International Standard Units:

Step 1: Find the mass of solids in the activated sludge basin:

$$\text{Activated sludge basin, kg} = \frac{\left[(\text{Concentration, }\frac{\text{mg}}{\text{L}})(\text{Volume, m}^3)\right]}{1000}$$

$$\text{Activated sludge basin, kg} = \frac{\left[(2800 \frac{\text{mg}}{\text{L}})(7570 \text{ m}^3)\right]}{1000}$$

$$= 21\ 196 \text{ kg}$$

Step 2: Find the mass of solids in the clarifier blanket:

$$\text{Clarifier blanket, kg} = \frac{\left[(\text{Concentration, } \frac{\text{mg}}{\text{L}})(\text{Volume, m}^3)\right]}{1000}$$

$$\text{Clarifier blanket, kg} = \frac{\left[(600 \frac{\text{mg}}{\text{L}})(1514 \text{ m}^3)\right]}{1000}$$

$$\text{Clarifier blanket, kg} = 908.4$$

Step 3: Find the mass of solids wasted:

$$\text{WAS, } \frac{\text{kg}}{\text{d}} = \frac{\left[(7400 \frac{\text{mg}}{\text{L}})(303 \frac{\text{m}^3}{\text{d}})\right]}{1000}$$

$$\text{WAS, } \frac{\text{kg}}{\text{d}} = 2242 \text{ kg/d}$$

Step 4: Solve for MCRT:

$$\text{Mean cell residence time} = \frac{\text{Activated sludge basin mass} + \text{Clarifier mass}}{\text{Waste activated sludge mass} + \text{Effluent solids mass}}$$

$$\text{Mean cell residence time} = \frac{21\ 196 \text{ kg} + 908.4 \text{ kg}}{2242 \frac{\text{kg}}{\text{d}} + 0}$$

$$\text{Mean cell residence time} = \frac{22\ 104.4 \text{ kg}}{2\ 242 \frac{\text{kg}}{\text{d}}}$$

$$\text{Mean cell residence time} = 9.86 \text{ days}$$

Solution in U.S. Customary Units:

Step 1: Find the mass of solids in the activated sludge basin:

$$\text{Activated sludge basin, lb} = (\text{Concentration, } \frac{\text{mg}}{\text{L}})(\text{Volume, mil. gal})(8.34 \frac{\text{lb}}{\text{mil. gal}})$$

$$\text{Activated sludge basin, lb} = (2800\ \frac{\text{mg}}{\text{L}})(2\ \text{mil. gal})(8.34 \frac{\text{lb}}{\text{mil. gal}})$$

Activated sludge basin = 46 704 lb

Step 2: Find the mass of solids in the clarifier blanket:

$$\text{Clarifier, lb} = (\text{Concentration, } \frac{\text{mg}}{\text{L}})(\text{Volume, mil. gal})(8.34 \frac{\text{lb}}{\text{mil. gal}})$$

$$\text{Clarifier, lb} = (600\ \frac{\text{mg}}{\text{L}})(0.4\ \text{mil. gal})(8.34 \frac{\text{lb}}{\text{mil. gal}})$$

Clarifier blanket mass = 2002 lb

Step 3: Find the mass of solids in the waste activated sludge:

$$\text{WAS, } \frac{\text{lb}}{\text{d}} = (\text{Concentration, } \frac{\text{mg}}{\text{L}})(\text{WAS flow, mgd})(8.34 \frac{\text{lb}}{\text{mil. gal}})$$

$$\text{WAS, } \frac{\text{lb}}{\text{d}} = (7400\ \frac{\text{mg}}{\text{L}})(0.08\ \text{mgd})(8.34 \frac{\text{lb}}{\text{mil. gal}})$$

$$\text{WAS, } \frac{\text{lb}}{\text{d}} = 4937$$

Step 4: Solve for MCRT:

$$\text{Mean cell residence time} = \frac{\text{Activated sludge basin mass} + \text{Clarifier mass}}{\text{Waste activated sludge mass} + \text{Effluent solids mass}}$$

$$\text{Mean cell residence time, days} = \frac{46\ 704\ \text{lb} + 2002\ \text{lb}}{4937 \frac{\text{lb}}{\text{d}} + 0}$$

$$\text{Mean cell residence time, days} = \frac{46\ 706\ \text{lb}}{4937\dfrac{\text{lb}}{\text{d}}}$$

$$\text{Mean cell residence time} = 9.86\ \text{days}$$

When the target MCRT is known, the calculation is done in reverse to find the total mass and volume that should be wasted each day. The operator determines that an MCRT of 9.86 days is too low and wishes to increase the MCRT to 12 days to maintain stable nitrification. Calculate the new wasting rate using the same information given in the previous example.

Solution in International Standard Units:

Step 1: Find the total mass to be wasted:

$$\text{Mean cell residence time} = \frac{\text{Activated sludge basin mass} + \text{Clarifier mass}}{\text{Waste activated sludge mass} + \text{Effluent solids mass}}$$

$$12\ \text{days} = \frac{21\ 196\ \text{kg} + 908.4\ \text{kg}}{\text{WAS, }\dfrac{\text{kg}}{\text{d}} + 0}$$

$$12\ \text{days} = \frac{22\ 104.4\ \text{kg}}{\text{WAS, }\dfrac{\text{kg}}{\text{d}}}$$

$$\text{WAS, }\frac{\text{kg}}{\text{d}} = \frac{22\ 104.4\ \text{kg}}{12\ \text{days}}$$

$$\text{WAS, }\frac{\text{kg}}{\text{d}} = 1842$$

Step 2: Find the volume of WAS to remove from the system:

$$\text{WAS, }\frac{\text{kg}}{\text{d}} = \frac{\left[(\text{Concentration, }\dfrac{\text{mg}}{\text{L}})(\text{WAS flow, }\dfrac{\text{m}^3}{\text{d}})\right]}{1000}$$

$$1842\ \frac{\text{kg}}{\text{d}} = \frac{\left[(7400\ \dfrac{\text{mg}}{\text{L}})(\text{WAS flow, }\dfrac{\text{m}^3}{\text{d}})\right]}{1000}$$

$$\text{WAS flow, } \frac{m^3}{d} = 248.9$$

Solution in U.S. Customary Units:

Step 1: Find the total mass to be wasted:

$$\text{Mean cell residence time} = \frac{\text{Activated sludge basin mass} + \text{Clarifier mass}}{\text{Waste activated sludge mass} + \text{Effluent solids mass}}$$

$$12 \text{ days} = \frac{46\,704 \text{ lb} + 2002 \text{ lb}}{\text{WAS, } \frac{\text{lb}}{\text{d}}}$$

$$12 \text{ days} = \frac{48\,706 \text{ lb}}{\text{WAS, } \frac{\text{lb}}{\text{d}}}$$

$$\text{WAS, } \frac{\text{lb}}{\text{d}} = 4059$$

Step 2: Find the volume of WAS to remove from the system:

$$\text{WAS, } \frac{\text{lb}}{\text{d}} = (\text{Concentration, } \frac{\text{mg}}{\text{L}})(\text{WAS flow, mgd})(8.34 \frac{\text{lb}}{\text{mil. gal}})$$

$$4059 \frac{\text{lb}}{\text{d}} = (7400 \frac{\text{mg}}{\text{L}})(\text{WAS flow, mgd})(8.34 \frac{\text{lb}}{\text{mil. gal}})$$

$$\text{WAS flow} = 0.066 \text{ mgd}$$

The U.S. Environmental Protection Agency (U.S. EPA) created a graphical tool in 1977 to assist operators with daily sludge age calculations. Figures 4.1 and 4.2 may be used in lieu of the calculations on the previous page(s) to estimate the wasting rate required to maintain a target sludge age. Keep in mind that calculations will be more accurate. This tool may be used with either International Standard or U.S. customary units of measurement. The graphs are based on the SRT calculation rather than the MCRT calculation. Solids retention time does not include solids in the secondary clarifier. Directions for using the tool are included in the figures. For daily use, consider copying and laminating Figure 4.1. Then, use a ruler and dry erase marker to find the wasting rate.

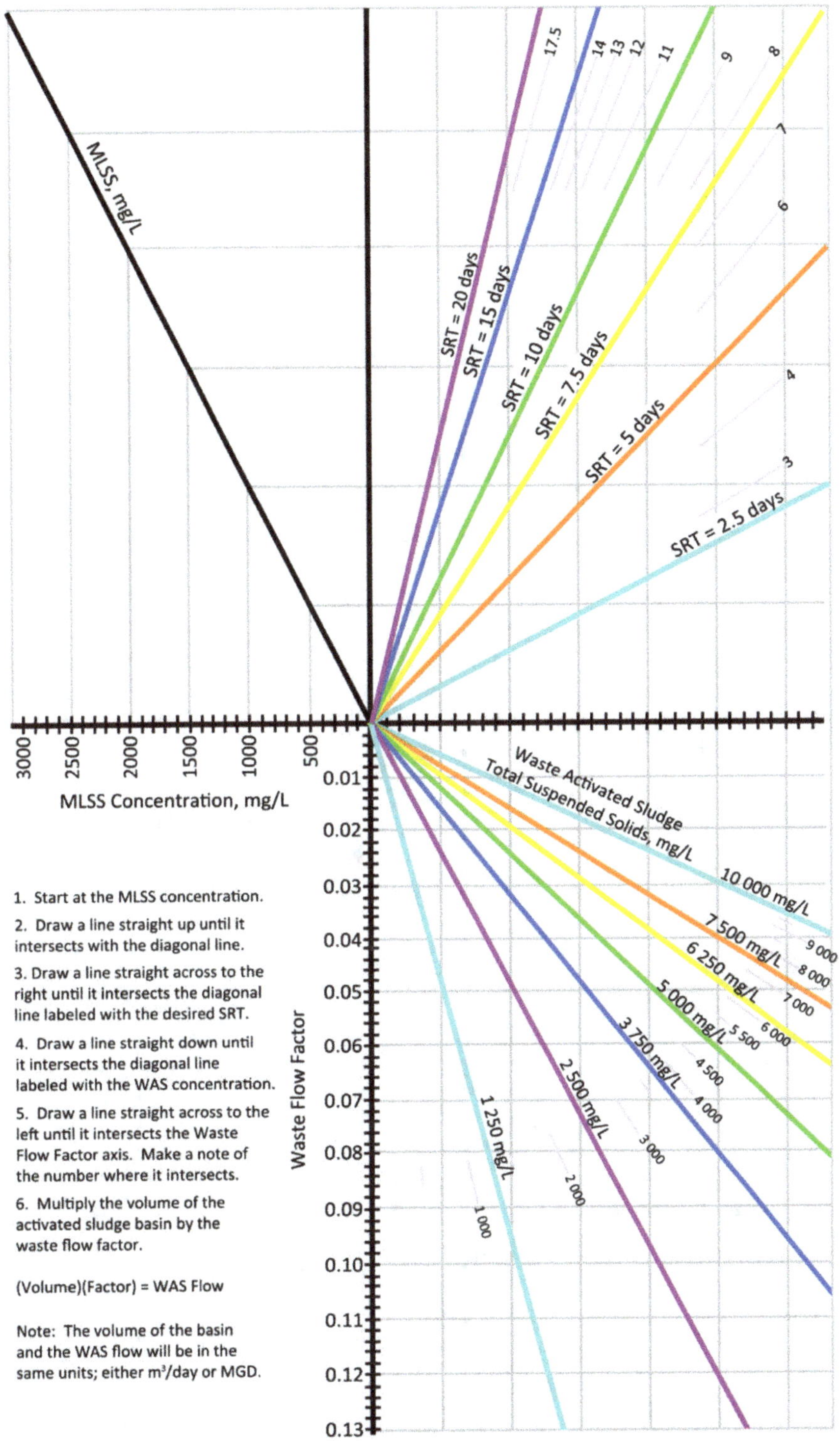

FIGURE 4.1 Graphical Tool for Estimating Wasting Rate for a Given Target SRT (Adapted from U.S. Environmental Protection Agency, 1977)

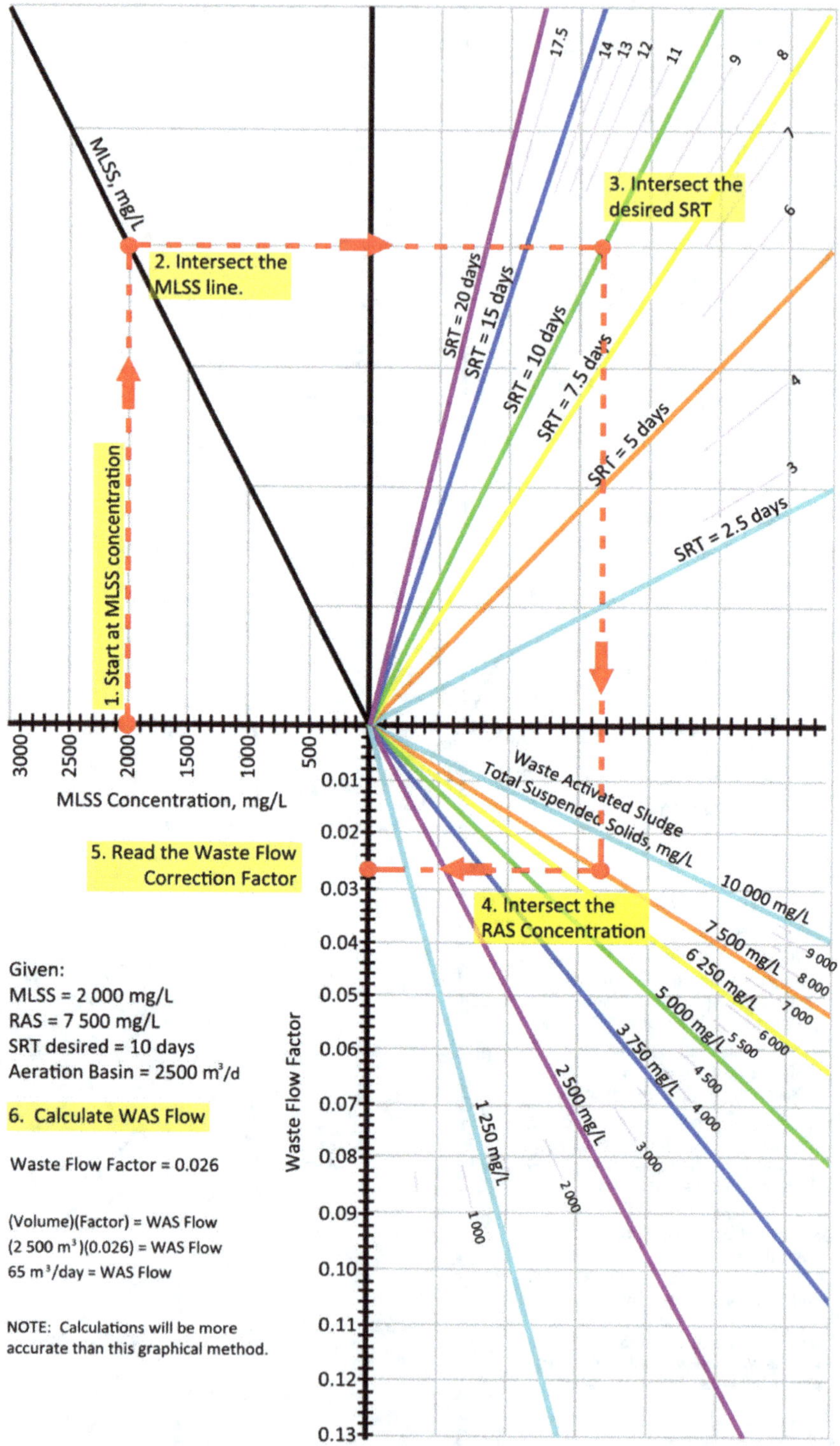

FIGURE 4.2 Example Calculation Using U.S. EPA Graphical Tool (Adapted from U.S. Environmental Protection Agency, 1977)

2.1.4 Advantages and Disadvantages

The MCRT, SRT, and $SRT_{aerobic}$ calculations are all based on the growth rates of microorganisms within the activated sludge process. Unlike the GSA, each of these equations accounts for the transformation of influent BOD into MLVSS during treatment, and they each provide a simple method for calculating the mass of waste solids to remove from the process each day. Instead of trying to predict the incoming load, the operator needs only maintain a constant percentage of solids removed from the process.

When BOD loading to the process increases or decreases, the number of new microorganisms produced will automatically increase or decrease according to the yield. The operator does not need to monitor or predict the influent load. If the same percentage of total mass is removed from the system each day, then the mass of solids remaining in the system will adjust to the influent load automatically and the influent BOD-to-MLVSS ratio (F:M) will remain relatively constant (Ekama, 2010). If the organic load to the process gradually increases, then the concentration of MLVSS will gradually increase to match the load. The converse is also true. Maintaining a constant sludge age can also reduce process upsets from sudden increases or spikes in organic loading rate.

Another advantage of using MCRT or SRT rather than the GSA or $SRT_{aerobic}$ is that much of the early research conducted on activated sludge processes was conducted using MCRT or SRT. For example, studies determining which sludge age ranges are likely to support the growth of different types of filamentous bacteria reference MCRT or SRT. Early researchers were not consistent about including solids in the clarifier blanket in their calculations. Continuing to use either MCRT or SRT allows for a comparison of current operating conditions with past research on filaments and helps operations staff determine when a filament bloom might be likely to occur.

Most domestic WRRFs experience diurnal variations in their influent flow, with the peak hour flow and load arriving between late morning and early afternoon. Exceptions include WRRFs with flow equalization basins or those with large service areas that damp out the peak hour flow. When peak flow enters the WRRF, it, along with RAS flow, pushes MLSS from activated sludge basins to secondary clarifiers. This can cause clarifier blankets to increase during portions of the day at some facilities. Because MCRT includes solids in all basins—whether they are aerated or not and in the clarifier blanket—grab samples can be collected from each location without regard to the time of day taken. Because all locations are monitored, the total sludge inventory will be accounted.

A potential disadvantage of using MCRT, SRT, or $SRT_{aerobic}$ for process control is that each of these calculations requires accurate and precise sampling and analysis of the MLSS in the activated sludge basins and WAS. Return activated sludge/WAS concentrations often change during the day, and it can be difficult to obtain representative samples from an intermittently operating WAS pump. Accurate measurement of MLSS and WAS concentrations requires training and specialized laboratory equipment such as an analytical balance, drying oven, vacuum pump, dessicator, and/or solids meter, which may be beyond the economic reach of some small systems. Centrifuge spin analysis may also be used to estimate MLSS and WAS concentrations; however, this method is less accurate. In the case of MCRT, it may not be possible to obtain a representative sample of the clarifier blanket if the clarifier is covered by panels or a dome. For systems that are unable to sample their clarifiers, SRT or $SRT_{aerobic}$ may be more appropriate.

Mean cell residence time/SRT control may be difficult to fully implement for systems that are not fully staffed. Ideally, operators will set a target MCRT/SRT and then collect and analyze samples each day for the MCRT/SRT calculation. Wasting rates should then be adjusted accordingly to maintain constant MCRT/SRT. Systems that are not fully staffed may not be able to collect the necessary data or perform the calculation daily. There are ways to compensate for less frequent MCRT/SRT calculations, including using a 3-, 5-, or 7-day running average MCRT/SRT. Using a running average helps to mitigate the effects of a single bad sampling event or a missing data point. A running average also provides a more realistic value for changes to wasting rates.

2.2 Solids Retention Time

2.2.1 Definition

The terms *SRT*, *MCRT*, and *sludge age* were used interchangeably throughout the early days of activated sludge and, in many cases, clarifier solids may or may not have been included in both calculations in different publications (Advanced Science Institute Series, 1983; Stall and Sherrard, 1978; U.S. Environmental Protection Agency, 1977). Today, MCRT is understood to include the solids in all activated sludge basins and the clarifier blanket, whereas SRT includes only the solids in the activated sludge basins. It is nearly impossible to pinpoint exactly when the terms *MCRT* and *SRT* took on their currently accepted definitions, but it appears that the first official definition of SRT was in *Process Design Manual for Upgrading Existing Wastewater Treatment Plants* (U.S. Environmental Protection Agency, 1974). This manual defined *SRT* as the mass of MLSS under aeration divided by the mass removed in the WAS and effluent each day. Here, solids are used in place of "mean cell" because all of the solids in the activated sludge basin are being included in the calculation, not just the volatile solids. This may seem like a fine point to make; however, as will be discussed later in this chapter, inert solids (sand, grit, eggshells, fibers, etc.) can compose a substantial percentage of the MLSS and they do not contribute to treatment. As late as 1977, there was considerable debate among researchers, engineers, and operators regarding whether solids in the clarifier blanket should be included in the sludge age calculation (U.S. Environmental Protection Agency, 1977). This debate continues today. There are multiple arguments for and against including the clarifier solids in the calculations, as shown in Table 4.3. There is a move in the wastewater community to use SRT rather than MCRT, particularly among design engineers. Many operator certification exams and state programs continue to use MCRT.

2.2.2 Calculation Method

The calculation method for SRT is almost identical to the calculation for MCRT. The only difference is that solids in the clarifier blanket are no longer included.

$$\text{Solids retention time} = \frac{\text{Activated sludge basin mass}}{\text{Waste activated sludge mass} + \text{Effluent solids mass}} \tag{4.6}$$

Mass may be calculated as either kilograms or pounds. In a well-functioning activated sludge process, the mass of effluent solids should be small compared to the mass of solids present in the activated sludge basin, clarifier blanket, and waste activated sludge. Often, effluent solids can be ignored and excluded from the SRT calculation. Refer back to Section 2.1.3 for example calculations.

2.2.3 Advantages and Disadvantages

The advantages of using SRT for process control are the same as those listed for MCRT control in Section 2.1.4 of this chapter. One additional advantage of using SRT over MCRT is that there is no need to collect and analyze samples of clarifier blankets. This reduces the total amount of time spent on sampling and analysis, which can be helpful for smaller facilities with limited staff.

A disadvantage of SRT control over MCRT control is that sampling may not account for a majority of the solids inventory. If an operator goes out to take a grab sample at 7:00 a.m. every morning and uses that value to calculate total mass of MLSS in the activated sludge basins, they can be confident that a majority of the MLSS has been accounted for by that sample. There should be a very small amount of MLSS in the clarifier blankets

TABLE 4.3 Inclusion of Clarifier Blankets Solids in the Sludge Age Calculation (Reprinted with permission from Indigo Water Group)

Arguments against inclusion	Arguments for inclusion
Collecting a representative sample from a clarifier blanket is not always possible. Clarifiers with covers or those in buildings with low ceilings can be difficult to access with a long core sampler.	Clarifier blanket depth can be estimated with a handheld TSS meter or by lowering a modified core sampler into the clarifier on a string. Once the blanket depth is known, the mass of solids in the clarifier can be calculated if the blanket concentration is assumed to be equal to the RAS concentration.
Microorganisms in the clarifier blanket do not have access to food and oxygen. They do not contribute to treatment while in the clarifier.	Microorganisms in the clarifier blanket are only in the blanket for a short period of time each cycle. They are exposed to both food and oxygen most of the time and do contribute to overall treatment capacity.
Assuming that clarifier blankets are maintained below 0.6 m (2 ft), the percentage of solids in the clarifier is insignificant compared to the total inventory. Ignoring these solids will not significantly change the results of the sludge age calculation. When blanket levels are relatively constant, the error in the sludge age calculation will also be constant.	For facilities that carry deeper sludge blankets in the clarifier, experience poor sludge settleability, or have large clarifiers relative to their activated sludge basins, the portion of solids in the clarifier blanket can be significant. Ignoring the solids in these situations can have a dramatic effect on the results of the sludge age calculation for some facilities. To be clear, deep sludge blankets in clarifiers are not recommended because this practice can degrade effluent quality and affect MLSS settling behavior.
Design engineers use process kinetics to design and size activated sludge basins. These calculations only include solids in the basins, not those in the clarifier.	Kinetics equations used to size activated sludge basins predict the total mass of solids produced. This is the yield discussed in Chapter 2. Although some of those solids may be in the clarifier temporarily, they are still included in design equations, even though, often, engineers assume that 100% of the sludge inventory remains in activated sludge basins.
Loading parameters (kg BOD/m^3 or lb BOD/1000 cu ft) for the activated sludge process were originally developed using only the MLSS in the activated sludge basin and not the MLSS in the clarifiers.	Again, kinetics dictate the mass of MLSS produced per mass of BOD or COD received. Some of the MLSS may temporarily be in the clarifier, but it would have been produced in response to the influent load.

because peak flow, for most WRRFs, will not arrive until later in the day. Recall that influent flow and RAS flow push MLSS from the activated sludge basins to the clarifiers. On a day when multiple interruptions occur—for example, an early morning safety meeting, an operator calls in sick, and a water main break requires an emergency response—the 7:00 a.m. grab sample may not be collected until after peak hour flow starts entering the WRRF. Now, the concentration of MLSS in the activated sludge basin may be slightly lower because there are more solids (potentially) sitting in the clarifier blanket. When the operator does the SRT calculation, the mass to waste will naturally be lower, and sludge age will increase for that day. While the differences in the MLSS concentration and SRT calculation result should not be large, they can create more variation in process control charts and may cause the operator to make unnecessary process changes. If, on the other hand, the operator collects grab samples from the activated sludge basins and clarifiers at about the same time (i.e., a few minutes apart), then 100% of the sludge inventory has been accounted for, regardless of the time of day the sample is collected or the influent flow regimen. When clarifiers are operated to keep the sludge blanket below 0.6 m (2 ft) at all times, differences between the MCRT and SRT calculation results should be small, regardless of the time of day that samples are collected for most facilities. An exception to this general rule is if facilities have large clarifiers relative to their activated sludge basins. Composite samplers may be used to collect samples of MLSS, RAS, and WAS, which can help eliminate inaccuracies associated with grab samples.

2.3 Solids Retention Time$_{Aerobic}$

Design engineers use $SRT_{aerobic}$ to determine the size of aerated basins for facilities that must remove ammonia. Bacteria responsible for converting ammonia to nitrate (nitrifying bacteria) grow much slower than bacteria responsible for removing BOD, especially at colder water temperatures. Nitrifiers are strict aerobes and cannot function without an adequate supply of dissolved oxygen. Biochemical oxygen demand removal, on the other hand, can take place under anaerobic, anoxic, and aerated conditions. The characteristics of nitrifying bacteria focus the design and operation of nitrifying activated sludge processes on the amount of aerobic activated sludge basin volume or time. Some WRRFs use $SRT_{aerobic}$ for process control, while others calculate both total SRT and $SRT_{aerobic}$.

2.3.1 Definition

The $SRT_{aerobic}$ is defined as total SRT multiplied by the ratio of the total activated sludge basin volume to the aerated volume (Argaman and Brenner,

1986). The aerated volume is further defined as any basin or zone containing dissolved oxygen at concentrations greater than 0.2 mg/L. The $SRT_{aerobic}$ calculation does not include any MLSS that may be in anaerobic zones or basins, anoxic zones or basins, or in the clarifier blanket. For an SBR-type activated sludge process, the $SRT_{aerobic}$ includes only the amount of time that the dissolved oxygen concentration is higher than 0.3 mg/L. The concentration of 0.3 mg/L was selected as the cutoff point between aerobic and anoxic conditions because this is the concentration of dissolved oxygen found to inhibit denitrification (Knowles, 1982). Chapter 2 provides more in-depth definitions of anaerobic, anoxic, and aerated conditions.

2.3.2 Calculation Method

For an activated sludge process with multiple basins or zones,

$$\text{Solids residence time}_{\text{aerobic}} = (\text{Solids residence time}_{\text{total}})\left(\frac{\text{Volume}_{\text{aerobic}}}{\text{Volume}_{\text{total}}}\right) \quad (4.7)$$

For an SBR,

$$\text{Solids residence time}_{\text{aerobic}} = (\text{Solids residence time}_{\text{total}})\left(\frac{\text{Time}_{\text{aerobic}}}{\text{Time}_{\text{total}}}\right) \quad (4.8)$$

2.3.3 Calculation Example

Find the $SRT_{aerobic}$ for a WRRF with the following characteristics:

Total activated sludge basin volume = 7570 m^3 (2 mil. gal)
Aerated basin volume = 4921 m^3 (1.3 mil. gal)
Anoxic basin volume = 2650 m^3 (0.7 mil. gal)
MLSS concentration = 2800 mg/L
WAS flowrate = 303 m^3/d (0.08 mgd)
WAS concentration = 7400 mg/L
Effluent TSS concentration = negligible

Solution in International Standard Units:

Step 1: Find the mass of solids in the activated sludge basin:

$$\text{Activated sludge basin, kg} = \frac{\left[(\text{Concentration}, \frac{\text{mg}}{\text{L}})(\text{Volume, m}^3)\right]}{1000}$$

$$\text{Activated sludge basin, kg} = \frac{\left[(2800\ \frac{\text{mg}}{\text{L}})(7570\ \text{m}^3)\right]}{1000}$$

Activated sludge basin mass = 21 196 kg

Step 2: Find the mass of solids wasted:

$$\text{WAS, } \frac{\text{kg}}{\text{d}} = \frac{\left[(\text{Concentration, } \frac{\text{mg}}{\text{L}})(\text{WAS flow, } \frac{\text{m}^3}{\text{d}})\right]}{1000}$$

$$\text{WAS, } \frac{\text{kg}}{\text{d}} = \frac{\left[(7400 \frac{\text{mg}}{\text{L}})(\text{WAS flow, } \frac{\text{m}^3}{\text{d}})\right]}{1000}$$

$$\text{WAS, } \frac{\text{kg}}{\text{d}} = 2242$$

Step 3: Solve for SRT:

$$\text{Solids retention time} = \frac{\text{Activated sludge basin mass}}{\text{Waste activated sludge mass} + \text{Effluent solids mass}}$$

$$\text{Solids retention time, days} = \frac{21\ 196\ \text{kg}}{2242 \frac{\text{kg}}{\text{d}} + 0}$$

$$\text{Solids retention time, days} = \frac{21\ 196\ \text{kg}}{2242 \frac{\text{kg}}{\text{d}}}$$

Solids retention time = 9.45 days

Step 4: Find the $\text{SRT}_{\text{aerobic}}$:

$$\text{Solids retention time}_{\text{aerobic}} = (\text{Solids retention time}_{\text{total}})\left(\frac{\text{Volume}_{\text{aerobic}}}{\text{Volume}_{\text{total}}}\right)$$

$$\text{Solids retention time}_{\text{aerobic}} = (9.45\ \text{days})\left(\frac{4921\ \text{m}^3}{7570\ \text{m}^3}\right)$$

$\text{Solids retention time}_{\text{aerobic}}$ = 6.14 days

Solution in U.S. Customary Units:

Step 1: Find the mass of solids in the activated sludge basin:

$$\text{Activated sludge basin, lb} = (\text{Concentration, } \frac{\text{mg}}{\text{L}})(\text{Volume, mil. gal})(8.34\frac{\text{lb}}{\text{mil. gal}})$$

$$\text{Activated sludge basin, lb} = (2800\ \frac{\text{mg}}{\text{L}})(2\text{ mil. gal})(8.34\frac{\text{lb}}{\text{mil. gal}})$$

Activated sludge basin mass = 46 704 lb

Step 2: Find the mass of solids in the waste activated sludge:

$$\text{WAS, }\frac{\text{lb}}{\text{d}} = (\text{Concentration, }\frac{\text{mg}}{\text{L}})(\text{WAS flow, mgd})(8.34\frac{\text{lb}}{\text{mil. gal}})$$

$$\text{WAS, }\frac{\text{lb}}{\text{d}} = (7400\frac{\text{mg}}{\text{L}})(0.08\text{ mil. gal})(8.34\frac{\text{lb}}{\text{mil. gal}})$$

$$\text{WAS, }\frac{\text{lb}}{\text{d}} = 4937$$

Step 3: Solve for SRT:

$$\text{Solids retention time} = \frac{\text{Activated sludge basin mass}}{\text{Waste activated sludge mass} + \text{Effluent solids mass}}$$

$$\text{Solids retention time, days} = \frac{46\ 704\text{ lb}}{4937\frac{\text{lb}}{\text{d}} + 0}$$

$$\text{Solids retention time, days} = \frac{46\ 704\text{ lb}}{4937\frac{\text{lb}}{\text{d}}}$$

Solids retention time = 9.46 days

Step 4: Find the $\text{SRT}_{\text{aerobic}}$:

$$\text{Solids retention time}_{\text{aerobic}} = (\text{Solids retention time}_{\text{total}})\left(\frac{\text{Volume}_{\text{aerobic}}}{\text{Volume}_{\text{total}}}\right)$$

$$\text{Solids retention time}_{\text{aerobic}} = (9.46 \text{ days})\left(\frac{1.3 \text{ mil. gal}}{2 \text{ mil. gal}}\right)$$

$$\text{Solids retention time}_{\text{aerobic}} = 6.15 \text{ days}$$

The SRT calculation shown here uses the same WRRF data as the MCRT calculation in Section 2.1.3. In this example, the SRT was 9.45 days, and the MCRT was 9.86 days. Excluding solids in the clarifier blanket made little difference in the sludge age calculation because the percentage of sludge in the clarifier blanket was small.

2.3.4 Advantages and Disadvantages

There are no real disadvantages of $SRT_{aerobic}$ other than those already given in Section 2.2.3 on SRT. The advantages of using $SRT_{aerobic}$ for process control are the same as those listed for MCRT and SRT control in Sections 2.1.4 and 2.2.3 of this chapter and in Table 4.3. Water resource recovery facilities that are required to remove ammonia may calculate $SRT_{aerobic}$ to ensure that there is enough aerated volume and/or time to support full and stable nitrification. For activated sludge facilities that were designed with separate basins or zones for anaerobic, anoxic, and aerated conditions, the $SRT_{aerobic}$ calculation was done by the design engineer, and the ratio of $SRT_{aerobic}$ to SRT_{total} is fixed. While the operator is unable to change the ratio in this case, he or she is able to indirectly control the $SRT_{aerobic}$ by controlling SRT_{total}. For activated sludge facilities that practice on/off aeration and for SBRs, the ratio of $SRT_{aerobic}$ to SRT_{total} is controlled in time rather than in space. In these situations, the operator has more direct control over the $SRT_{aerobic}$ and can adjust it to almost any percentage of the total. For additional discussion on minimum $SRT_{aerobic}$ required for nitrification, refer to Section 5.0 of this chapter that discusses selecting a target sludge age.

2.4 Dynamic Solids Retention Time

2.4.1 Definition

All of the different sludge age calculations discussed so far assume that steady-state conditions exist in the activated sludge process. In other words, grab samples collected for the calculations are representative of process conditions, regardless of what time of day they were collected, and that changes are not taking place in the process. Steady state also assumes that the process has completely equilibrated to past changes. These assumptions rarely hold true under field conditions because the activated sludge process is constantly changing and adapting as influent flows, loads, and other environmental conditions are constantly changing. This is especially true for industrial WRRFs, municipal facilities that receive large quantities of industrial wastewater, and WRRFs that have high peak-hour flows compared to their average daily flows. Another potential issue with the standard MCRT/SRT calculations is that facilities typically use 3-, 5-, or 7-day moving averages to smooth away variations in daily MCRT/SRT calculations when making process control decisions. As a result, the MCRT/SRT in use is an approximation of the true age of the sludge.

Takács et al. (2008) proposed a new method for calculating sludge age called *dynamic solids retention time* (DSRT). Dynamic solids retention time attempts to compensate for non-steady-state conditions by determining the true SRT. A key component of DSRT is that it compares today's sludge age to yesterday's sludge age. The underlying assumption behind DSRT is that the true SRT is a combination of the SRT of the solids that were already in the system (yesterday's sludge) and the SRT of the new solids produced in the system today (today's sludge).

2.4.2 Calculation Method

Dynamic solids retention time is calculated as follows:

$$\mathrm{SRT}_{\mathrm{today}} = (\mathrm{SRT}_{\mathrm{yesterday}} + 1)\left(\frac{\mathrm{SRT}_{\mathrm{yesterday}} F_{\mathrm{p}}}{M_{\mathrm{today}}}\right) \tag{4.9}$$

Where

$\mathrm{SRT}_{\mathrm{today}}$ = actual SRT for a given day,
$\mathrm{SRT}_{\mathrm{yesterday}}$ = previous days SRT (days),
M_{today} = mass of solids in the system (kg [lb]), and
F_p = Sludge production in the system (kg/d [lb/d]).

The equation has two components. The sludge that was in the system yesterday and is still in the system will have a sludge age equal to yesterday's sludge age plus one day. This is the first term. The second term says that yesterday's sludge age will be increased or decreased according to how many new solids are produced and remain in the process. Sludge production can be estimated from calculations; however, accurate estimates require the operator to know influent BOD, TSS, and total VSS concentrations at a minimum. When possible, sludge production should be estimated using the quantity of sludge hauled from the facility. Refer to literature by Takács et al. (2008) for a detailed, explanation of calculating DSRT.

2.4.3 Advantages and Disadvantages

The principal advantage of the DSRT approach is that it eliminates the need to use running-average SRT calculations. Dynamic solids retention time also does a better job at estimating sludge age under certain conditions, such as when wasting does not occur for a period of time. If no wasting occurred for a day or two, the traditional MCRT/SRT calculations give unrealistically high numbers that can skew the running average. The disadvantages of using DSRT are that it requires a lot of detailed information about the influent loads to the WRRF and the amount of solids produced each day and that the calculations are fairly complicated. Dynamic solids retention time may be useful for WRRFs that experience large swings in flows and loads from day to day but is of limited use to WRRFs operating under more stable conditions.

2.5 Return Activated Sludge/Waste Activated Sludge Concentrations

Obtaining accurate estimates of RAS and WAS concentrations can be difficult because sampling locations are not representative, pumps may operate intermittently, grab samples do not reflect changing conditions, composite sampling may change the solids characteristics, and the RAS flow to influent flow ratio may be changing. Fortunately, if the influent and RAS flows are both known, it is possible to calculate the RAS concentration from the MLSS concentration. Increasing the RAS flowrate decreases the amount of time that solids are allowed to remain in the clarifier. Less settling time means that RAS will be less concentrated. Decreasing the RAS flowrate increases the amount of time that solids are allowed to remain in the clarifier. More settling time means that RAS will be more concentrated, but only to a point. Compaction is limited by MLSS characteristics.

Many WRRFs are able to set the RAS flow as a fixed percentage of influent flow. This practice is called *flow pacing*. In this situation, the RAS concentration will remain relatively constant over the course of a day as long as the percentage of solids in the clarifier remains low relative to the percentage of solids in the activated sludge basin; that is, the clarifier is not building a blanket. Flow pacing can be difficult to maintain during certain times of the day because of pumping limitations. In some cases, RAS pump(s) cannot be turned down far enough and, as a result, the percentage of RAS to influent flow increases. This causes the RAS concentration to decrease. In other cases, RAS pumping capacity becomes the limiting factor. For example, if influent flow to a facility peaks at 2000 m^3/d (0.53 mgd) but RAS pump(s) have a capacity of only 1000 m^3/d (0.26 mgd), then the percentage of RAS that may be returned during peak hour influent flows will only be 50%. In this situation, the clarifier may begin to accumulate solids and build a blanket. When influent flows decrease, the RAS flow percentage will increase again and solids will be removed from the clarifier more quickly, which draws the blanket (if any) back down.

Some WRRFs do not have the ability to flow pace their RAS pump(s). In this situation, the operator may decide to either (1) manually adjust the RAS flow multiple times over the course of a day or (2) select a fixed flowrate for the entire day. When the RAS flow is fixed, solids will enter the clarifier faster during some parts of the day and slower during others but will be removed at the same rate. As a result, the RAS concentration will vary over the course of the day. The clarifier solids loading rate and other clarifier-related process control variables are discussed in Chapter 6.

2.5.1 Calculation Method

The maximum RAS concentration achievable by the clarifier can be estimated from settleometer test results. As indicated previously, settleometer tests are typically conducted using a 1000 mL graduated cylinder. For the test, a sample of MLSS is collected from the end of the activated sludge basin, poured into the 1000 mL graduated cylinder and allowed to settle for 30 minutes. The settleometer mimics how the MLSS settles in the clarifier. A blanket should form in the bottom of the settleometer, as shown in Figure 4.3. After 30 minutes, the volume of the settled sludge is recorded. If the MLSS is allowed to remain in the clarifier for 30 minutes, then it should

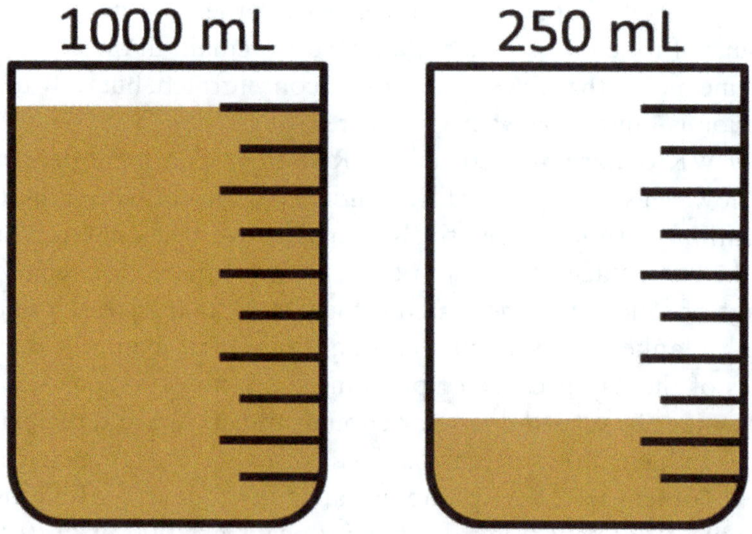

FIGURE 4.3 Settleometer Test (Reprinted with permission from Indigo Water Group)

compact to the same degree as the MLSS in the settleometer. If the MLSS concentration is known, the settled sludge concentration is easily calculated using the following formula:

$$SSC_{30} = \left(\frac{\text{Initial settleometer volume, mL}}{SSV_{30}}\right) \times \text{MLSS, mg/L} \qquad (4.10)$$

Where

SSC_{30} = settled sludge concentration after 30 minutes and
SSV_{30} = settled sludge volume after 30 minutes.

Often, the MLSS will be in the clarifier for fewer than 30 minutes and will not achieve maximum compaction. This is especially true for WRRFs that must remove ammonia. The actual WRRF RAS concentration may be estimated from the following equation (influent flow and RAS flow must be in the same units):

$$X_{RAS} = \left(\frac{Q_{influent}}{Q_{RAS}} + 1\right) \times X_{MLSS} \qquad (4.11)$$

Where

X_{RAS} = RAS concentration,
$Q_{influent}$ = influent flowrate,
Q_{RAS} = RAS flowrate, and
X_{MLSS} = MLSS concentration.

2.5.2 Calculation Example

The influent flow to a WRRF is 600 m^3/d (0.16 mgd) at 9:00 a.m. and increases to 1500 m^3/d (0.40 mgd) at 2:00 p.m. The MLSS concentration is 2800 mg/L. Assuming that the blanket in the clarifier remains below 0.6 m (2 ft) over the course of the day, calculate the RAS concentration at both 9:00 a.m. and 2:00 p.m. Assume that the RAS pump rate is fixed at 900 m^3/d. The settled sludge volume at 30 minutes is 420 mL.

Step 1: Find the RAS concentration at 9:00 a.m.:

$$X_{RAS} = \left(\frac{Q_{influent}}{Q_{RAS}} + 1\right) \times X_{MLSS}$$

$$X_{RAS} = \left(\frac{600\ \text{m}^3/\text{d}}{900\ \text{m}^3/\text{d}} + 1\right) \times 2800\ \text{mg/L}$$

$$X_{RAS} = (0.67 + 1) \times 2800\ \text{mg/L}$$

$$X_{RAS} = 4667\ \text{mg/L}$$

Step 2: Find the RAS concentration at 2:00 p.m.:

$$X_{RAS} = \left(\frac{Q_{\text{influent}}}{Q_{RAS}} + 1\right) \times X_{MLSS}$$

$$X_{RAS} = \left(\frac{1500\ \text{m}^3/\text{d}}{900\ \text{m}^3/\text{d}} + 1\right) \times 2800\ \text{mg/L}$$

$$X_{RAS} = (1.7 + 1) \times 2800\ \text{mg/L}$$

$$X_{RAS} = 7467\ \text{mg/L}$$

Step 3: Compare the calculated RAS concentrations to the settleometer:

$$SSC_{30} = \left(\frac{\text{Initial settleometer volume, mL}}{SSV_{30}}\right) \times \text{MLSS, mg/L}$$

$$SSC_{30} = \left(\frac{1000\ \text{mL}}{420\ \text{mL}}\right) \times 2800\ \text{mg/L}$$

$$SSC_{30} = 6667\ \text{mg/L}$$

In this example, the settleometer result indicates that the maximum RAS concentration achievable by the clarifier is 6667 mg/L. The RAS ratio calculation estimated a higher RAS concentration at 7467 mg/L. Because the sludge is not settling and compacting well, the RAS concentration at 2:00 p.m. will be 6667 mg/L and not 7467 mg/L, assuming the MLSS remains in the clarifier for 30 minutes. Experienced operators will observe times when a slow-settling sludge continues to compact in the settleometer after 30 minutes; however, operators should be mindful that turning the RAS down to exceedingly low levels in an attempt to further thicken and compact the solids in the clarifier blanket is counterproductive and can cause effluent quality to deteriorate.

2.5.3 Sampling Locations for Return Activated Sludge/Waste Activated Sludge

Ideally, samples will be collected from well-mixed locations that are representative of the system as a whole. Splitter boxes, joiner boxes, flumes, and pump discharges all make excellent sampling locations for determining solids concentrations because they are well mixed. Figure 4.4 shows two possible configurations for RAS and WAS withdrawal from a clarifier. In the first scenario, the RAS and WAS pumps are pulling from the same pipeline. There may be a hopper or pit at the bottom of the clarifier where the pipeline enters the tank. The RAS pump operates continuously and the RAS concentration can be predicted using Equation 4.11. The WAS pump operates intermittently, but because it pulls from the same pipeline as the RAS pump, WAS and RAS concentrations will be the same.

In the second scenario, the RAS and WAS pumps do not share a pipeline. The RAS pipeline withdraws from the clarifier at a higher elevation than the WAS pipeline. The WAS concentration is often higher than the RAS concentration because the MLSS has had more time to settle and compact before reaching the WAS withdrawal point. It becomes more difficult to estimate the WAS concentration using Equation 4.11. This disparity can be exaggerated by intermittent operation of the WAS pump. If the WAS pump only operates for 10 or 15 minutes out of each hour, then grit and heavier particles tend to accumulate in the WAS hopper while lighter floc is removed and recycled through the RAS line. If a grab sample is collected from the WAS pump immediately after it starts up, these heavier materials may fool

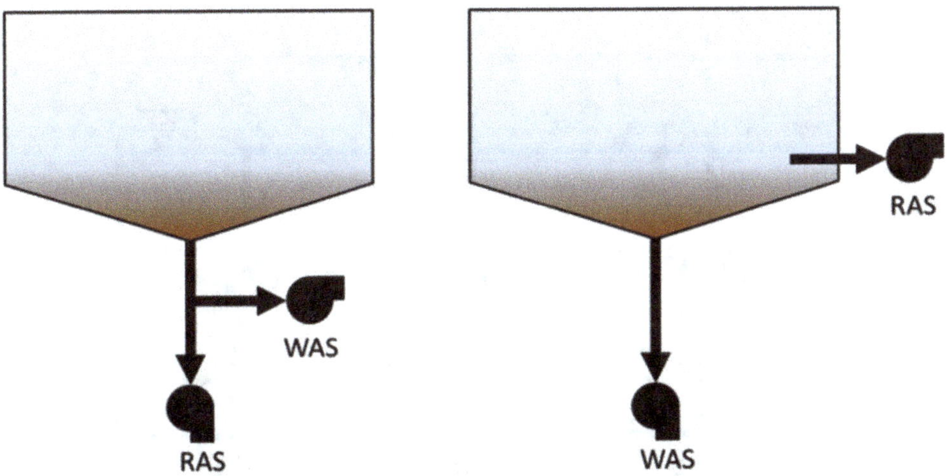

FIGURE 4.4 Return Activated Sludge and WAS Withdrawal Points (Reprinted with permission from Indigo Water Group)

an operator into believing that the WAS concentration is much higher than it is the case. As soon as the hopper has been emptied, the WAS concentration will become approximately the same as the RAS concentration. For this reason, a series of grab samples should be collected at the beginning, middle, and end of a pumping cycle to measure the true WAS concentration. If the WAS concentration estimate is incorrect, then the MCRT/SRT calculations will also be incorrect.

Some activated sludge processes waste directly from the activated sludge basin rather than from the bottom of the clarifier. Wasting from the activated sludge basin simplifies process control because the operator sets the wasting rate as a fixed percentage of the activated sludge basin volume. Careful measurement of MLSS and RAS/WAS concentrations is no longer needed. Disadvantages of this approach are that a much larger volume of MLSS must be removed each day and that a prethickening step will likely be needed before sending wasted MLSS to a solids-handling process such as a digester.

Figure 4.5 illustrates an RAS/WAS pump gallery, where all three pumps are manifolded together. For this particular WRRF, any combination of clarifiers can be placed into service with one or more RAS pumps. Because

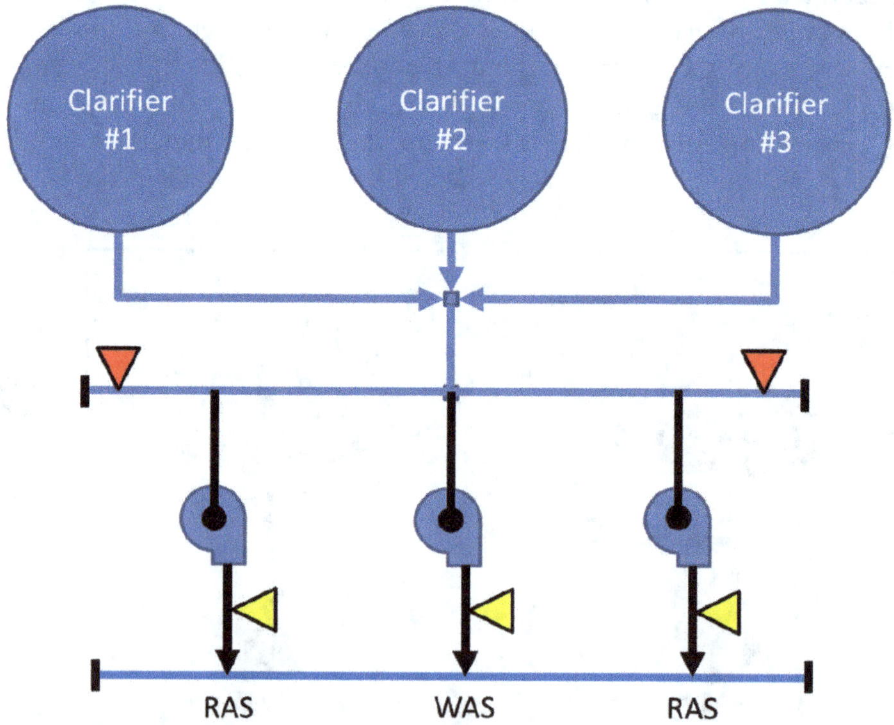

FIGURE 4.5 Return Activated Sludge/WAS Pump Gallery Diagram (Reprinted with permission from Indigo Water Group)

all of the pumps are manifolded together, a sample collected from any one pump should be comparable to any other pump. The WAS and RAS concentrations will be equal.

For the example gallery shown as Figure 4.5, the design engineer provided two sample taps, shown as red triangles in figure, that were placed at opposite ends of the manifold. Operations staff diligently flushed the line before collecting samples for process control. Mixed liquor suspended solids concentrations measured at either of these sample taps were extremely erratic and were often lower in concentration than the MLSS in the activated sludge basins. It was clear that sampling points were not representative. It was theorized that the MLSS was settling out at the ends of the pipe. Never knowing the true WAS concentration made it impossible for the operators to use MCRT/SRT control.

Facility operators installed three new sampling taps, shown as yellow triangles in Figure 4.5. By placing the sample taps downstream of the pumps, they ensured that RAS would be well mixed before sample collection. Measured RAS concentrations have been consistent since the sampling points were moved. Return activated sludge results agree with the concentrations predicted using the RAS-to-influent flow ratio calculation discussed in Section 2.5.1.

2.6 When to Include Effluent Total Suspended Solids

Clean Water Act set secondary treatment standards for 5-day BOD (BOD_5) and TSS removal. For mechanical facilities like activated sludge processes, final effluent concentrations for BOD_5 and TSS cannot exceed 30 mg/L for the 30-day average concentration or 45 mg/L for the 7-day average concentration (Secondary Treatment, 1993). These limits represent the expected level of effluent quality attainable. Many, if not most, activated sludge processes perform better than this and routinely discharge BOD_5 and TSS concentrations less than 10 mg/L. For a WRRF that is meeting secondary treatment standards, the mass of TSS leaving in the final effluent should be small compared to the mass in the activated sludge basin(s) and clarifier(s). Table 4.4 compares the effect of including, or not including, effluent TSS in both SRT and MCRT calculations for a conventional activated sludge system. Facility data are given in Section 2.1.3 of this chapter. To calculate the mass of TSS leaving in the effluent, an effluent flow is needed. For this example, an effluent flow of 30 280 m^3/d (8 mgd) was chosen, giving an HRT in the activated sludge basin of 6 hours at average daily flow.

Using the above data, the results indicate that when the effluent TSS is below 15 mg/L, the effect of effluent TSS on the sludge age calculations is relatively small at 1.7 days. In this case, effluent TSS could be safely left

TABLE 4.4 Effect of Effluent TSS on SRT and MCRT Calculations (Reprinted with permission from Indigo Water Group)

Effluent TSS concentration and mass	MCRT, days		SRT, days	
	Effluent TSS included	Effluent TSS not included	Effluent TSS included	Effluent TSS not included
15 mg/L 454.2 kg (1001 lb)	8.2	9.9	7.9	9.5
30 mg/L 908.4 kg (2001.6 lb)	7.0	9.9	6.7	9.5
45 mg/L 1362.6 kg (3002.4 lb)	6.1	9.9	5.9	9.5
60 mg/L 1816.8 kg (4003.2 lb)	5.5	9.9	5.2	9.5
100 mg/L 3028.0 kg (6672.0 lb)	4.2	9.9	4.2	9.9

out of the SRT or MCRT calculation for most facilities. Section 5.0 of this chapter discusses minimum sludge ages required to foster the growth of nitrifier populations (ammonia oxidizing bacteria and nitrite oxidizing bacteria) and maintain stable nitrification at different water temperatures. It is possible for a facility to nitrify well at an SRT of 9.5 days but be unable to nitrify at an SRT of 7.9 days. In this situation, the effluent TSS concentration could have a huge effect on whether the WRRF is able to continue meeting its discharge permit limits. When determining whether to include effluent TSS in the MCRT or SRT calculation, operators must take permit limits into consideration as well as the need to select a sludge age that provides an adequate buffer for errors in sampling, analysis, and calculations. Operators may also choose to leave effluent TSS out of their calculation and consistently operate with a slightly higher sludge age to compensate. With this strategy, facilities with effluent ammonia limits should keep careful watch on effluent ammonia concentrations. If they increase even slightly, the sludge age may need to be increased. As effluent TSS increases and the facility is no longer in compliance, effluent TSS has a larger and larger effect on SRT and MCRT calculations. As sludge age increases beyond 10 days, the effect of effluent TSS increases because less sludge is being removed from the system overall and the mass of solids leaving in the final effluent may be nearly the same as the mass of solids leaving in the WAS.

Larger facilities often have daily sampling requirements and will collect a sample for effluent TSS every day. If data are available, they should be included in the MCRT or SRT calculation for greater accuracy. Including the effluent TSS routinely will provide a more accurate record over time, especially during periods of process upset. Some facilities may wish to exclude effluent TSS from their daily process calculations for regulatory reasons. Typically, if samples of the final effluent are collected and analyzed more frequently than is required by the discharge permit, sample results must be included in the monthly monitoring report to the state or U.S. EPA. Many states exempt samples collected solely for process control purposes if they are taken in a different manner or location than the permit requires and/or if they are analyzed by a non-U.S. EPA-approved method. A sample collected from secondary clarifier effluent may be interpreted by a regulatory agency as representative of the final effluent, even though it is collected upstream of disinfection because secondary treatment is complete. Collecting grab samples for process control can complicate permit reporting because they may not be representative of average day operation, but a regulatory agency may wish to have the results included in the monthly monitoring report. Be sure to check with your regulatory agency and report all sample results as required.

Some activated sludge facilities have settling ponds or tertiary filters that remove solids from the process without recycling them back to activated sludge basins. It may be necessary to include these solids in the SRT and MCRT calculations as well.

2.7 Accounting for Solids Recycle from Downstream Processes

Many WRRFs send WAS to thickening processes such as gravity thickeners, rotary drum thickeners, and dissolved air flotation thickeners. However, thickeners are not 100% efficient at capturing solids. Typically, excess liquid and uncaptured solids are returned to facility headworks downstream of flow monitoring or sometimes directly to the activated sludge basin. Either way, returned solids effectively reduce the mass of WAS removed from the system and can affect the MCRT/SRT (Johnson et al., 2000). If returned solids are still active and involved in treatment, then they may need to be accounted for in the MCRT/SRT equation. This is especially true when thickening processes are not functioning well and solids capture is low. Solids that have been treated with chemicals or passed through digestion should not be considered active. Johnson et al. (2000) proposed a method for compensating for returned, active solids using the following formula:

$$\text{MCRT/SRT} = \frac{\text{Mass of MLSS in the system}}{\text{Mass wasted} + \text{Mass in effluent} + \text{Mass returned to the process}} \quad (4.12)$$

2.8 Potential Effects of Foaming Events on Sludge Age

During a foaming event, it is possible for a significant portion of the sludge inventory to be entrained in the foam. Foam solids are not easily quantified and, typically, are not included as part of the total sludge inventory. As a result, during a foaming event, the sludge age may be higher than sludge age calculations indicate. In extreme foaming events, foam may overflow activated sludge basins and end up outside of the basins on the ground and walkways. The loss of solids decreases sludge age; however, because this lost foam is almost impossible to measure, it may not be accounted for in sludge age calculations. In this case, the sludge age may be much lower than sludge age calculations indicate.

2.9 Wasting Frequency

The SRT or MCRT should be calculated at least once per day and wasting rates should be adjusted accordingly to maintain a constant or near constant SRT/MCRT. Wasting over longer periods of time, 10 minutes out of every hour, for example, is preferable because this practice maintains a more constant SRT/MCRT, which, in turn, maintains sludge settleability. Mass wasting of solids should be avoided because this does not control SRT/MCRT and can cause sudden changes in sludge settleability. As a general rule, SRT/MCRT should not change by more than approximately 15% of the previous day's values (this is kilograms [pounds] removed, not cubic meters [gallons]).

Wasting frequency and duration at some facilities may be limited by staffing or pumping capacity. For example, if the maximum WAS pumping rate is 5 m^3/min and the operator needs to waste 4000 m^3/d to maintain a desired SRT, then wasting will have to be spread out over 800 minutes (13.3 hours). Package plants are small, typically skid-mounted or trailer-type activated sludge facilities that, often, are used to treat wastewater from small areas such as an individual business, trailer park, camp, or subdivision. Often, package plants lack automatic controls and may be staffed only a few days per week. Fortunately, many package plants do not have ammonia limits in their discharge permits, so tight control over SRT is less important than it is for larger facilities. Still, the SRT must be controlled well enough for the MLSS to flocculate, settle, and compact well. Wasting should be done at as high a frequency as possible, with daily being ideal and several days each week being less ideal.

In the early days of activated sludge, extended aeration activated sludge processes were allowed to "self-waste" into the final effluent. This operational "technique" is documented in several U.S. EPA manuals and operator training guides. Self-wasting occurs when the MLSS concentration increases to the point at which the clarifier is no longer capable of settling those solids. The blanket gradually builds in the clarifier and, eventually, MLSS flows out over the weirs and into the final effluent. This is referred to as "blowing the blanket." Self-wasting will continue until the MLSS concentration decreases to a point at which the clarifier becomes capable of settling once again. This cycle of self-wasting will continue until operational changes are made to reduce the MLSS concentration through intentional wasting. Next, this cycle is repeated. A self-wasting activated sludge process would never be allowed today because operating in this way does not meet secondary treatment standards for BOD_5 and TSS. Facilities that are staffed only 1 or 2 days per week may be self-wasting when the operator is not there. Owners should be encouraged to staff package plants and other small systems adequately so that intentional wasting can be done several days each week, at a minimum.

2.9.1 Example No. 1—Programmable Wasting

Influent flow and load to a WRRF is relatively constant, with few changes from day to day. Operations staff is targeting an SRT of 5 days because this SRT has historically produced the best settling MLSS and the lowest turbidity effluent. The SRT is calculated every day, Monday through Friday. The operator then sets the wasting pump to run for a certain number of minutes out of each hour over a 24-hour period. The WAS pump is controlled by a computer. The number of minutes is adjusted each day to match the WAS volume from the SRT calculation. The WRRF is not staffed on the weekends. On Saturday and Sunday, the operator uses the average wasting rate for the week because they will not be there to adjust the cycles.

2.9.2 Example No. 2—Manual Wasting

A smaller WRRF does not have the ability to program its wasting pump and is only staffed during the day. Influent flow and load are fairly constant from day to day. The operator knows from past experience that the facility will need to waste between 300 and 800 m^3 (0.08 and 0.21 mil. gal) each day. Wasting for a few minutes of every hour would require the facility operator to manually turn the pump on and off each time. Instead, the operator elects to waste from the system twice each day. In the morning, they collect samples so they can perform the SRT calculation. At the same time, they start the WAS pump and waste 300 m^3 (0.08 mil. gal) while they analyze process control samples and perform the SRT calculation. The SRT calculation indicates that an additional 375 m^3 (0.99 mil. gal) of WAS should be removed to maintain the target SRT. Just before going home at the end of the day, the operator starts the WAS pump and removes the additional WAS.

2.10 When to Use Mean Cell Residence Time, Solids Retention Time, or Solids Retention Time$_{Aerobic}$

Whether a facility chooses to use SRT$_{aerobic}$, SRT, or MCRT for process control depends, in part, on how their process is configured and whether ammonia removal is required by the discharge permit. An SBR with an effluent ammonia limit will find SRT$_{aerobic}$ useful for controlling the amount of air-on time in a treatment cycle, as will facilities with the ability to independently control airflow to different portions of activated sludge basins or operate some basins as either anoxic or aerobic. As shown in Figure 4.6, for a facility with small activated sludge basins and large clarifiers, MCRT may be more appropriate because, even when the clarifier sludge blanket is below 0.6 m (2 ft) of depth, a significant portion of the MLSS may be in the clarifier. Facilities with large activated sludge basins, such as oxidation ditches, may prefer to use SRT because it requires less sampling and analysis and, even when the clarifiers contain several meters (several feet) of sludge blanket, the majority of the MLSS will be in the activated sludge basin.

2.10.1 *Example No. 1—Selection of Solids Retention Time*

An extended aeration oxidation ditch activated sludge facility was designed for an HRT of 24 hours. The influent flow is 9462.5 m^3 (2.5 mgd) and the oxidation ditch volume is 9462.5 m^3 (2.5 mil. gal). There are two secondary

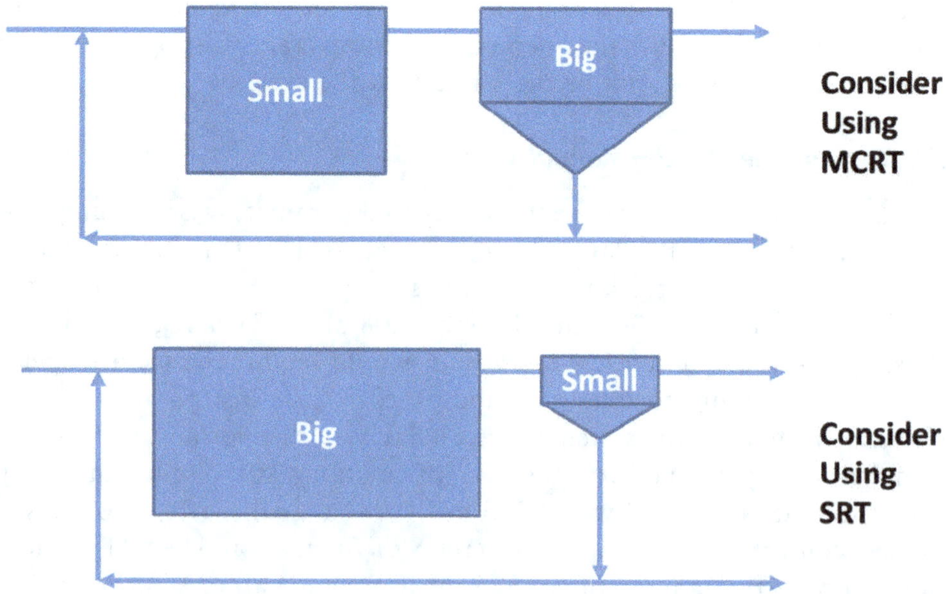

FIGURE 4.6 Activated Sludge Basin Versus Clarifier Size (Reprinted with permission from Indigo Water Group)

clarifiers that are 18.3 m (60 ft) in diameter and 3.65 m (12 ft) deep. The MLSS concentration is 3000 mg/L and the RAS concentration is 7800 mg/L. The clarifier blanket depth is 1.5 m (5 ft). Calculate the percentage of MLSS in the oxidation ditch and clarifier. Assume that the RAS concentration and the sludge blanket concentration are approximately equal.

Step 1: Find the mass of solids in the activated sludge basin:

$$\text{Activated sludge basin, kg} = \frac{\left[(\text{Concentration, }\frac{\text{mg}}{\text{L}})(\text{Volume, m}^3)\right]}{1000}$$

$$\text{Activated sludge basin, kg} = \frac{\left[(3000\frac{\text{mg}}{\text{L}})(9462.5\text{ m}^3)\right]}{1000}$$

$$\text{Activated sludge basin} = 28\ 388\text{ kg}$$

Step 2: Find the volume of the clarifier:

$$\text{Volume} = \Pi/4\ (\text{Diameter}^2)(\text{Depth})(2\text{ Clarifiers})$$
$$\text{Volume} = 0.785(18.3\text{ m})^2(3.65\text{ m})(2\text{ Clarifiers})$$
$$\text{Volume} = 1919\text{ m}^3$$

Step 3: Find the mass of solids in the clarifier blanket:

$$\text{Clarifier, kg} = \left[\frac{(\text{Concentration, }\frac{\text{mg}}{\text{L}})(\text{Volume, m}^3)}{1000}\right]\left(\frac{\text{Blanket depth}}{\text{Clarifier depth}}\right)$$

$$\text{Clarifier, kg} = \left[\frac{(7800,\ \frac{\text{mg}}{\text{L}})(1919\text{ m}^3)}{1000}\right]\left(\frac{1.5\text{ m}}{3.65\text{ m}}\right)$$

$$\text{Clarifier} = 6152\text{ kg}$$

Step 4: Find the percentage of solids in the clarifier blanket:

$$\text{Percent} = \frac{\text{Part}}{\text{Whole}} \times 100$$

$$\text{Percent} = \frac{6152 \text{ kg}}{28\ 387.5 \text{ kg} + 6152 \text{ kg}} \times 100$$

$$\text{Percentage of solids in the clarifier blanket} = 17.8\%$$

Because the activated sludge basin is large compared to the clarifiers, most of the MLSS will be in the basin, even when the clarifier blanket is deep. These solids will not have a significant effect on SRT or MCRT calculation. Keep in mind that the blanket depth should be maintained at less than 0.6 m (2 ft) at all times. For this situation, a facility may choose to use SRT for process control.

2.10.2 Example No. 2—Selection of Mean Cell Residence Time

A conventional activated sludge facility was designed for an HRT of 6 hours. Influent flow is 9462.5 m^3 (2.5 mgd). The activated sludge basin holds 2366 m^3 (0.625 mil. gal). There are two secondary clarifiers that are 24.4 m (80 ft) in diameter and 3.65 m (12 ft) deep. The MLSS concentration is 3000 mg/L and the RAS concentration is 7800 mg/L. The clarifier blanket depth is 0.6 m (2 ft). In this situation, the total mass of MLSS in the activated sludge basin is 7097 kg and the mass of MLSS in the clarifier blankets is 4374 kg. Even though the blanket depth is only 0.6 m (2 ft) in each of the clarifiers, 38% of the total sludge inventory is in the clarifier. For this situation, a facility should use MCRT for process control.

2.10.3 Example No. 3—Selection of Either Solids Retention Time or Mean Cell Residence Time with Mixed Liquor Volatile Suspended Solids

An activated sludge facility receives alum sludge from its water treatment facility. The water facility discharge is somewhat irregular, with no discharge at all on some days and large volumes discharged on other days. The MLSS concentration in the activated sludge basins can suddenly increase when alum sludge is entering the facility. As a result, the operations staff has a difficult time calculating sludge age. In this case, the operator should select either SRT or MCRT for process control but use MLVSS rather than MLSS in the calculation. The alum sludge does not contribute to the growth of microorganisms and passes through the process unchanged. The total MLSS measurement includes this inert material in the sludge age calculation, which could result in accidental overwasting and loss of nitrification. A WRRF receiving water treatment facility residuals may also want to base adjustments to the sludge wasting rate on the 5- or 7-day running average SRT or MCRT.

2.10.4 Example No. 4—Selection of Solids Retention Time$_{Aerobic}$

An activated sludge facility has multiple basins in each treatment train. The first basin is anaerobic, the second is anoxic, the third is a swing basin that can be either anoxic or aerated, and the last basin is fully aerated. The operator must decide whether to operate the third basin as an additional anoxic basin or if it should be fully aerated. In this scenario, the operator should calculate the $SRT_{aerobic}$ to determine if there is sufficient aerobic sludge age to support full nitrification when the swing basin is operated in anoxic mode. The $SRT_{aerobic}$ should also be calculated for all SBRs and for facilities practicing on–off aeration.

3.0 FOOD-TO-MICROORGANISM RATIO

3.1 Definition

The *F:M* is defined as the mass of BOD or COD added to the activated sludge process each day divided by the mass of MLVSS in the process. *It is important to note that the F:M equation always uses MLVSS rather than MLSS.* The volatile portion of the MLSS is a good approximation of the mass of live microorganisms in the process but also includes dead cells and organic debris. The F:M is literally the amount of "food" available per microorganism in the process. The F:M is also the inverse of the modified GSA equation (Section 2.0). With the modified GSA equation, sludge age is calculated by taking the mass of solids in the activated sludge basin and dividing by the mass of influent BOD. The F:M calculation simply flips the modified GSA on its head.

Activated sludge operational modes are defined, in part, by the range of F:M, as shown in Table 4.1. The F:M for conventional activated sludge processes is accepted to be between 0.2 and 0.4 kg of BOD_5/kg of MLVSS·d (0.2 and 0.4 lb of BOD_5/lb of MLVSS·d) and between 0.05 and 0.15 kg of BOD_5/kg of MLVSS·d (0.05 and 0.15 lb of BOD_5/lb of MLVSS·d) for extended aeration systems (U.S. Environmental Protection Agency, 1974, 1978). Typically, conventional activated sludge processes are designed to remove BOD and TSS but not ammonia, whereas extended aeration activated sludge processes are designed for complete ammonia removal and may also remove nitrate.

3.2 Calculation Method

The F:M is calculated according to the following formula:

$$\text{Food-to-microorganism ratio} = \frac{F}{M} = \frac{\text{kg or lb of BOD}_5\text{ applied per day}}{\text{kg or lb of MLVSS inventory}} \tag{4.13}$$

International Standard units for calculating mass:

$$\text{kg} = \frac{\left[(\text{Concentration, }\frac{\text{mg}}{\text{L}})(\text{Volume, m}^3)\right]}{1000}$$

To calculate kilograms per day, use the flow in cubic meters per day instead of volume in cubic meters. Alternatively, use dimensional analysis:

$$\frac{X\text{ mg}}{\text{L}}\left|\frac{1\text{ g}}{1000\text{ mg}}\right|\left|\frac{1\text{ kg}}{1000\text{ g}}\right|\left|\frac{1000\text{ L}}{1\text{ m}^3}\right|\left|\frac{X\text{ m}^3}{\text{d}}\right| = \text{kg/d}$$

U.S. customary units for calculating mass:

$$\text{Mass, lb} = (\text{Concentration, }\frac{\text{mg}}{\text{L}})(\text{Volume, mil. gal})(8.34\frac{\text{lb}}{\text{mil. gal}})$$

To calculate pounds per day, use the flow in million gallons per day instead of volume in million gallons. Alternatively, use dimensional analysis to arrive at lb/d:

$$\frac{X\text{ mg}}{\text{L}}\left|\frac{1\text{ g}}{1000\text{ mg}}\right|\left|\frac{1\text{ kg}}{1000\text{ g}}\right|\left|\frac{2.204\text{ lb}}{1\text{ kg}}\right|\left|\frac{3.785\text{ L}}{\text{gal}}\right|\left|\frac{1\,000\,000\text{ gal}}{1\text{ mgd}}\right|\left|\frac{X\text{ mil. gal}}{1\text{ d}}\right| = \text{lb/d}$$

Multiplying and dividing all the coefficients across the dimensional analysis, you arrive at the factor 8.34 applied in the equation above.

3.3 Calculation Example

A WRRF consists of screening, grit removal, primary clarification, conventional activated sludge, and disinfection. The influent flow is 4000 m^3/d (1.06 mgd) and it has a BOD_5 concentration of 300 mg/L. The primary clarifier removes 30% of the influent BOD_5 and 40% of the influent TSS. The activated sludge basins have a combined volume of 1600 m^3 (0.42 mg) and an MLSS concentration of 3200 mg/L. The MLSS is 80% volatile. Calculate the F:M.

Solution in International Standard Units:

Step 1: Find the mass of BOD_5 in the influent:

$$\frac{300\text{ mg}}{\text{L}}\left|\frac{1\text{ g}}{1000\text{ mg}}\right|\left|\frac{1\text{ kg}}{1000\text{ g}}\right|\left|\frac{1000\text{ L}}{1\text{ m}^3}\right|\left|\frac{4000\text{ m}^3}{\text{d}}\right| = 1200\text{ kg/d}$$

Step 2: Find the mass of BOD_5 going to the activated sludge process. If the primary clarifier is removing 30% of the influent BOD, then 70% is going to the activated sludge process.

$$\begin{aligned}
(\text{Total mass})(\text{Percentage}) &= \text{Mass to activated sludge process}\\
(1200\text{ kg/d})(100\% - 30\%) &= \text{Mass to activated sludge process}\\
(1200\text{ kg/d})(0.7) &= \text{Mass to activated sludge process}\\
840\text{ kg/d} &= \text{Mass to activated sludge process}
\end{aligned}$$

Step 3: Find the mass of MLSS in the activated sludge process:

$$\frac{3200\text{ mg}}{\text{L}}\left|\frac{1\text{ g}}{1000\text{ mg}}\right|\left|\frac{1\text{ kg}}{1000\text{ g}}\right|\left|\frac{1000\text{ L}}{1\text{ m}^3}\right|\left|1600\text{ m}^3\right| = 5120\text{ kg}$$

Step 4: Find the mass of MLVSS in the activated sludge process:

$$\begin{aligned}
(\text{Total mass})(\text{Percentage}) &= \text{Volatile mass}\\
(5120\text{ kg})(80\%,\text{ or }0.8) &= 4096\text{ kg}
\end{aligned}$$

Step 5: Calculate the F:M:

$$\text{Food-to-microorganism ratio} = \frac{\text{F}}{\text{M}} = \frac{\text{kg or lb of BOD}_5\text{ applied per day}}{\text{kg or lb of MLVSS inventory}}$$

$$\text{Food-to-microorganism ratio} = \frac{\text{F}}{\text{M}} = \frac{840\text{ kg BOD applied per day}}{4096\text{ kg MLVSS inventory}}$$

$$\text{Food-to-microorganism ratio} = \frac{\text{F}}{\text{M}} = 0.21\text{ per day}$$

Based on the F:M alone, this is likely a conventional activated sludge process.

3.4 Advantages and Disadvantages

Using the F:M for process control suffers many disadvantages. The most obvious drawbacks to F:M are that it puts the operator in a position of both monitoring and predicting the influent load and does not provide a clear mechanism for determining how much MLSS should be removed from the process each day by wasting. As will be discussed in Section 4.0 of this chapter, F:M is dependent on MCRT/SRT. A WRRF may select a target MCRT/SRT designed to bring the F:M within a desired range. If the F:M is too high or too low, the MLSS may not settle well (U.S. Environmental Protection Agency, 1977). Although many WRRFs calculate F:M regularly, F:M should not be used as the sole or primary method of process control (Ekama, 2010).

Early activated sludge process control focused on trying to maintain enough sludge mass or inventory to effectively treat the incoming organic load. *Operation of Municipal Wastewater Treatment Plants* (Water Environment Federation, 1996) explained it this way:

> A good way to understand the importance of [maintaining the optimal amount of activated sludge solids in the system to treat the BOD in the influent wastewater] is to consider the amount of BOD in the untreated wastewater as "food", and the activated sludge solids as "microorganisms" that will eat the food in one sitting. If there are not enough microorganisms available to eat all the food in one sitting, there will be leftovers. Leftovers means that effluent BOD is higher than desired.

We now know that the number of microorganisms available is determined by the influent organic load and the yield (Chapter 2). We also know that the mass of MLSS in the system is controlled by the wasting rate, which is determined by the desired sludge age. Setting the sludge age automatically sets the F:M (Section 4.0). Finally, if the sludge age is long enough to support flocculation and compaction so that solids separation can occur in the secondary clarifier, there should be a large enough population of microorganisms in the process to completely remove available BOD.

The F:M can be useful for predicting denitrification rates, as described in Chapter 8. The rate, or speed, of denitrification is directly proportional to the amount of available BOD and F:M. In an activated sludge process with both preanoxic and postanoxic zones like the one shown in Figure 4.7,

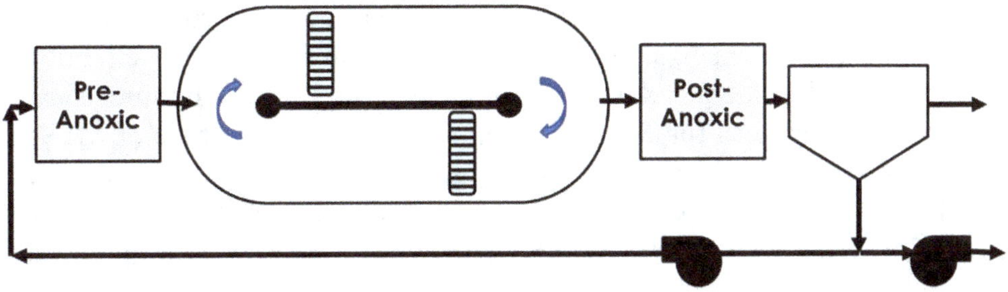

FIGURE 4.7 Pre- and Post-Anoxic Zones (Reprinted with permission from Indigo Water Group)

denitrification rates may be as much as 10 times faster in the preanoxic zone than the postanoxic zone because of the availability of excess BOD.

4.0 RELATIONSHIPS AMONG PROCESS VARIABLES

Sludge age should be the primary control variable for all activated sludge systems (Ekama, 2010; Jenkins and Garrison, 1968; Pretorius et al., 2016; U.S. Environmental Protection Agency, 1977). Once a target sludge age has been selected, it will dictate the wasting rate, MLSS concentration, MLVSS concentration, F:M and MLSS growth rate, microbiology, and sludge settleability. It is not possible to control more than one of these variables at a time. Burchett and Tchobanoglous (1974) noted that "it is evident that the new growth of microorganisms per day in the activated sludge system is related to the F:M ratio and the MCRT. Second, and more important, it may be concluded that if one of the two variables is controlled at a given level, the other will seek and find its associated level." Figure 4.8 is a simplistic

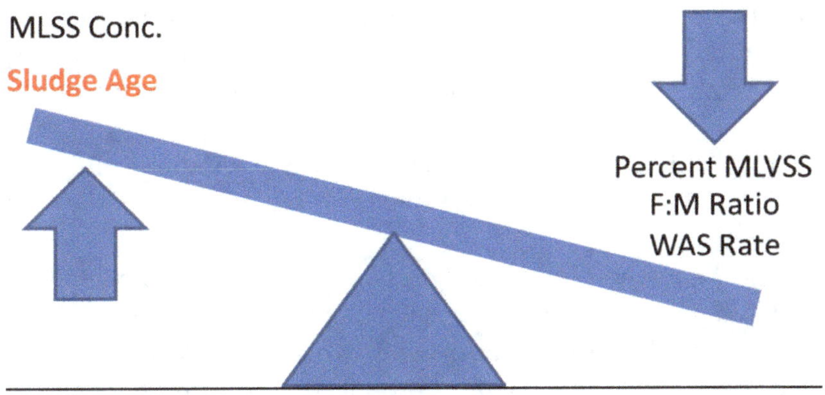

FIGURE 4.8 Relationships Among Activated Sludge Process Control Variables (Reprinted with permission from Indigo Water Group)

representation of the relationships among these important process variables. In reality, the relationships between the variables are more nuanced, but the basic relationships depicted here remain true. In Figure 4.8, the sludge age is increasing along with the MLSS concentration. The percentage of MLVSS, F:M, and wasting rate are all decreasing. If the operator decided to lower the sludge age, then the MLSS would go down with it and all of the variables on the right side (percentage MLVSS, F:M, and WAS rate) would increase.

4.1 Why Do Mixed Liquor Suspended Solids Increase with Increasing Solids Retention Time?

As sludge age is increased, fewer kilograms (pounds) of MLSS are removed from the system in WAS each day. This is the definition of sludge age. For a 10-day sludge age, one-tenth (10%) of the solids are removed each day. For a 5-day sludge age, one-fifth (20%) of the solids are removed each day. Removing fewer kilograms (pounds) of MLSS causes the MLSS concentration (milligrams per liter) in the activated sludge basins to increase.

4.2 Why Does the Ratio of Mixed Liquor Volatile Suspended Solids to Mixed Liquor Suspended Solids Decrease with Increasing Solids Retention Time?

The MLSS consist of both organic and inorganic (inert) material. The live microorganisms constitute the "active" portion of the MLSS and are the fraction of MLSS that provides treatment. The MLVSS represent the active microorganisms. We assume that all organic solids in the MLSS are live, active microorganisms, even though dead microorganisms and organic debris are also measured in the MLVSS test. The MLSS also contain inert material that came in with influent wastewater. Sand, grit, and eggshells cannot be broken down by microorganisms. This material passes through the process unchanged. The total MLSS added each day will be the sum of new bacteria produced (yield) plus non-biodegradable organics (leftovers from endogenous respiration) and inert material, as follows:

$$\text{Total MLSS added} = \text{New cells grown} + \text{Existing cells} + \text{Inert material} \quad (4.14)$$

Figure 4.9 illustrates how biodegradable materials are used and changed through biological treatment. Biochemical oxygen demand entering the system is used by the bacteria for both cellular maintenance and to build more bacteria. The mass of new bacteria produced is fixed by the yield (0.4 kg MLVSS/kg COD or 0.67 kg MLVSS/kg BOD_5), as discussed in Chapter 2. At the same time, the amount of BOD required to maintain existing bacteria

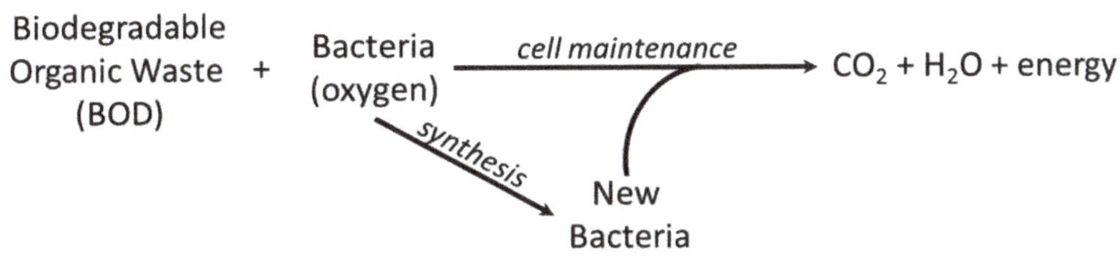

FIGURE 4.9 Conversion of BOD to MLSS

increases as the number of active bacteria increases. In other words, a larger fraction of influent BOD is going toward maintenance of existing cells and a smaller percentage of total cells in the system will have enough energy to reproduce. As the bacterial mass grows, it becomes limited by the amount of influent BOD, the F:M ratio decreases and the growth rate of viable bacteria, as represented by the MLVSS, as a whole decreases.

As sludge ages and F:M decreases, there is less food available for the greater mass of bacteria. With the food supply limited, one bacteria will be lost through endogenous respiration as they consume their own cell mass and exopolymer in an attempt to survive. Some bacteria will lyse or break open and become food for others. Still others will be lost through predation as higher lifeforms such as stalked ciliates become more prevalent in older sludge ages. The percentage of MLVSS to MLSS concentration and the relative number of viable bacteria will decrease as sludge age increases, as shown in Figure 4.10. The MLVSS concentration does not tell the whole story. Both dead bacteria and organic debris are also volatile. By one estimate, approximately 23% of biological solids (MLVSS) produced are relatively inert and, thus, accumulate in the system (McCarty and Brodersen, 1962). It may appear as though the number of live bacteria is increasing with longer and longer sludge ages; but, because the yield is fixed, what is really happening is that inert material is building up in the activated sludge basins. It is not any more possible to build new bacteria without influent BOD than it is for people to gain weight without eating.

Sand, grit, eggshells, and some organic compounds cannot be broken down by bacteria. This inert fraction of influent wastewater builds up in the activated sludge basins at the same rate each day. Think about adding 100 kg (100 lb) of sand to the activated sludge process on day one. On day two, 100 kg (100 lb) more are added. On day three, another 100 kg (100 lb) are added. If the sludge age is 20 days, this means that each day an additional 100 kg (100 lb) of inert material will be added and only 5% or 5 kg will be removed. After 20 days, 1900 kg of inert material will have accumulated in the process. As this continues, with increasing sludge age, the percent of inert material versus total suspended solids increases in essentially

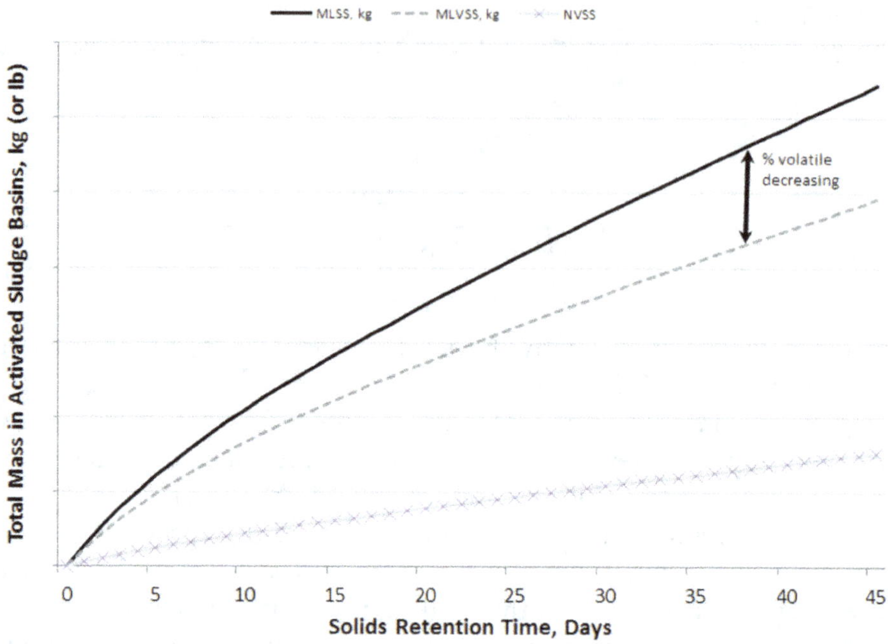

FIGURE 4.10 Changes to MLSS, MLVSS, and Nonvolatile Suspended Solids (NVSS) With Increasing SRT (Reprinted with permission from Indigo Water Group)

a straight line, as shown in Figure 4.10. At the same time, as sludge age increases, the MLVSS concentration begins to level off because microorganisms grow slower when less food is available (lower F:M). The combination of effects causes the total solids concentration (MLSS) to increase while the percentage of volatile solids (MLVSS) decreases. For wastewater that contain larger amounts of inert material, the effect will be more dramatic. A healthy activated sludge system that is not adding chemicals will typically have an MLVSS-to-MLSS ratio in the 0.75 to 0.85 range (U.S. Environmental Protection Agency, 1977). The MLSS in an aerobic or anaerobic digester, on the other hand, may only be 60% volatile. The percentage of volatile solids in the digester is lower because bacteria are breaking down by endogenous respiration, but inert solids pass through the digester unchanged.

4.3 Is Maintaining a Constant Mixed Liquor Suspended Solids Concentration the Same as Constant Sludge Age?

The MLSS concentration increases with increasing sludge age, but, without more information, it is impossible to predict the MLSS concentration just from sludge age. For many wastewater facilities, the influent load will be different on the weekends than on weekdays and may vary seasonally as well. The organic load can go up and down as can the amount of inert material. To keep a constant MLSS concentration under changing influent conditions, operators must increase and decrease the amount of solids they are removing from the system each day. If the percentage of mass removed each day changes, then sludge age is also changing. For some systems with consistent influent flows and loads, maintaining constant MLSS can approximate constant sludge age.

4.3.1 Example No. 1—Constant Solids Retention Time

The Post Point Wastewater Treatment Plant in Bellingham, Washington, uses in-basin TSS probes to monitor MLSS and WAS concentrations. The WAS flow is automatically adjusted to maintain an operator-selected SRT. At one point, the operators noticed that the MLSS concentration had dropped below its typical range. The operators assumed that the in-basin MLSS probe had malfunctioned or was out of calibration. After reviewing influent data, they discovered that the influent BOD load had decreased. As a result, the MLSS concentration had automatically decreased to maintain a constant SRT (Pretorius et al., 2016).

4.3.2 Example No. 2—Constant Mixed Liquor Suspended Solids and Dropping Solids Retention Time

The fictitious Metro City WRRF receives alum residuals from its water treatment facility. The mass of alum residuals varies considerably from day to day and can cause the MLSS concentration to change by as much as 500 mg/L in a single day. The WRRF uses constant MLSS for process control rather than constant SRT. On Monday, the MLSS concentration had increased to 3200 mg/L. The operator increased the wasting rate to drop the MLSS concentration back to their target of 3000 mg/L. On Tuesday, the water treatment facility discharged an unknown quantity of alum residuals that caused the MLSS concentration to increase to 3700 mg/L. The operator increased the wasting rate again to bring the MLSS concentration back to 3000 mg/L. On Monday, the percentage of MLVSS was 80% but, on Tuesday, had fallen to 76% VSS because of the alum addition. On Wednesday, the operator collected samples for effluent ammonia and was surprised to discover that the effluent ammonia concentration had increased from less than 1 mg/L to 22 mg/L. Why did this happen? A minimum SRT is required to maintain stable nitrification. When the operator increased wasting on Tuesday, they removed too much of the biological solids and dropped the SRT below the required minimum. The MLSS concentration stayed constant, but the SRT dropped dramatically.

4.3.3 *Example No. 3—Constant Mixed Liquor Suspended Solids and Seasonal Solids Retention Time*

A resort community has more visitors in the winter than in the summer. Each visitor contributes to the amount of load received by the WRRF. The WRRF uses constant MLSS concentration for process control and has a target MLSS concentration of 2500 mg/L. Past experience has shown that this concentration of MLSS produces a well-settling sludge and good effluent quality. Because the organic load is higher in the winter, more MLSS are produced and the operator must waste more to keep the MLSS concentration at the target of 2500 mg/L. In the summer, when the organic load decreases, the operator wastes less as less sludge is produced. While the MLSS concentration is kept constant year-round, the SRT is not constant. It is higher in the summer and lower in the winter. This WRRF often loses the ability to remove ammonia for short periods of time during the winter when the SRT drops too low.

4.4 Can I Operate with Constant Mixed Liquor Suspended Solids Instead of Constant Sludge Age?

A constant MLSS concentration or constant MLSS mass approach is appealing because it is simple to understand, reduces the amount of sampling and process control required, and is less dependent on accurate sampling of waste activated sludge concentration. Many activated sludge systems operate effectively using constant MLSS. When influent loading and water temperature are consistent, maintaining a constant MLSS can approximate maintaining a constant SRT. For this to be true, both of the following conditions must be true:

- Influent BOD (kilograms [pounds]) is constant and
- Influent TSS (kilograms [pounds]) is constant.

A disadvantage of this control method is its secondary emphasis on SRT. Caution must be used with this control technique, in which widely varying BOD concentrations are experienced. For example, a situation may exist in which secondary influent BOD to a particular activated sludge process suddenly increases by 50% and stabilizes at this level. Because of the increased food supply, the amount of activated sludge produced would also increase, and the MLSS levels in the activated sludge basin would climb. Observing this, the operator who is holding a fixed MLSS level naturally would increase the wasting rate, trying to cut down the solids concentration to its typical range. By reducing the solids concentration to a "normal" value, the operator is decreasing the SRT and increasing the F:M ratio. This could cause serious problems. Overwasting can cause a loss of nitrification, change sludge-settling characteristics, and affect overall effluent quality. The constant MLSS approach should not be used by WRRFs that are required to remove ammonia.

4.5 Why Does the Food-to-Microorganism Ratio Decrease with Increasing Solids Retention Time?

If the MLSS concentration is increasing and the amount of BOD entering the facility remains approximately the same, the amount of food per microorganism will decrease. Yield and decay are discussed in Chapter 2.

Solids retention time is related to F:M by the following equation (U.S. Environmental Protection Agency, 1974):

$$\text{SRT, days} = \frac{1}{(Y)\left(\frac{F}{M}\right)(\text{BOD removal efficiency}) - k_d} \tag{4.15}$$

Where

Y = yield,
F:M = food-to-microorganism ratio, and
k_d = decay coefficient.

As SRT increases, F:M will decrease proportionally. The converse is also true. Referring back to Table 4.1, for an SRT in the range of 20 to 30 days, the F:M will be considerably lower (0.05 to 0.15 kg BOD_5/kg MLVSS·d) than that for a system with an SRT in the 5- to 15-day range (0.2 to 0.4 kg BOD_5/kg MLVSS·d). The activated sludge process should be controlled by setting the target SRT or MCRT and allowing the F:M to self-adjust. It is not possible to control both of these variables at the same time. Selection of an MCRT or SRT is discussed in Section 5.0 of this chapter.

4.6 The Design Condition

Almost every WRRF operations and maintenance manual or drawing set contains a chart listing the design parameters for MCRT or SRT, MLSS concentration, and F:M. This is the design point. Many operators frustrate themselves trying to meet all three conditions simultaneously or they may attempt to maintain a fixed mass (kilograms [pounds]) of MLSS in the activated sludge basins each day, regardless of changing influent wastewater characteristics. In-facility operating conditions will never exactly match the design point for several reasons. First, the design point can only be reached when the WRRF is operating at its design flow and load. Most discharge permits require that a facility be in the planning stages for the next expansion, when either the maximum monthly flow or load exceeds 80% of design capacity, and to be under construction when they reach 95%. Because WRRFs are designed for peak loads and flows, they will have excess capacity under average operating conditions. Second, few WRRFs monitor influent for volatile and nonvolatile suspended solids. Without this information, it is impossible for the design engineer to predict total MLSS concentration. Imagine two different activated sludge processes. The first receives 200 kg/d (440 lb/d) of TSS that is 85% volatile. This means there are 30 kg (66 lb) of nonvolatile (inert) solids building up in the basins each day. The other process receives 200 kg (440 lb) of TSS that is 50% volatile. This means there are 100 kg (220 lb) of nonvolatile (inert) solids building up in the basins each day. Even if both of these WWRFs maintain the same sludge age, their MLSS concentrations will be different. The design engineer made a reasonable estimate of how much inert material was in the influent wastewater. If the estimate does not match the actual concentration in the influent exactly, the MLSS concentration will not match the design point exactly either. The MLSS concentration can also be influenced by recycle streams from solids-handling processes, chemical addition, and receipt of water treatment facility residuals. Facilities that practice chemical phosphorus removal generally use alum or ferric in large doses downstream of activated sludge basins. These chemicals are often recycled as part of filter backwash or other processes and end up in the activated sludge basins.

4.7 Summary of Process Control Strategies

Table 4.5 summarizes the various process control strategies currently in use, along with the advantages and disadvantages of each. As noted by Pretorius et al. (2016), operators sometimes abandon MCRT/SRT control of their activated sludge processes in favor of less reliable control parameters, such as attempting to maintain a constant MLSS concentration or F:M. Reasons for using constant MLSS or constant F:M in place of MCRT/SRT control include adherence to older process control strategies, staffing limitations, difficulty in obtaining representative samples, lack of on-site laboratory facilities, a lack of understanding of the advantages of MCRT/SRT control, and successful operation using either constant MLSS or constant F:M. However, MCRT/SRT control offers the best control of the activated sludge process and should be used whenever practicable (Ekama, 2010; Jenkins and Garrison, 1968; Pretorius et al., 2016; U.S. Environmental Protection Agency, 1977). Constant MLSS and constant F:M control strategies should not be used at WRRFs that are required to remove ammonia.

TABLE 4.5 Advantages and Disadvantages of Common Process Control Strategies (Reprinted with permission from Indigo Water Group)

Process control method	Advantages	Disadvantages
GSA	None	Does not account for the biological nature of activated sludge. Operator must predict and match changing influent conditions. Sensitive to small changes in wasting rate. Chronic over- or under-wasting of solids is likely. Fails to control nitrification.
Sludge age, including MCRT, SRT, and $SRT_{aerobic}$	Widely used with reliable results. Based on growth kinetics. Accounts for transformation of influent BOD and TSS into biological solids (MLSS). Automatically adjusts MLSS concentration to match incoming BOD load, thereby simultaneously controlling MLSS concentration and F:M. Operator does not need to predict or match influent conditions. Clear mechanism for calculating wasting rates. Target sludge age controls the presence or absence of different types of microorganisms in the process. Self-stabilizing. Produces excellent sludge quality when the target MCRT/SRT is within a reasonable range. When kept in the correct range for a given water temperature, maintains stable nitrification and minimizes filament growth.	Requires accurate measurement of MLSS and WAS solids. May be difficult to obtain representative samples needed for calculations. Staffing limitations in small systems may make sludge age control difficult to fully implement. Control charts are recommended with calculation of running averages to minimize the influence of bad data points or days when wasting is not performed.
MCRT	Accounts for the total sludge inventory by including both activated sludge basins and clarifier solids.	Requires more data collection than either SRT or $SRT_{aerobic}$.

(continued)

TABLE 4.5 Advantages and Disadvantages of Common Process Control Strategies (Reprinted with permission from Indigo Water Group) (*Continued*)

Process control method	Advantages	Disadvantages
MCRT (*continued*)	Preferred for systems with large clarifiers and smaller activated sludge basins. Historic data for sludge ages that promote filament growth are based on MCRT and SRT.	
SRT_{total}	Less data collection required than for MCRT. Preferred for systems with small clarifiers and large activated sludge basins. Historic data for sludge ages that promote filament growth are based on MCRT and SRT.	Assumes solids are not stored in the secondary clarifier blanket. For systems with large clarifiers relative to the activated sludge basins, this may not be true even when blankets are below 0.6 m (2 ft). Historic data for predicting the growth of filamentous bacteria is based on MCRT or SRT_{total} and not $SRT_{aerobic}$. Calculating $SRT_{aerobic}$ only may result in excessively long sludge ages, growth of filaments, and subsequent deterioration of sludge settleability.
$SRT_{aerobic}$	Used by design engineers for basin sizing. Ensures that there is adequate aerobic basin time or space to support stable nitrification. May be used to adjust anoxic versus aerated zone size within basins. May be used to adjust on/off cycle times for sequencing batch reactors.	
Constant MLSS (mass or concentration)	Simple to understand. Minimum amount of laboratory data required. Capable of producing good settling sludge and effluent quality when influent loads are consistent. Can approximate SRT control for some systems. Best suited for WRRFs that are not required to nitrify.	Constant MLSS does not hold sludge age constant, which may result in variable sludge quality (poor settleability) and uncontrolled loss of nitrification. No clear mechanism for calculating wasting rates. If the influent load is not constant, the SRT and F:M can fluctuate. Filamentous bacteria may be easily retained when the unmonitored SRT exceeds the SRT required for filament growth. More prone to foaming and bulking.

(*continued*)

TABLE 4.5 Advantages and Disadvantages of Common Process Control Strategies (Reprinted with permission from Indigo Water Group) (*Continued*)

Process control method	Advantages	Disadvantages
Constant F:M	Some argue that constant F:M has an advantage in WRRFs, such as those that receive large quantities of industrial wastewater, where influent characteristics can change rapidly. Should produce a consistently settling sludge. Early work in activated sludge microbiology is based on F:M.	Requires accurate measurement of MLSS, MLVSS, WAS, and influent BOD. Operator must predict and match changing influent conditions. Necessarily uses 5-day old data in the case of BOD_5 or 3+ hour old data in the case of COD. Reactive—responds to past conditions Requires a significant amount of laboratory work for influent and process monitoring. No clear mechanism for calculating wasting rates. Chasing F:M may drastically alter sludge age. May not maintain stable nitrification.

5.0 SELECTING A TARGET MEAN CELL RESIDENCE TIME, SOLIDS RETENTION TIME, OR SOLIDS RETENTION TIME$_{AEROBIC}$

Conventional activated sludge processes typically have MCRTs or SRTs between 5 and 15 days, whereas extended aeration activated sludge processes have MCRTs or SRTs as long as 30 days. Selecting the right MCRT, SRT, or $SRT_{aerobic}$ for a facility should be based on achieving the best possible sludge-settling characteristics while meeting other process goals, such as low effluent ammonia, nitrate, and phosphorus limits. If the sludge is too young or too old, process goals will not be met.

5.1 Flocculate, Settle, and Compact

The optimum MCRT, SRT, or $SRT_{aerobic}$ for a given WRRF will be the one that produces an MLSS that flocculates, settles, compacts, and meets the limitations of the discharge permit (Wahlberg, 2016). The ability of MLSS to flocculate, settle, and compact is measured by conducting daily settleometer tests and calculating the sludge volume index (SVI). If a sludge flocculates well, it will be capable of agglomerating loose bacteria and particulate matter to leave a crystal-clear supernatant, as shown in Figure 4.11(a). If a sludge does not flocculate well, the supernatant may be turbid or cloudy, as shown in Figure 4.11(b). A turbid supernatant may not meet the 30-mg/L TSS limit in most discharge permits and will be more difficult to disinfect.

Research has shown that, for bioflocculation to occur, a minimum SRT of 2 days is needed, and that, at SRTs between 0.25 and 2 days, free bacteria dominate rather than flocs (Maharajh, 2010). Free bacteria and smaller floc particles do not settle. Many activated sludge design manuals set the minimum SRT for various system types at 3 days (Metcalf & Eddy, Inc./AECOM, 2013; Water Environment Federation, 2023). Figure 4.12 illustrates some early work by Sayigh and Malina, Jr. (1978) correlating BOD removal to SRT. Better than 85% BOD removal was achieved for SRTs greater than 2 days when the water temperature was at least 10 °C. For colder water, the minimum SRT required for good flocculation was longer than 3 days.

FIGURE 4.11 Evidence of MLSS Flocculating Properties (Reprinted with permission from Indigo Water Group)

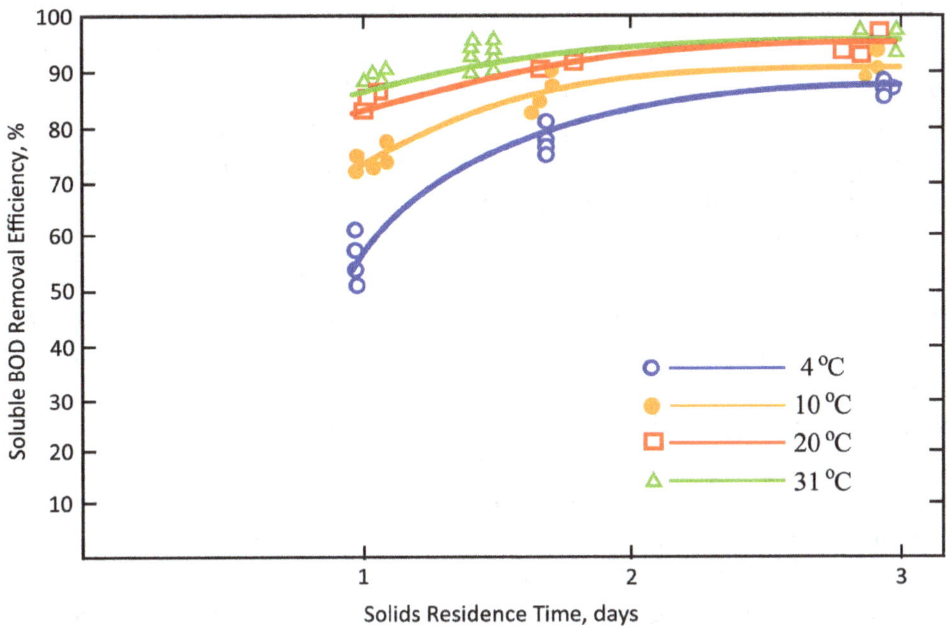

FIGURE 4.12 Soluble BOD Removal Efficiency Versus SRT (Sayigh and Malina, Jr., 1978)

5.1.1 Young Sludge Bulking

When a new activated sludge process is started up, it is common to see white foam with poor settling. As the sludge gets a bit older, MLSS flocs begin to take shape and become heavy and dense enough to begin to settle. However, young sludge floc is often open and lacy in structure because of the fast growth rate of floc formers. As a result, young sludge floc can be light and fluffy, which causes it to settle slowly. In older manuals, this condition is sometimes referred to as young sludge "bulking," with the floc described as "straggler floc." Bulking is any MLSS with an SVI greater than 200 mL/g. West (1973a) described straggler floc as "small, almost transparent, very light, fluffy, buoyant sludge particles 3.2 to 6.4 mm (one-eighth to one-quarter inch) in diameter." Once the SRT reaches 2 to 3 days, the floc structure becomes denser and settleability improves. In summary, for an activated sludge process to flocculate, settle, compact, and meet permit limits for BOD_5 and TSS, the SRT must be at least 2 days.

5.1.2 Old Sludge Bulking

Excessively long SRTs can also result in bulking sludge either because of the formation of "pin floc" or an abundance of filamentous bacteria. *Pin floc* is small and compact. Pin floc may be observed floating suspended without obvious movement upward or downward throughout a moderately turbid secondary clarifier. In the settleometer test, pin floc is characterized by rapidly settling, discrete sludge particles in place of a well-flocculated blanket. Pin floc is associated with high SRT, low F:M operation. High SRT, low F:M operation can cause floc formers to go into endogenous respiration, where they consume their external polymers and cell mass to survive. The loss of exopolymer can break up floc particles.

High SRTs encourage the growth of certain types of filamentous bacteria (Jenkins et al., 2003). Old sludge filaments include *Microthrix parvicella*, Type 0041, Type 0675, Type 1851, and Type 0803 (Jenkins et al., 2003). Note that filaments can proliferate under a number of different environmental conditions, as described in Chapter 3, and that many are able to proliferate at relatively short SRTs (Jenkins et al., 2003). Filamentous bacteria can cause the floc structure to be more open and lacy and less dense. In extreme circumstances, filaments can extend so far beyond the floc particle that they connect floc particles together. This is called *bridging*. An example of bridging is shown in Figure 4.13. When filaments are present in abundance, MLSS particles will settle more slowly and will not compact as well at the bottom of the clarifier. The goal becomes to keep the MCRT or SRT long

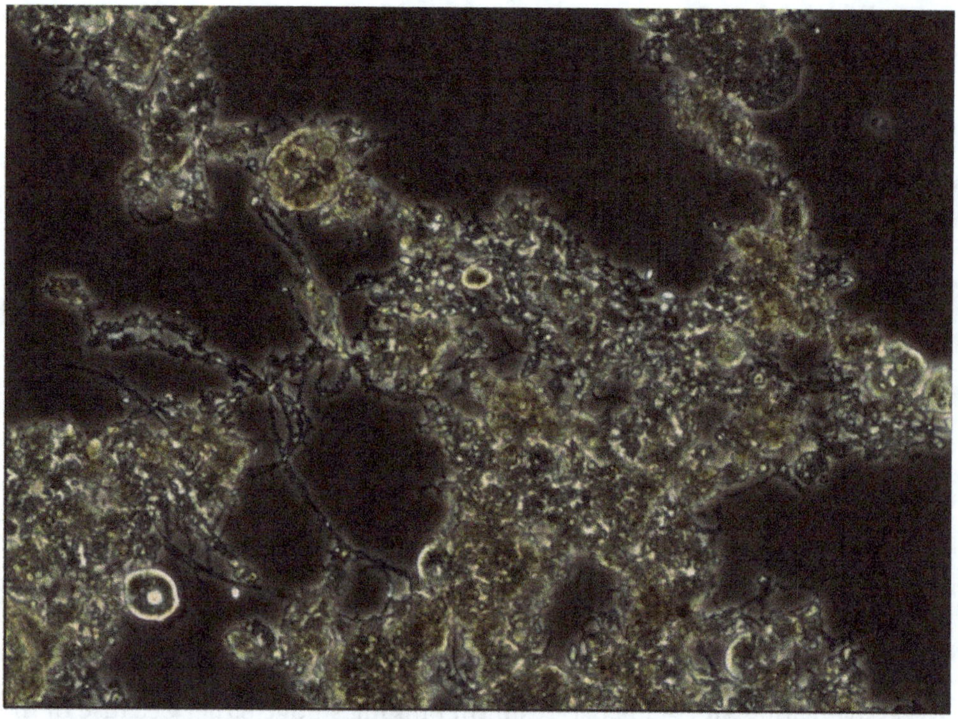

FIGURE 4.13 Filamentous Bridging Between Floc Particles (Reprinted with permission from Indigo Water Group)

enough to flocculate but short enough to prevent the growth of old sludge filaments and the formation of pin floc.

5.1.3 Sludge Settleability

The settled sludge volume after 5 minutes of settling in the settleometer test indicates how fast the floc is settling. For the first 5 minutes, the floc is essentially in free fall, with few collisions between floc particles to slow flocs down. This is the free-settling zone (Eckenfelder and Melbinger, 1957). As discussed in Chapter 3 and later in Chapters 5 and 6, the 5-minute settled sludge volume (SSV_5) is affected by water temperature, floc particle size and structure, and the presence or absence of filamentous bacteria. Many facilities see a strong seasonal pattern in which the SSV_5 is higher in the winter and lower in the summer simply because of changes in water temperature. Floc particles are close to the same density as water. When water gets colder, it becomes more dense and the difference in density between the floc and the surrounding water is less pronounced. As a result, settling velocity decreases. Settling velocity can also be affected by solids concentration. As MLSS concentration increases, settling velocity decreases because the density of fluid increases and there are more collisions between particles (Eckenfelder and Melbinger, 1957). This can be observed by comparing how samples of MLSS and RAS perform in the settleometer. The RAS will always settle at a slower rate (i.e., higher SSV_5) than MLSS. If filaments are present and extend beyond the edges of floc particles or, worse, create bridges between floc particles, this can create drag, which will slow down the settling velocity. If stalked ciliates and other predator organisms are present, they will scavenge free bacteria and help to produce a clear supernatant. Developing a sludge that flocculates and settles to leave behind a clear supernatant requires that the operator control the types and numbers of microorganisms present by adjusting sludge age.

5.1.4 Sludge Compaction

The settled sludge volume after 30 minutes of settling in the settleometer indicates how well floc is compacting. For slow-settling sludges, it may take longer than 30 minutes for compaction to begin. A series of U.S EPA pamphlets produced in the 1970s recommended leaving the MLSS in the clarifier for 60 minutes or longer to achieve maximum sludge compaction (West, 1973a, 1973b; 1978). Leaving MLSS in the clarifier for extended periods of time is no longer recommended because it can contribute to high effluent suspended solids from deflocculation and denitrification in the clarifier blanket (Ekama, 2010). Methods for determining the optimum RAS pumping rate are discussed in Chapter 5.

The SVI calculation takes the 30-minute SSV and divides it by the MLSS concentration in milligrams per liter. The calculation result is in of milliliters per gram. In other words, how much space does 1 g of solids occupy after being allowed to settle and compact for 30 minutes. Typically, the lower the SVI, the better the sludge quality. However, for the SVI calculation to be meaningful, the MLSS concentration must be within a reasonable range; otherwise, results cannot be compared in a meaningful way to literature values. A detailed discussion of the SVI calculation can be found in Chapter 6, Section 9.0. It has long been accepted that an SVI below 70 mL/g can leave behind a turbid supernatant and that SVIs greater than 200 mL/g represent bulking sludge that is not compacting well (Palm et al., 1980; Trygar, 2010). Often, water resource recovery facilities that practice enhanced biological phosphorus removal (EBPR) use chemically enhanced primary treatment, receive water facility residuals, and/or add other flocculants to their treatment process report SVIs lower than 70 mL/g. Typically, how well a sludge compacts is dictated by the presence or absence of filamentous bacteria. Higher concentrations of filaments will prevent MLSS particles from compressing together, much in the same way that it is difficult to stuff unwound Christmas tree lights back into a box for storage. For the sludge to compact well, the operator must control the number of filaments, many of which can be controlled by maintaining an appropriate sludge age.

5.2 Maintain Mixed Liquor Suspended Solids Concentrations with a Reasonable Range

Activated sludge MLSS concentrations are limited on the low end by bioflocculation and on the high end by both (1) the ability to transfer enough oxygen to activated sludge basins to support microorganisms and (2) the ability of secondary clarifiers to flocculate, settle, and compact those solids. For most systems, this effectively limits the MLSS concentration to a range of 1000 to 4000 mg/L, although some types of activated sludge systems can operate successfully outside of this range. For MLSS to flocculate, there needs to be at least a minimum number of collisions between floc particles. Collisions help floc particles combine into larger particles and sweep up particulates from influent wastewater. Typically, an MLSS concentration of at least 1000 mg/L is needed for enough collisions to occur. If the MLSS concentration falls too low, an operator can reduce the number of activated sludge basins in service.

As MCRT/SRT increases, so does the MLSS concentration. As MLSS concentration increases, it becomes more and more difficult to dissolve oxygen into the wastewater, which increases energy use. Eventually, a point is reached where so much air is being blown into the activated sludge basin that the resulting turbulence breaks flocs apart, MLSS settleability deteriorates, and effluent turbidity increases. High MLSS concentrations increase the solids loading rate to the secondary clarifier. If the solids loading rate gets too high, the clarifier will not be able to settle solids. Solids loading to the secondary clarifiers is discussed in detail in Chapter 5.

5.3 Washout Mean Cell Residence Time/Solids Retention Time—Influencing Biology and Settleability

Activated sludge is a diverse community of bacteria, protozoans, and metazoans, as described in Chapter 3. Each of these organisms has a minimum time requirement to grow and reproduce. Most bacteria in the system reproduce rapidly and have generation times of 1 to 3 hours, with a few types reproducing as rapidly as once every 10 minutes under ideal conditions (Brock and Madigan, 1991). For these bacteria, the MCRT/SRT must simply be long enough for the MLSS to flocculate to keep them in the system. Some bacteria, like the nitrifying bacteria and some filamentous bacteria, take much longer to reproduce. Some ciliates reproduce once every 4 to 5 hours (Water Environment Federation, 2017). Like all ecological systems, higher predators and more complex organisms take longer to reproduce and can only develop after their prey are well established.

Washout MCRT or SRT is the minimum MCRT/SRT required to keep a particular organism in the system. Microorganisms are constantly entering the facility in influent. In fact, many microorganisms in the activated sludge process are soil organisms (Water Environment Federation, 2017). If MCRT/SRT is too low, these organisms will be removed from the system as fast as they enter, and a stable population will not develop. Although the term *washout* is used, microorganisms are removed from the system through wasting. An operator performing a microscopic examination of the facility's activated sludge may see the occasional nematode, for example, but unless the MCRT/SRT is old enough to keep that nematode in the system long enough for it to reproduce, the numbers of nematodes will not increase. A stable population of nematodes will not be established. This is true for every organism that can live, thrive, and survive in the activated sludge process, including nitrifying bacteria, filamentous bacteria, and macrofauna discussed in Chapter 3. Target MCRT/SRT selected by the operator will determine which microorganisms have an opportunity to develop stable populations. Other environmental conditions will also affect the biology of the activated sludge and organisms present, including organic and nitrogen loading rates, dissolved oxygen concentrations, pH, alkalinity, and temperature. If other required conditions are not met (e.g., food supply, oxygen), then even having a long enough sludge age may not produce a stable population of a desired organism. In practice, select a sludge age that is 2 to 3 times higher than the washout sludge age to keep a desired organism in MLSS.

The following are some estimated washout SRTs for different microorganisms at 20 °C:

- *Nocardioforms* = 1.5 days (Jenkins et al., 2003),
- Ammonia-oxidizing bacteria (AOB) = 1.6 days (U.S. Environmental Protection Agency, 2010),
- Phosphate-accumulating organisms (PAOs) = 3 to 4 days (U.S. Environmental Protection Agency, 2010),
- Nitrite-oxidizing bacteria (NOB) = 2 days (U.S. Environmental Protection Agency, 2010),
- *Sphaerotilus natans* = 3 to 7 days (Jenkins et al., 2003), and
- *M. parvicella* = greater than 10 days (Jenkins et al., 2003).

What these washout SRTs tell us is that if an activated sludge process has a total SRT of 6 days and the in-basin water temperature is 20 °C, we would expect to find stable populations of AOB and NOB, but not *S. natans* or *M. parvicella*. If the process had an anaerobic zone and the correct ratio of influent BOD to phosphorus, we would also expect to find PAOs. Solids retention time, by itself, does not guarantee a stable population if other necessary conditions are not also met. Similarly, *Nocardioforms* may or may not be present. Some filaments can be controlled simply by maintaining an SRT below the washout SRT for that filament. Old sludge filaments include *M. parvicella*, Type 0041, Type 0675, Type 1851, and Type 0803 (Jenkins et al., 2003). *Microthrix parvicella* is known for causing foaming and bulking episodes, especially during winter operating conditions, in WRRFs that must remove ammonia year-round. Figure 4.14 shows ranges of MCRTs for different filaments. Note that, for many filaments, like *Nocardioforms* and *S. natans*, the washout SRT is so low that it is not possible to waste them out of the process.

The estimated washout SRTs provided earlier in this section for AOB and NOB are much longer than those for the majority of floc-forming, BOD-consuming bacteria in the process. The fact that floc-forming heterotrophs grow faster than autotrophic AOB and NOB bacteria means that the conversion of ammonia to nitrate is slower than uptake of BOD. Low effluent ammonia concentrations mean that effluent BOD concentrations should also be low. The washout SRT for PAOs is similar to the NOB, which means that minimum SRTs required for phosphorus removal are similar to those required for ammonia removal. In general, an SRT of 3 to 4 days at 20 °C

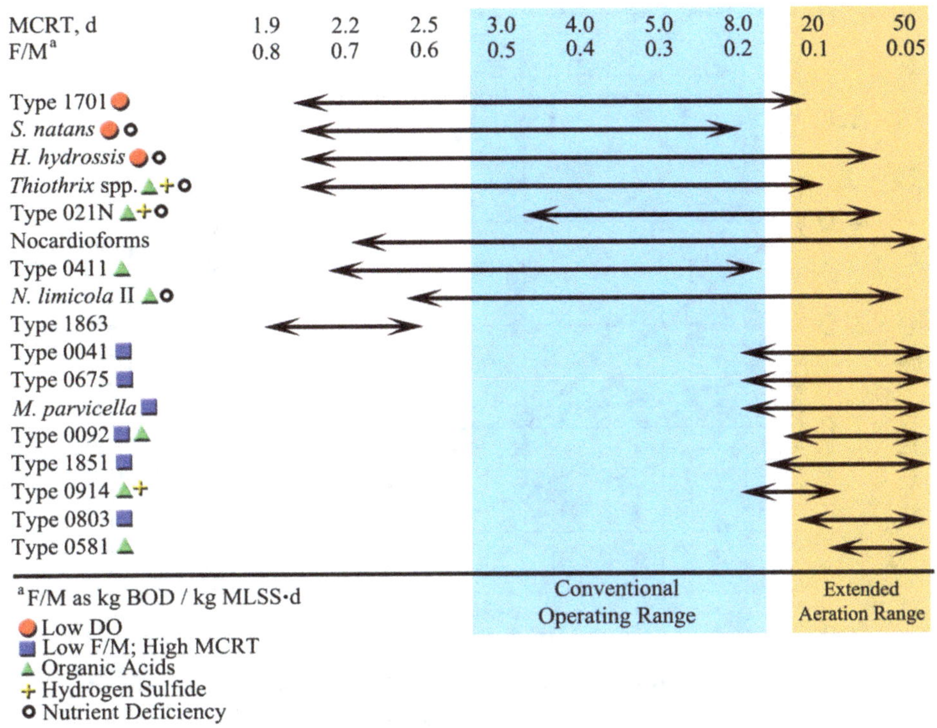

FIGURE 4.14 Washout SRTs for Various Filaments (*H. hydrossis* = *Haliscomenobacter hydrossis*; *N. limicola* II = *Nostocoida limicola* II; DO = dissolved oxygen) (Modified from Jenkins et al. [2004])

is sufficient to support stable EBPR (U.S. Environmental Protection Agency, 2010). Several researchers have proposed eliminating effluent monitoring for BOD and TSS in WRRFs that are required to monitor for effluent ammonia or that practice EBPR (Muirhead et al., 2009).

Washout SRT is affected by temperature. For every 10 °C drop in water temperature, the growth rate of bacteria decreases by 50% and washout SRT doubles. Similarly, growth rates for floc-forming and filament-forming bacteria are affected. U.S. EPA (2010) created Figure 4.15 using a combination of calculated theoretical washout SRTs and experimental data by Hellinga et al. (1998). Data labeled as "H" is experimental and data labeled as "M" is calculated from AOB and NOB growth rates. At a water temperature of 20 °C, the washout SRT for AOB is approximately 1.6 days and the washout SRT for NOB is approximately 2 days. To maintain a stable population and avoid accidental loss of these bacteria because of accidental overwasting, the target SRT would need to be 2 to 3 times as long or between 4 and 6 days. U.S. EPA (2010) recommends a safety factor between 40% and 80% of the theoretical minimum SRT. If water temperature drops to 10 °C, the

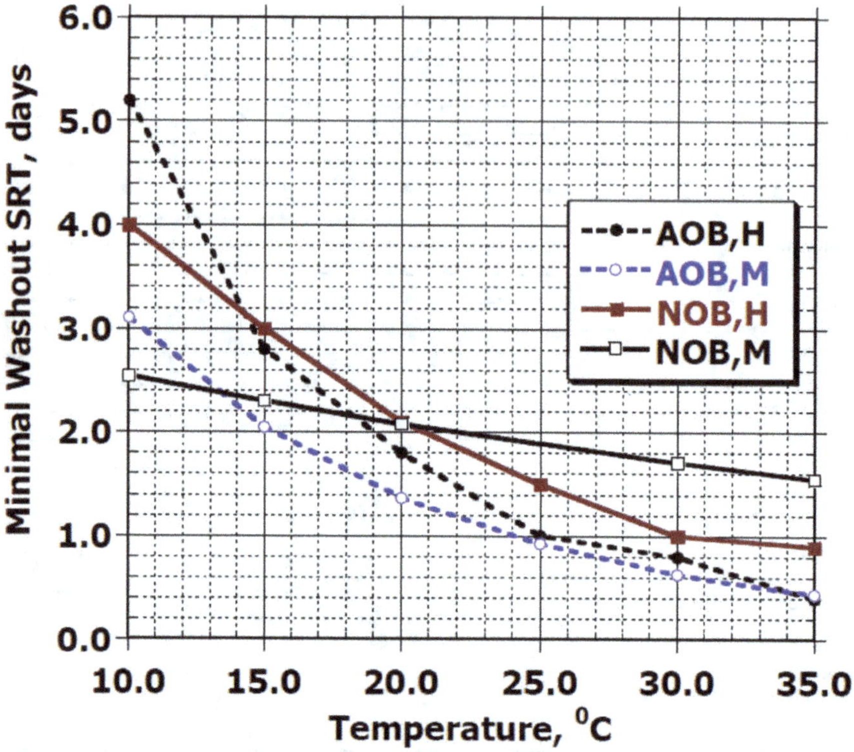

FIGURE 4.15 Washout SRTs for AOB and NOB at Different Water Temperatures (U.S. Environmental Protection Agency, 2010)

washout SRT for AOB increases to 3 to 5 days, which means a target SRT of between 10 and 15 days is prudent to maintain stable nitrification. For a WRRF with winter water temperatures of 10 °C and summer water temperatures of 20 °C, the target SRT will need to be between 6 and 15 days to maintain nitrification. The PAOs are similarly affected by changes in water temperature. Keep in mind that these minimum SRTs assume that there are not any other environmental conditions such as low dissolved oxygen, insufficient alkalinity, or low pH that could cause AOB and NOB to grow slower. Under adverse conditions, even longer SRTs will be needed to maintain these bacteria in the system.

5.4 Meeting Effluent Ammonia Limits

The "right" sludge age for a facility depends on whether ammonia removal is required to meet permit limits. For WRRFs that are not required to nitrify, the MCRT/SRT only needs to be long enough to support bioflocculation and produce a low turbidity effluent. Typically, the lower the MCRT/SRT, the less expensive it will be to operate the facility as fewer kilograms (pounds) of MLSS will be under aeration. The longer the MLSS is aerated, the more energy is required to operate the blowers. On the other hand, total sludge production decreases slightly with increasing sludge age as a greater portion of incoming BOD is used for cell maintenance instead of growth. For WRRFs that are required to nitrify, the $SRT_{aerobic}$ must be long enough to support stable nitrification but short enough to maintain good sludge settleability. Because the washout $SRT_{aerobic}$ for nitrifying bacteria depends on water temperature, many WRRFs that remove ammonia will need to adjust their target $SRT_{aerobic}$ seasonally.

The total MCRT/SRT must be long enough to retain desired microorganisms (nitrifiers) and short enough to flush out certain undesirable, old sludge filamentous bacteria like *M. parvicella* (Bartle and Revis, 2014). *Microthrix parvicella* can cause foaming and settleability issues in activated sludge. *Microthrix parvicella* commonly occur in low-loaded domestic WRRFs in which the F:M is less than 0.2 kg BOD_5/kg MLVSS, the anoxic zones comprise more than 40% of the total activated sludge basin volume, water temperatures fall below 15 °C, and the total SRT is greater than 10 days. Figure 4.16 illustrates how the SRT ranges where there is risk for growth of two *M. parvicella* strains (4B and RN1) (pink area between red and magenta lines) relate to typical SRTs required to support nitrification

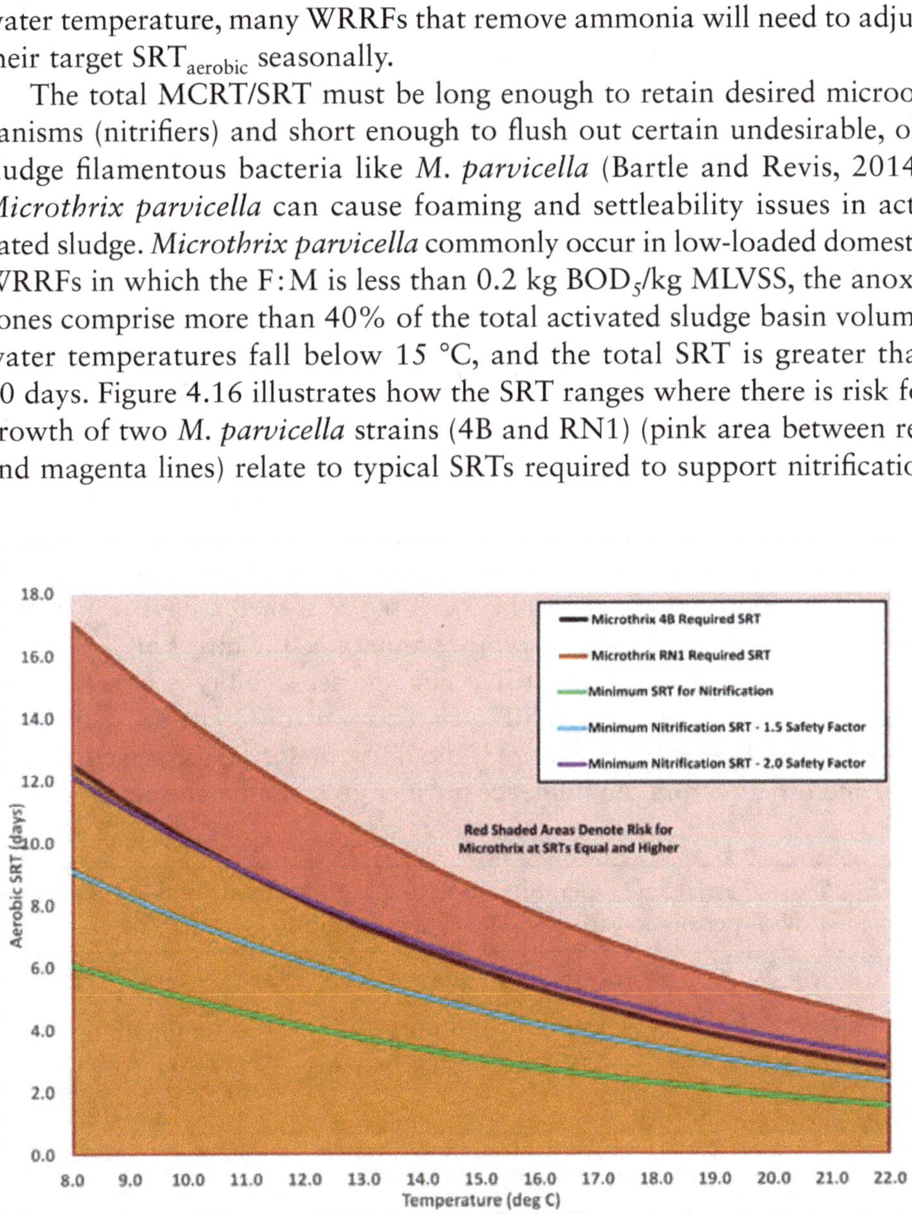

FIGURE 4.16 *Microthrix parvicella*/Nitrification Versus Temperature (Reprinted with permission from Bill Marten)

(orange area) across a range of temperatures (Marten & Lynne, 2011). As the figure shows, there is a degree of overlap at higher SRTs for nitrification that results in many WRRFs that are required to remove ammonia and are experiencing bulking and foaming events at certain temperatures.

The minimum nitrification SRT lines on Figure 4.16 were created by taking minimum washout SRTs for nitrifying bacteria (Figure 4.15) and multiplying by a safety factor of 1.5 and a safety factor of 2 to get the two upper lines. These safety factors are commonly used for WRRF design. The orange area between these lines shows where the SRT is theoretically long enough to support the growth of AOB and NOB. The red area between the SRT lines for the two *M. Parvicella* strains is where conditions may support the growth of these filaments in an activated sludge process that may experience filamentous bulking. The area of the graph below the *Microthrix* 4B required SRT line indicates a safe operating zone, where stable nitrification is supported and *M. parvicella* can be flushed out.

As shown in the graph, at a water temperature of 12 °C, the target SRT is below 8 days to flush out *M. parvicella* while maintaining nitrification. In this region of the graph, a typical SVI is anticipated to be below 200 mL/g. At water temperatures below 12 °C in the activated sludge basin, the SRT can be longer than 8 days but should be maintained below the SRT required for *M. parvicella* growth. An operator at this facility would want to set a target SRT of fewer than 8 days and then monitor the final effluent for ammonia on a daily basis. If effluent ammonia started to increase, even slightly, the SRT would need to be increased to 8 or 8.5 days or longer. If this facility was not required to remove ammonia, the target SRT would be set as low as possible to prevent foaming and settleability issues. Table 4.6 suggests some safe minimum SRTs to nitrify at different in-basin water temperatures. Keep in mind that pH, alkalinity, dissolved oxygen concentration, and other factors can influence nitrifier growth rate and washout SRT.

TABLE 4.6 Suggested Safe Operating SRT Ranges to Maintain Stable Nitrification (Reprinted with permission from Indigo Water Group)

	Washout SRT (days)			
Temperature	AOB	NOB	*Microthrix parvicella* foaming	Safe minimum operating SRT to nitrify (days)
10 °C	5.0	4.0	~12	12 to 16
15 °C	2.8	3.0	~20	6 to 9
20 °C	1.8	2.0	>25	4 to 6
25 °C	1.0	1.5	—	3 to 4.5

Consider Figure 4.15 and Table 4.6 as starting points for selecting a target MCRT or SRT for your facility.

5.4.1 Example No. 1—Seasonal Solids Retention Time Control

An oxidation ditch activated sludge process experiences changes in water temperature from 14 °C in January to 23 °C in July. The WRRF has effluent ammonia limits of 1.2 mg/L most months. The operator of this facility records the water temperature in the ditch every day and then selects a target SRT based on water temperature. As a result, the SRT is gradually increased and decreased over the year to compensate for seasonal changes. Before implementation of seasonal SRT control, this facility experienced severe bulking and foaming episodes during the winter, with SVIs of more than 600 mL/g. After implementation of seasonal SRT control, the facility has operated for three successive winters without bulking or foaming.

5.4.2 Example No. 2—Solids Retention Time Experimentation

A three-stage, plug flow activated sludge process discharges its treated effluent to a storage reservoir. Discharges from the reservoir must meet stringent ammonia limits, but the reservoir can store more than 180 days of effluent. This gives the operators at this facility considerable flexibility to experiment with SRT. If the SRT is dropped too low for a short period of time and nitrification is lost, any ammonia discharged will be diluted below permit limits by the reservoir. The activated sludge process has baffles separating anaerobic, anoxic, and oxic zones that trap foam in the basins. As a result, the WRRF can have outbreaks of severe *M. parvicella* foaming at relatively low SRTs. The chief facility operator has been pushing the SRT progressively lower to determine the lower boundary of safe operation. During the summer, when water temperatures were 22 °C, the WRRF was able to maintain nitrification with an SRT of only 5 days. During the winter, when water temperatures dropped to 12 °C, it was able to maintain nitrification with an SRT of 11.5 days. If the SRT fell to 11 days, effluent ammonia concentrations would suddenly increase.

5.5 Summary on Selecting a Target Mean Cell Residence Time, Solids Retention Time, or Solids Retention Time$_{Aerobic}$

The following are things to keep in mind when selecting a target sludge age:

- A minimum sludge age is needed to support bioflocculation, even when nitrogen removal is not a requirement. If the sludge age is too low, MLSS flocs that form will not be dense enough to settle in the clarifier. Use settleometer results to determine best achievable settling characteristics at the lowest sludge age.

- Longer sludge ages are necessary when ammonia removal is required by the discharge permit. Target sludge ages should be based on a combination of water temperature, sludge settleability, and effluent ammonia concentrations.
- Suboptimal conditions—low pH, low dissolved oxygen, and insufficient alkalinity—affect the growth rate of nitrifying bacteria and may require longer SRTs.
- As sludge age and MLSS concentrations increase, aeration costs increase.
- Longer sludge ages will encourage growth of undesirable organisms such as *M. parvicella,* which may result in foaming and settling issues. Stay below the washout sludge age for these organisms whenever possible. Note that not all filaments are associated with longer SRTs.
- Excessively long sludge ages will cause MLSS to go into endogenous respiration. Floc size will decrease as exopolymer is consumed as food. Settling may be fast with turbid effluent.
- Check operating data. Is there a past sludge age that produces the best SVI?

6.0 SETTING A TARGET DISSOLVED OXYGEN CONCENTRATION

Selection of a target dissolved oxygen concentration in aerated zones or basins should be based on (1) maintaining enough mixing energy to keep MLSS in suspension, (2) providing enough oxygen to remove BOD_5, (3) preventing the growth of low dissolved oxygen filaments, (4) supporting nitrification if required by the discharge permit, and (5) minimizing operational costs by saving energy. The following sections discuss recommended dissolved oxygen concentrations for achieving these goals.

6.1 Preventing the Growth of Low Dissolved Oxygen Filaments

For decades, operators have been taught to target a minimum dissolved oxygen concentration of 2 mg/L in activated sludge basins. U.S. EPA recommended keeping dissolved oxygen concentrations between 2 and 4 mg/L in their 1973 operations pamphlets (West, 1973a, 1973b). This concentration range was thought to be the safe operating range that would prevent the growth of low dissolved oxygen filaments and minimize sludge bulking.

Because of the way they are operated today, many systems operate successfully at much lower dissolved oxygen concentrations without experiencing low dissolved oxygen filament bulking. This section will describe conditions necessary for low dissolved oxygen filaments to proliferate.

For microorganisms in the activated sludge process to grow at their maximum rate, they need access to both oxygen and food. Recall from Chapter 2 that bacteria, at the most basic level, are similar to combustion engines. They require both fuel and oxygen to run. The structure of floc particles creates a situation in which the oxygen concentration inside the floc will always be lower than the oxygen concentration outside of the floc. Figure 4.17 is a simple schematic of a floc particle. The microorganisms on the outside of the floc can easily access oxygen and fuel. As long as concentrations of both are available in excess (i.e., microorganisms have access to more than they can immediately use), microorganisms will grow at their maximum rate, that is, growth is not limited, and excess oxygen and fuel will pass deeper into the floc. Microorganisms on the inside of the floc do not have free access. They can only access oxygen and fuel that penetrates the floc. That means oxygen and fuel have to make it past bacteria on the outside of the floc. A gradient is established at which oxygen and fuel concentrations are highest on the outside of the floc and are gradually decreasing through to the center of the floc. For larger flocs, the interior of the floc will be anoxic rather than aerobic, as shown in Figure 4.17. The lack of resources inside the floc causes microorganisms there to grow slower. They cannot achieve their

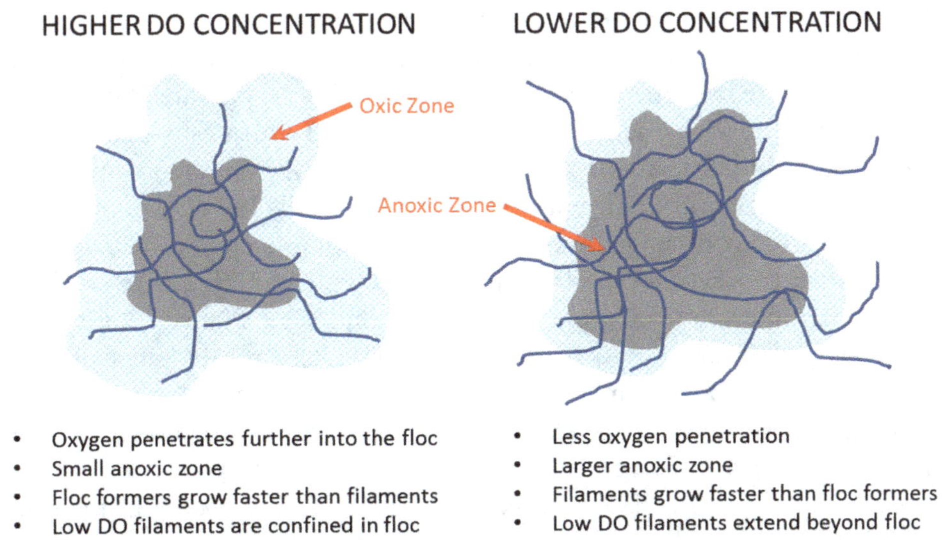

FIGURE 4.17 Growth of Low Dissolved Oxygen Filaments (DO = dissolved oxygen)

maximum growth rates and growth is limited. Lau et al. (1984) concluded that the rapid growth of floc formers under high dissolved oxygen conditions prevented both food and dissolved oxygen from reaching the interior of floc particles. In other words, the apparent decrease in the growth rate of *S. natans* is caused by the competition for food between floc formers and *S. natans* and not because dissolved oxygen is inhibitory.

In Chapter 3, the competition theory between filament formers and floc formers was discussed. Essentially, filament formers can achieve their maximum growth rates at much lower oxygen and fuel concentrations than floc formers. When dissolved oxygen concentrations are low, low dissolved oxygen filaments like *S. natans* and *Haliscomenobacter hydrossis* can keep growing at their maximum rate, but floc formers slow down. If the oxygen concentration is low enough, low dissolved oxygen filaments will grow faster than floc formers. The result is filaments that extend beyond the edges of the floc, as shown in Figures 4.13 and 4.17. Something similar happens with old sludge filaments. Those filaments are better able to compete under low F:M conditions that go hand in hand with old sludge. Floc formers slow down, but filaments keep chugging along at their slow, consistent rate. The million-dollar question for operators is, "How low can the dissolved oxygen concentration go before low dissolved oxygen filamentous bulking occurs?" The answer to this question has huge implications for overall operating costs.

The 2- to 4-mg/L dissolved oxygen recommendation from U.S. EPA was based on data collected from multiple full-scale activated sludge facilities, most of which were operated in conventional mode for F:M. It was not until 1980 that the first controlled experiments were conducted to determine the effect of dissolved oxygen concentration on sludge settleability. Palm et al. (1980) conducted a series of experiments in which they operated laboratory-scale, continuous-flow activated sludge systems. Two side-by-side systems were operated at the same F:M but different dissolved oxygen concentrations. Dissolved oxygen concentrations were gradually lowered in one of the systems until low dissolved oxygen sludge bulking occurred. *Sphaerotilus natans* was identified as the filament responsible for bulking. The results of these experiments are shown in Figures 4.18 and 4.19. Figure 4.18 is based on COD data. Figure 4.19 is based on BOD data. Chemical oxygen demand was converted to BOD using a conversion factor established for wastewater used in these experiments (Palm et al., 1980). Blocks in red show when low dissolved oxygen bulking occurred and *S. natans* was present. Blocks in green show when low dissolved oxygen bulking did not occur at the same F:M. The larger shaded areas in Figures 4.18 and 4.19 are the safe operating zones where low dissolved oxygen bulking does not occur. Palm et al. (1980) were able to demonstrate the following:

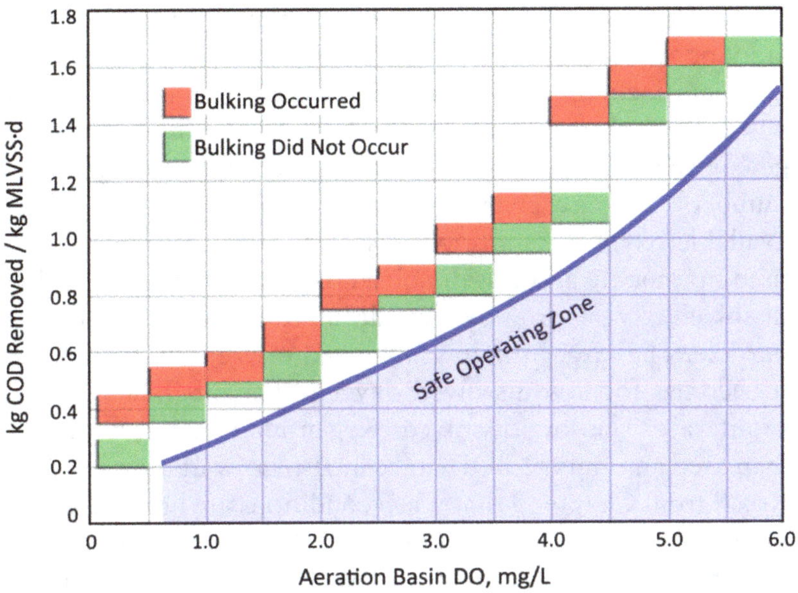

FIGURE 4.18 Incidence of Low Dissolved Oxygen Bulking Versus F:M Calculated Using COD (DO = dissolved oxygen) (Adapted from Palm et al. [1980])

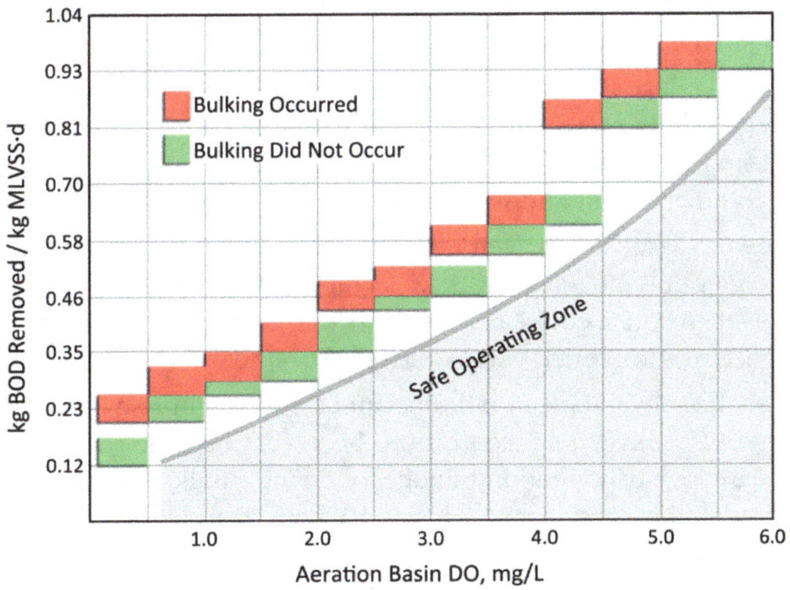

FIGURE 4.19 Incidence of Low Dissolved Oxygen Bulking Versus F:M Calculated Using BOD_5 (DO = dissolved oxygen)(Modified from Palm et al. [1980]); Figure 4.19 was created by applying a correction factor to data presented in Figure 4.18 to convert COD to BOD; the correction factor, included in the paper by Palm et al. (1980), assumes a COD-to-BOD ratio of 1.6 for domestic wastewater; the COD-to-BOD ratio can vary from one WRRF to another, especially if there are large industrial dischargers in the service area; experiments by Palm et al. (1980) used primary effluent (settled domestic wastewater) rather than raw wastewater

- The amount of dissolved oxygen uptake is directly related to the amount of COD or BOD_5 removed. Recall from Chapter 2 that when nutrients, including dissolved oxygen, are not limited, microorganisms will grow at their maximum rate. High F:M and high dissolved oxygen on the outer portion of the floc encourage rapid growth. Higher numbers of rapidly growing bacteria result in a lot of competition for available oxygen. As a result, dissolved oxygen at the floc surface is used up quickly and less dissolved oxygen penetrates the floc. A low dissolved oxygen or anoxic condition results in the center of the floc particle. The larger the low dissolved oxygen zone, the more likely it becomes for low dissolved oxygen filaments like *H. hydrossis* to proliferate. The low dissolved oxygen filaments also have access to increased amounts of BOD and can also grow at their maximum rate. Recall from Chapter 3 that when conditions are limiting, filaments are often able to outcompete and outproduce floc formers. This is why higher F:M requires higher dissolved oxygen concentrations.
- Some filaments can grow faster than floc formers under low dissolved oxygen conditions. *Sphaerotilus natans* was the dominant low dissolved oxygen filament in these experiments.
- Higher dissolved oxygen concentrations are needed at higher F:Ms to prevent bulking.
- Dissolved oxygen of 1.5 to 2 mg/L is sufficient to prevent low dissolved oxygen bulking when the F:M is less than 0.35 kg BOD_5/kg MLVSS (0.5 to 0.6 kg COD/kg MLVSS). This agrees with U.S. EPA's earlier recommendation and with research by Wilen and Balmer (1999).
- Lower dissolved oxygen concentrations can prevent low dissolved oxygen bulking at lower F:M. Figures 4.18 and 4.19 imply that dissolved oxygen could be taken all the way to zero, but the lowest concentration tested varied between 0.1 and 0.5 mg/L during the test.
- Low dissolved oxygen bulking can be cured simply by increasing the dissolved oxygen concentration; however, the time required to produce a stable, nonbulking sludge following a bulking episode is two to three sludge ages. If the SRT was 10 days, then the system will require 20 to 30 days of operation at a higher dissolved oxygen concentration for bulking to subside. It takes time to flush filaments from the system.

6.2 Maximizing Nitrification Rates

Traditionally, as processes for nitrogen removal were being developed, it was thought that nitrifying bacteria, AOB and NOB, required higher dissolved oxygen concentrations than other floc-forming, carbon-oxidating bacteria. A dissolved oxygen concentration of 2 mg/L was recommended to support maximum nitrification rates (ASIS, 1983). This was a general view that formed from focus on a specific population of bacteria known to perform nitrification under conditions limited to conventional activated sludge operation. Over the last few decades, as field observations and subsequent investigations of phenomena such as sustained low dissolved oxygen operation and simultaneous nitrification and denitrification became more common, our understanding of complex factors involved in biological nutrient removal increased. These factors include things like the complex dynamics of bacterial populations and their adaptability, and the ability of a population of microorganisms contributing to ammonia oxidation to shift toward one with higher affinity for oxygen under sustained low dissolved oxygen (less than 0.5 mg/L). The observation of other phenomena such as microenvironments within activated sludge floc (Figure 4.17) hypothesized from initial observations of its occurrence in biofilms, the more recent discovery of complete ammonia oxidation (*commamox*) and anaerobic ammonia oxidation (*anammox*), the increasing incorporation of fixed film media to augment suspended growth, as well as the successful implementation of ammonia-based aeration control, have broadened the spectrum of what can be considered optimum conditions for nitrification and denitrification. From the operator's standpoint, optimum conditions will depend on what variation of the activated sludge process one is operating, the particular effluent quality requirements for the WRRF, and monitoring and control parameters available. For operation of a conventional process geared primarily toward BOD removal, traditional recommendations will likely be acceptable. For nutrient removal in general, and nitrification in particular, a finer degree of monitoring and control, particularly of inorganic nitrogen and dissolved oxygen, will be necessary for maximizing effectiveness. For a more detailed discussion of nitrification kinetics and process control variables, see Chapter 8.

7.0 PRACTICE QUESTIONS

1. The accepted difference between SRT and MCRT is
   a. SRT includes solids in the clarifier
   b. MCRT only includes solids in aerated zones
   c. MCRT includes solids in the clarifier
   d. SRT only includes solids in anoxic zones

2. Gould sludge age was developed in the 1920s and was based on the idea that
   a. Activated sludge is primarily a physical–chemical treatment process
   b. Activated sludge is primarily a biological treatment process
   c. Influent soluble BOD is critical to activated sludge performance
   d. Influent TSS is transformed to MLSS in the activated sludge basin

3. The modified GSA equation substitutes influent BOD for influent TSS. The inverse of this relationship is known today as
   a. SRT
   b. MCRT
   c. F:M
   d. $SRT_{aerobic}$

4. The most advantageous method for activated sludge process control is
   a. Sludge age
   b. Constant MLSS mass
   c. Constant F:M
   d. RAS control

5. Under what condition will maintaining a constant MLSS concentration approximate constant sludge age?
   a. Volume of WAS removed each day is constant
   b. RAS is a fixed percentage of influent flow
   c. Effluent solids remain below 30 mg/L
   d. Influent loads are constant

6. The desired MCRT is 10 days. If MLSS is removed directly from the activated sludge basin, what percentage of MLSS must be removed each day?
   a. 5%
   b. 10%
   c. 20%
   d. 25%

7. Using Figure 4.1, estimate the desired wasting rate given the following information: The MLSS concentration is 2000 mg/L. Target

sludge age is 7.5 days. The WAS concentration is 6250 mg/L. The activated sludge basin holds 2000 m^3 (0.53 mil. gal).

a. 0.044 mgd
b. 160 m^3/d
c. 0.088 mgd
d. 88 m^3/d

8. The target SRT is 12 days. Currently, the MLSS concentration is 2800 mg/L. Tourist season begins this weekend and the influent BOD load is expected to increase by 25% in the next few days. If the target SRT is maintained, what will happen to the MLSS concentration?

a. It will increase
b. It will decrease
c. It will remain about the same
d. It will increase by 25%

9. An activated sludge process has extremely large basins with a combined HRT of 28 hours. The clarifiers are relatively small and have an HRT of only 2 hours. The operator maintains the blankets below 1 m (3 ft) of depth. Which of the following calculations for sludge age is most appropriate for this facility?

a. GSA
b. MCRT
c. SRT
d. $SRTa_{erobic}$

10. A conventional activated sludge process has an HRT of 6 hours. There are three clarifiers online with a combined detention time of 4 hours. Which of the following sludge age calculations is most appropriate for this system?

a. GSA
b. MCRT
c. SRT
d. $SRT_{aerobic}$

11. Solids in the clarifier blanket should be included in the sludge age calculation when

a. The MLSS is settling and compacting well
b. During peak wet weather events

c. Clarifier sludge blankets are deep
d. Clarifiers are small compared to basins

12. Effluent TSS are often excluded from sludge age calculations. Why is this acceptable practice?
    a. The number of required sampling locations is decreased
    b. The solids in the final effluent are not part of the activated sludge process
    c. Often, the discharge permit does not require sampling for TSS
    d. Typically, effluent TSS is too small to affect the calculation

13. One advantage of using MCRT over SRT is
    a. Additional samples are required
    b. One hundred percent of the sludge inventory is accounted for
    c. Difficulty in sampling clarifier sludge blankets
    d. Enables sludge storage in clarifiers

14. The RAS flow is currently set at 50% of influent flow. If the RAS flow is increased to 70%, the operator should expect to see
    a. Reduced overall treatment time
    b. Rat holing of the clarifier sludge blanket
    c. Decreased solids loading to the clarifier
    d. Decrease in RAS concentration

15. Mixed liquor suspended solids with a concentration of 1800 mg/L were placed into a settleometer and allowed to settle for 30 minutes. The initial sludge volume was 2000 mL and the 30-minute sludge volume was 280 mL. Find the settled sludge concentration.
    a. 6429 mg/L
    b. 7143 mg/L
    c. 12 857 mg/L
    d. 14 286 mg/L

16. The RAS flow is set at 70% of influent flow. If the MLSS concentration is 2600 mg/L, what is the anticipated RAS concentration, assuming that solids are not being stored in the clarifier blanket?
    a. 6300 mg/L
    b. 3700 mg/L
    c. 7500 mg/L
    d. 9000 mg/L

17. The operator knows from the settleometer test that the settled sludge concentration at 30 minutes is 10 000 mg/L. The initial MLSS concentration was 2500 mg/L. If the operator wants to achieve the thickest RAS concentration for this sludge but does not want to leave the RAS in the clarifier any longer than required, what should the RAS flowrate be when the influent flow is 420 m^3/d (0.11 mgd)?
    a. 140 m^3/d (0.037 mgd)
    b. 250 m^3/d (0.066 mgd)
    c. 280 m^3/d (0.074 mgd)
    d. 510 m^3/d (0.134 mgd)

18. An activated sludge facility uses large quantities of ferric chloride in a downstream process to precipitate phosphorus. The precipitated phosphorus and iron, now iron sludge, is removed with tertiary filters. When the filters are cleaned, iron sludge is recycled back to the activated sludge process. How might the operator compensate for changes in total MLSS concentration caused by these nonbiological solids?
    a. Increase blower output to increase available oxygen
    b. Decrease sludge age to remove solids faster
    c. Switch to constant mass for process control
    d. Substitute MLVSS for MLSS in the SRT calculation

19. All of the following may affect the accuracy of an SRT calculation EXCEPT
    a. Accuracy of WAS flowmeter
    b. Lack of a blanket in the clarifier
    c. Foaming in the activated sludge basin
    d. High-effluent suspended solids

20. The SRT is currently at 6 days. The target SRT is 8 days. The operator should
    a. Increase the WAS
    b. Decrease the WAS
    c. Increase the RAS
    d. Decrease the RAS

21. The SRT is currently 12 days. If the SRT is decreased to 8 days, what will happen to the F:M?
    a. It will increase by 50%
    b. It will remain the same

c. It will increase
d. It will decrease

22. An activated sludge process is experiencing extreme foaming and settleability is poor. Reviewing records, the chief facility operator discovers that the sludge age has increased over the last week to 16 days. The operator knows from experience that the foaming subsides as long as the SRT is well controlled at fewer than 12 days. They make necessary adjustments to bring the SRT to fewer than 12 days. How long will it take for the process to completely recover and stabilize at the new sludge age?
    a. Less than 12 days
    b. Between 24 and 36 days
    c. Between 36 and 50 days
    d. Greater than 75 days

23. Solids retention time calculations and wasting rate adjustments (if needed) should be done
    a. Daily
    b. Weekly
    c. Every 2 weeks
    d. Monthly

Answer key:

1. c, 2. a, 3. c, 4. a, 5. d, 6. b, 7. d, 8. a, 9. c, 10. b, 11. c, 12. d, 13. b, 14. d, 15. c, 16. a, 17. a, 18. d, 19. b, 20. b, 21. c, 22. b, 23. a.

8.0 REFERENCES

Advanced Science Institutes Series. (1983). Sludge characteristics and behavior: NATO ASI series, series E: Applied sciences—No. 66; Carberry, J. B., Englande, A. J., Jr. (Eds.). Martinus Nijhoff Publishers.

Argaman, Y., & Brenner, A. (1986). Single-sludge nitrogen removal: Modeling and experimental results. *Journal of the Water Pollution Control Federation*, *58*(8), 853–860.

Bartle, H., & Revis, C. (2014, September 29–October 1). Controlling *Microthrix parvicella* with foam control and ammonia-SRT. *Proceedings of the 87th Annual Water Environment Federation Technical Exhibition and Conference* [CD-ROM], New Orleans, LA, United States. Water Environment Federation.

Brock, T. D., & Madigan, M. T. (1991). *Biology of Microorganisms* (6th ed.). Prentice-Hall.

Burchett, M. E., & Tchobanoglous, G. (1974). Facilities for controlling the activated sludge process by mean cell residence time. *Journal of the Water Pollution Control Federation, 46*(5), 973–979.

Eckenfelder, W., & Melbinger, N. (1957). Settling and compaction characteristics of biological sludges: I. General considerations. *Journal of the Water Pollution Control Federation, 29*(10), 1114–1122.

Ekama, G. (2010). The role and control of sludge age in biological nutrient removal activated sludge systems. *Water Science and Technology, 61*(7), 1645–1652.

EnviroSim Associates, Ltd. (2005). *Influent specifier (raw) 2_2.xls.* Spreadsheet for fractionating wastewater into components required for Biowin modeling. EnviroSim Associates, Ltd.

Garrett, M. T., Jr. (1958). Hydraulic control of activated sludge growth rate. *Sewage and Industrial Wastes, 30*(3), 253–261.

Garrett, M. T., Jr., & Sawyer, C. N. (1952). Kinetics of removal of soluble BOD by activated sludge. *Proceedings of the 7th Industrial Waste Conference,* Purdue University, West Lafayette, IN.

Gould, R. H. (1953, October 6–9). Sewage disposal problems in the world's largest city [Paper presentation]. 25th Anniversary Meeting of the Federation of Sewage and Industrial Wastes Associations, New York.

Hellinga, C., Schellen, A. A. J. C., Mulder, J. W., & Loosedrecht, M. C. M. (1998). The SHARON process: An innovative method for nitrogen removal from ammonium-rich wastewater. *Water Research, 37*(9), 135–142.

Jenkins, D., & Garrison, W. E. (1968). Control of activated sludge by mean cell residence time. *Journal of the Water Pollution Control Federation, 40*(11), 1905–1919.

Jenkins, D., Richard, M. G., & Daigger, G. T. (2003). *Manual on the causes and control of activated sludge bulking, foaming, and other solids separation problems* (3rd ed.). Lewis Publishers.

Johnson, W. S., Leveque, E. G., Narayanan, B., & Buhr, H. O. (2000, October 15–18). The effect of in-plant recycles on control of an activated sludge process. *Proceedings of the 73rd Annual Water Environment Federation Technical Exposition and Conference* [CD-ROM], Anaheim, CA, United States. Water Environment Federation.

Knowles, R. (1982). Denitrification. *Microbiological Reviews, 46*(1), 43–70.

Lau, A., Strom, P., & Jenkins, D. (1984). The competitive growth of floc-forming and filamentous bacteria: A model for activated sludge bulking. *Journal of the Water Pollution Control Federation, 56*(1), 52–61.

Maharajh, N. (2010). Effect of feed rate and solids retention time (SRT) on effluent quality and sludge characteristics in activated sludge systems using sequencing batch reactors. Master's thesis, Virginia Polytechnic Institute and State University.

Marten & Lynne. (2011). *Controlling activated sludge bulking and foaming: From theory to practice*. [Online presentation.]

McCarty, P., & Brodersen, C. (1962). Theory of extended aeration activated sludge. *Journal of the Water Pollution Control Federation, 34*(11), 1095–1103.

Metcalf & Eddy, Inc./AECOM. (2013). Microbial growth kinetics. In *Wastewater engineering: Treatment and resource recovery*. McGraw-Hill.

Muirhead, W. M., Baur, R., Darr, M., Farmer, G., Jiminez, J., Schutz, W., & Taylor, R. (2009, October 10–14). Thirty-six years after the Clean Water Act—Time for a new regulatory framework. *Proceedings of the Water Environment Federation Nutrient Removal Conference* [CD-ROM], Orlando, FL, United States. Water Environment Federation.

National Research Council Subcommittee. (1946). Sewage treatment at military installations. *Sewage Works Journal, 18*(5), 794.

Palm, J. C., Jenkins, D., & Parker, A. D. (1980). Relationship between organic loading, dissolved oxygen concentration, and sludge settleability in the completely-mixed activated sludge process. *Journal of the Water Pollution Control Federation, 52*(10), 2484–2506.

Pearson, E. A. (1966). Kinetics of biological treatment. In *Special lecture series, advances in water quality development.* University of Texas.

Pretorius, C., Appleton, R., Walker, S., Jorgensen, E., Stevensen, B., & Bateman, A. L. (2016, September 24–28). Why SRT control is not more widely practiced and three ways to fix it: Making SRT control a reliable tool for operators to use. *Proceedings of the 89th Annual Water Environment Federation Technical Exhibition and Conference* [CD-ROM], New Orleans, LA. United States. Water Environment Federation.

Sayigh, B. A., & Malina, J. F., Jr. (1978). Temperature effects on the activated sludge process. *Journal of the Water Pollution Control Federation, 50*(4), 678–687.

Secondary treatment, 40 CFR § 133.102 (1993).

Stall, R. T., & Sherrard, J. H. (1978). Evaluation of control parameters for the activated sludge process. *Journal of the Water Pollution Control Federation, 50*(3) 450–457.

Takács, I., Stricker, A. E., Achleitner, S., Barrie, A., Rauch, W., & Murthy, S. (2008, October 18–22). Do you know your sludge age. *Proceedings of the 81st Annual Water Environment Technical Exhibition and Conference* [CD-ROM], Chicago, IL. Water Environment Federation.

Torpey, W. N., & Chasick, A. H. (1955). Principles of activated sludge operation. *Sewage and Industrial Wastes, 27*(11), 1217–1233.

Trygar, R. (2010). Sludge volume index is a valuable measure of sludge settleability characteristics and can be monitored to help prevent process problems. *Treatment Plant Operator*, March.

U.S. Environmental Protection Agency. (1974). *Process design manual for upgrading existing wastewater treatment plants.* EPA-625/1-71-004A.

U.S. Environmental Protection Agency. (1977). *Aerobic biological wastewater treatment facilities process control manual.* EPA III-A 524-77.

U.S. Environmental Protection Agency. (1978). *Design of wastewater treatment facilities major systems.* EPA-430/9-79-008.

U.S. Environmental Protection Agency. (2010). *Nutrient control design manual.* EPA-600/R-10-100.

Wahlberg, E. J. (2016). Personal communication.

Water Environment Federation. (1996). *Operation of municipal wastewater treatment plants* (5th ed., Manual of Practice No. 11).

Water Environment Federation. (2017). *Wastewater biology: The microlife* (3rd ed.).

Water Environment Federation. (2023) *Standards for design of water resource recovery facilities* (7th ed., WEF 8).

West, A. W. (1973a). *Operational control procedures for the activated sludge process—Part I observations and part II control tests.* U.S. Environmental Protection Agency, Office of Water Program Operations.

West, A. W. (1973b). *Operational control procedures for the activated sludge process—Part IIIA calculation procedures.* U.S. Environmental Protection Agency, Office of Enforcement and General Counsel.

West, A. W. (1978). *Updated summary of the operational control procedures for the activated sludge process.* U.S. Environmental Protection Agency, Operational Technology Branch National Training and Operational Technology Center.

Wilen, B.-M., & Balmer, P. (1999). The effect of dissolved oxygen concentration on the structure, size, and size distribution of activated sludge flocs. *Water Research, 33*(2), 391–400.

9.0 SUGGESTED READINGS

Hatfield, W. D. (1958) Activated Sludge. *Sewage and Industrial Wastes, 30*(6), 689–693.

Illinois Environmental Protection Agency. (1995). Formula sheets from wastewater operator certification. http://www.epa.state.il.us/water/operator-cert/waste-water/publications/form-sht.pdf

Keinath, T. K., & Cashion, B. S. (1980). *Control strategies for the activated sludge process*. EPA-600/2-80-131. U.S. Environmental Protection Agency.

Kraus, L. S. (1965). Operating practices for activated sludge plants. *Journal of the Water Pollution Control Federation, 37*(5), 713–721.

Manser, R., Gujer, W., & Siegrist, H. (2005). Consequences of mass transfer effects on the kinetics of nitrifiers. *Water Research, 39*(19), 4633–4642.

Wanner, J. (2014). IWA conference: Activated sludge—100 years and counting. http://www.iwa100as.org/history.php#

Water Environment Federation. (2024). *Standards for operation of Nutrient Removal Facilities* (2nd ed., WEF 37).

5

Process Control Methods for the Secondary Clarifier

1.0 INTRODUCTION TO SECONDARY CLARIFIERS

The purpose of clarifiers is to facilitate the flocculation and settling of particles entering the clarifier. Primary, secondary, and tertiary clarifiers all have the same purpose but are different because they each address different types of particles. Primary clarifiers remove screened and degritted suspended and colloidal solids that are organic and inorganic (gritlike), typically occurring at concentrations of 100 to 500 mg/L; they may also be chemically modified, as in chemically enhanced primary systems. Secondary clarifiers remove organic suspended solids that collectively are regarded as mixed liquor and that are an order of magnitude higher in concentration than primary clarifier influent suspended solids (1000 to 4000 mg/L). Tertiary clarifiers remove suspended solids that remain after secondary clarification so that suspended solids entering tertiary clarifiers are relatively low in concentration (10 to 50 mg/L). Typically, solids are treated with chemicals to increase their size and density and, as a result, can be primarily inorganic. In all three cases, clarifiers provide a means to flocculate influent solids and allow them to settle by gravity.

1.1 Purpose of Secondary Clarifiers

A secondary clarifier is required to flocculate mixed liquor suspended solids (MLSS) that are generated in the upstream aeration basin; the process of flocculation gathers the particles in the mixed liquor, leaving behind a clear liquid that meets total suspended solids (TSS) permit requirements (as the name implies, it is a "clarifier"). A second, equally important purpose of secondary clarification is to settle the flocculated solids into a denser layer of solids that can be removed from the bottom of the clarifier. The settled solids are returned to the aeration basin to maintain a desired sludge inventory according to the targeted solids retention time (SRT) setpoint. This flow of solids is known as the *return activated sludge (RAS) recycle stream*. If the aeration basin were to be operated without an RAS recycle, the rate of growth of new organisms would not be fast enough to maintain the desired sludge inventory and, eventually, the MLSS would be washed out of the system. Therefore, this second purpose of secondary clarification is critical to maintaining the desired inventory of organisms in the aeration basin, and the process of flocculation is critical to ensuring that the mixed liquor settles quickly enough at the residence time in the clarifier.

To maintain a desired sludge inventory in the aeration basin, periodically, settled solids must be removed from the system. This flow of solids is known as the *waste activated sludge (WAS) stream*. The WAS is the flow that is used for solids handling.

1.2 Settling Concepts

The settling behavior of particles is best described by a relationship developed by a British scientist, George Stokes, in 1851, a relationship that today is known as Stoke's law. He showed that the settling velocity of particles, settling as discrete or single particles, is proportional to their density and the square of their diameters when they are settling under the influence of gravity. What this means is that the larger and denser the particles, the faster they will settle.

In wastewater treatment systems, conditions prevailing in different unit processes and wastewater characteristics create different types of particles that settle in different ways; the settling behavior of solids in wastewater treatment is classified into four regimes:

1. Class I (discrete nonflocculent) settling,
2. Class II (discrete flocculent) settling,
3. Class III (zone or hindered) settling, and
4. Class IV (compression) settling.

In Class I, discrete nonflocculent settling, there is no interaction between the particles and no tendency to adhere to one another (i.e., no flocculation). As a result, each particle settles at its own characteristic terminal settling velocity. This type of settling behavior is observed in grit-removal systems.

In Class II, discrete flocculent settling, the particles are fairly dilute but have a strong tendency to flocculate. Particles with higher densities settle faster and "catch up" to slower settling particles. This collision results in one larger particle. This type of settling occurs in primary clarifiers.

In Class III, zone settling, particles are so concentrated and flocculent that they interact with each other. As a result, all particles settle at the same velocity with the formation of a distinct interface. This is observed in secondary clarifiers.

In Class IV, compression settling, particles rest on particles that are resting on particles that are resting on the floor of the tank or container. The resulting compressive stress "squeezes" water out as the particles get closer and closer together. This settling behavior is observed mostly in gravity thickeners. However, it can occur in secondary clarifiers when operated with deep blankets (>1 m [3 ft]), a practice that is discouraged because of the fermentation or denitrification that can occur in such deep blankets.

Flocculation is a dominant force in the settling process that occurs in secondary clarifiers. However, the principle of settling as envisaged by Stokes remains the same in that the larger and denser the flocs, the more quickly they will settle in the secondary clarifier. It underscores the importance of inducing the growth of organisms that will flocculate well.

In the secondary clarifier, as flocs leave the center well, they will aggregate further as they fall through the clarifier until they form a solid mass. This behavior can be seen in a 2-L Mallory jar that is used for settling tests and measuring the sludge volume index (SVI). In this test, an interface is seen to form between the solids and clear supernatant above it that is typical of hindered or zone settling; the reason for this is that individual discrete particles are hindered from settling individually and interact with each other such that they settle as one entity.

The velocity with which the interface settles is referred to as the *zone settling velocity* (ZSV). This relationship is shown graphically in Figure 5.1. Initially, a graph of the position of the interface versus time will be linear and the ZSV is measured as the slope of the linear portion. As the particles get in each other's way, the settling velocity of the combined particles will slowly decrease as the particles fall farther down into the clarifier. This change in settling velocity is seen in the curvature of the graph generated by plotting the position of the interface over a period of time. The curvature illustrates the transition from zone settling to compression.

By the time the mixed liquor has settled to the bottom of the clarifier, it has increased in concentration by a factor of 2 or 3, depending on the degree of consolidation of the resulting sludge. This becomes the source of RAS and is removed using either sludge scrapers or vacuum-assisted withdrawal devices from the bottom of the clarifier.

2.0 FACTORS AFFECTING THE OPERATION OF SECONDARY CLARIFIERS

2.1 Influent Flow

Influent flow can affect clarifier operation and performance because of its inherent variability. Secondary clarifiers are connected hydraulically to influent flow and, thus, can respond to changes in flow, although the effect is

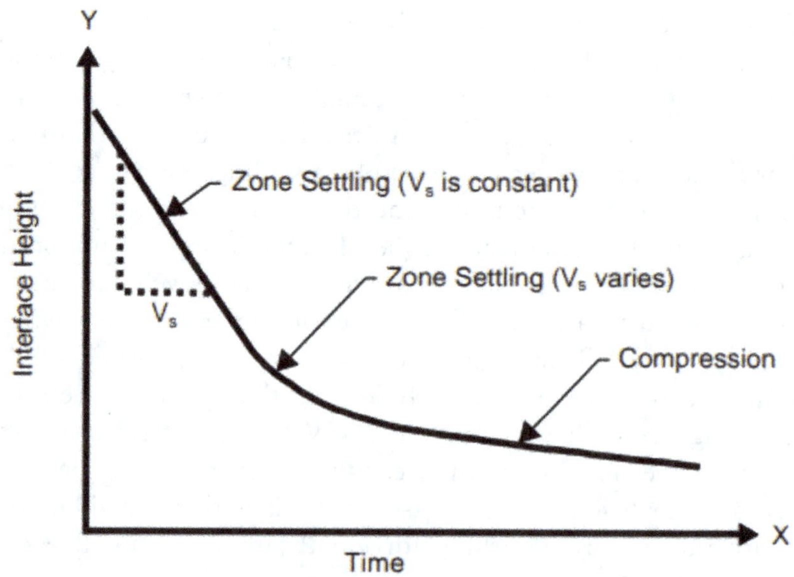

FIGURE 5.1 Zone Settling Velocity in a Secondary Clarifier

attenuated by the number of unit processes between the headworks and clarifiers and the residence time therein. The typical response to flow variations in a secondary clarifier is to elevate the sludge blanket; however, this can cause effluent TSS concentrations to increase. Extreme changes in flowrate can elevate the blanket until it crests the clarifier weir and causes significant solids loss.

Flow variability has two components: diurnal and meteorological. Diurnal flow changes contribute a flow peaking factor of approximately 1.5 and, typically, do not impose any significant hydraulic disturbance on clarifiers. Meteorological or seasonal changes can cause significant changes in influent flow according to rain frequency and intensity, which can be elevated in the summer or the winter depending on geography and climate. Also, the level of infiltration and inflow can elevate the flow peaking factor significantly, particularly in small communities. The net effect is that seasonal changes in flowrate can cause significant increases in effluent TSS concentrations.

The hydraulic effect of influent flow changes induced by the aforementioned range of operating conditions can be estimated by calculating the surface overflow rate (SOR) and observing clarifier responses to changes in SOR. Flow changes can also affect clarifier solids loading, the effect of which can be estimated by calculating the solids loading rate (SLR), as discussed in the following section.

2.2 Surface Overflow Rate

The SOR is a measure of the hydraulic loading to the secondary clarifier. It is calculated from the influent component of total flow to the clarifier, Q (in cubic meters per day [gallons per day]), divided by the surface area, A (in square meters [square feet]), available for the flow to pass through. (Note: this term does not include RAS flowrate, unlike in the definition of SLR.) This yields Q/A (in cubic meters per square meter per day [gallons per day per square foot]) as follows:

$$\text{Surface overflow rate} = \frac{Q\left(\dfrac{\text{m}^3}{\text{d}}\right)}{A(\text{m}^2)} \tag{5.1}$$

The value of A changes according to the number of clarifiers in service.

The clarifier SOR at average and peak flows can be calculated using the following example:

$$\text{Surface area per clarifier} = 729 \text{ m}^2$$

$$\text{For two clarifiers, total surface area} = 2 \times 729 = 1458 \text{ m}^2$$

$$\begin{aligned}\text{Average facility flow} &= 28\,400 \text{ m}^3\text{/d}\\ \text{Peak facility flow} &= 56\,800 \text{ m}^3\text{/d}\\ \text{Average SOR (from Equation 5.1)} &= 28\,400/1458 = 19.5 \text{ m}^3\text{/m}^2\cdot\text{d}\\ \text{Peak SOR (from Equation 5.1)} &= 56\,800/1458 = 38.9 \text{ m}^3\text{/m}^2\cdot\text{d}\end{aligned}$$

An industry range of values recommended for SOR is 33 to 62 $m^3/m^2{\cdot}d$ (800 to 1500 gpd/sq ft) *at peak hour flow conditions*. The practice has evolved with early observations that SOR exerted a significant influence on effluent TSS concentrations (Parker, 1983); it was also suggested that the clarifier depth was linked to SOR in exerting a combined influence on effluent TSS concentrations (Parker, 1983). The SOR will increase during wet weather events.

The SOR is equal to upward liquid velocity resulting from flow exiting a secondary clarifier. It is important because the sum of particle downward velocities resulting from solids settling and removal of RAS from the bottom of the clarifier must be greater than SOR for successful liquid–solids separation. It was not until much later that, after much field testing and observations, it became clear that this downward velocity is almost always greater than SOR. Rarely do clarifiers fail because the SOR is pushing solids out (clarification failure) over the clarifier weirs; clarifiers almost always fail because more solids are being put into them than can be settled to the bottom and removed in RAS (thickening failure) (Wahlberg, 1996b). Indeed, modeling of clarifiers and the documentation of full-scale clarifier experience has suggested that SOR alone is not a significant contributor to poor clarifier performance; rather, SLR has been identified as a more important criterion that controls secondary clarifier performance (Parker et al., 2001). In many cases, the observation that high effluent TSS concentrations have been induced by high SOR values has been tempered by the discovery that other design and operational constraints have contributed to poor clarifier performance, including hydraulic factors such as poor inlet/outlet design, inadequate RAS return capacity at peak flow conditions, inadequate detention time for flocculation in the center well, and denitrification caused by excess detention time of sludge in the blanket.

2.3 Solids Loading Rate

The SLR is a measure of total solids loading to the secondary clarifier. It is calculated from the mass of solids flowing to all of the clarifiers divided by the total surface area of the clarifiers, A. The mass of solids flowing to all of the clarifiers is estimated from the total flow to the clarifier (influent flow + RAS flow) (Q + RAS) multiplied by the MLSS concentration and multiplied by an International Standard unit conversion factor of 10^{-3} (8.34 lb/gal in U.S. customary units). This yields (Q + RAS) × (MLSS/A):

$$\text{Solids loading rate, kg/m}^2\cdot\text{d} = \frac{\left[Q\left(\frac{\text{m}^3}{\text{d}}\right) + \text{RAS}\left(\frac{\text{m}^3}{\text{d}}\right)\right]\cdot\text{MLSS}\left(\frac{\text{mg}}{\text{L}}\right)\cdot 10^{-3}}{A(\text{m}^2)} \quad (5.2)$$

The clarifier SLR at average flow for the previous example can be calculated as follows:

$$\begin{aligned}
\text{Surface area per clarifier} &= 729 \text{ m}^2 \\
\text{For two clarifiers, total surface area} &= 2 \times 729 = 1458 \text{ m}^2 \\
\text{Average facility flow} &= 28\,400 \text{ m}^3/\text{d} \\
\text{RAS flow} &= 14\,400 \text{ m}^3/\text{d} \\
\text{MLSS concentration} &= 2500 \text{ mg/L}
\end{aligned}$$

$$\text{SLR, kg/m}^2\cdot\text{d} = \frac{(28\,400 + 14\,400)\frac{\text{m}^3}{\text{d}}\cdot 2500\frac{\text{mg}}{\text{L}}\cdot 10^{-3}}{1458 \text{ m}^2} = 73.4 \text{ kg MLSS/m}^2\cdot\text{d}$$

An industry range of values for peak solids loading rate is 122 to 195 kg/m^2·d (25 to 40 lb/sq ft·d); this is determined at the design maximum day flowrate, the design maximum RAS return rate, and the design MLSS concentration.

Increasing influent flowrate increases SLR, which can constrain operation during wet weather flow conditions. This effect can be minimized by changing the aeration basin flow regime (e.g., by implementing step feed or sludge reaeration—also known as contact stabilization—modes of operation or some combination of the two). Both modes result in solids being stored at the influent end of the aeration basin. Mixed liquor downstream of the sludge reaeration zone or step feed points will be diluted according to the proportion of solids held back in the aeration basin, which is the level of dilution contributing to a reduction in SLR.

Whichever of the SOR- or SLR-based estimates of required clarifier surface area is the largest is what controls the design and operation of secondary clarifiers. In most cases, except in the case of low MLSS concentrations, it is SLR that controls the sizing of total clarifier surface area and, therefore, operability of the secondary clarifier system. This finding underscores the importance of monitoring clarifier SLR. Based on the aforementioned expression, ([Q + RAS]) × MLSS/A, the operator can address a rising solids loading problem by reducing the RAS rate; increasing the number of clarifiers in service or changing to a solids contact or step feed aeration basin feeding regime (if these options are available; this would have the effect of spreading solids loading across more clarifiers); or lowering the MLSS concentration by increasing the number of aeration basins in service, if that is an option (this would have the effect of diluting mixed liquor in a larger volume). Reducing RAS rate is likely to be a temporary solution because it will result in an increase in sludge blanket depth. This constraint points to the need for a more sophisticated tool that includes the effect of all controllable variables in clarifier operation; the state point analysis is such a tool.

2.4 Surface Area

In a facility with two secondary clarifiers or more, it is best to think of the total surface area available, that is, the number of clarifiers multiplied by the surface area of one clarifier. It is this total surface area that is an integral part of the definition of SOR and SLR. The ability to control critical SLR and SOR parameters by changing clarifier surface area (bringing clarifiers in and out of service) is an important process control tool for operators, as demonstrated in the previous section.

2.5 Mixed Liquor Suspended Solids Concentration

Other than by changing the number of aeration basins in service, the concentration of MLSS is controlled by the SRT setpoint. As discussed earlier in this manual, defining SRT requires control of the solids wasting rate, which then determines residual MLSS concentration in the aeration basins. Because the biochemical oxygen demand (BOD) loading rate varies, the growth of organisms responds by increasing or decreasing in tandem with the change in BOD loading. As a result, MLSS concentration will vary in accordance with BOD loading for a given SRT setpoint. As seen in the definition of SLR in Equation 5.2, MLSS concentration is an integral part of the definition of SLR. Therefore, changes in MLSS concentration can exert changes in SLR. The rate of change of MLSS concentration on a daily basis is not large and, in a diurnal changing environment, there will be little effect on clarifier SLR. However, if diurnal variability is large or the water resource recovery facility (WRRF) is close to its maximum design loading, there may be a deleterious effect on SRT. If significant changes in SRT are proposed, the operator should anticipate changes in MLSS concentration and plan accordingly. Changes in SRT will occur on a regular basis as a result of changing seasons; depending on the geographical region, this can mean lowering SRTs in the summer and elevating them in the winter. Therefore, the operator needs to consider when and how the number of clarifiers and aeration basins might be changed to regulate the change in MLSS concentration.

2.6 Return Activated Sludge Flow and Concentration

The RAS is needed to transfer sludge from secondary clarifiers to the aeration basins and maintain a desired sludge inventory (see Section 2.1). Typically, RAS rates are expressed as a percentage of flow to the aeration basins.

For low SRT systems with low MLSS concentrations, RAS flow is typically 25% to 50% of influent flow. For higher SRT systems with high MLSS concentrations, RAS flow is on the order of 75% to 120% of influent flow.

For a given sludge inventory, reducing the RAS rate has the effect of allowing settled sludge to accumulate and thicken, thereby increasing settled solids concentration. The response to this action is an increase in RAS concentration. There is, however, a limit to which RAS concentration can be increased; indeed, it can lead to an increase in sludge blanket depth, with the potential concomitant problems of fermentation or denitrification occurring. The converse is seen with an increasing RAS rate; less time is available for sludge consolidation at the bottom of the clarifier and RAS concentration will decline.

Waste activated sludge is typically withdrawn from the RAS flow to control SRT. An elevated RAS concentration provides a WAS stream that is at the same higher concentration. Sending this thicker flow to WAS thickeners will improve the efficiency of thickeners, producing a thickened sludge with a higher concentration than would be achieved for a WAS feed at lower solids concentration (as would occur if the RAS flow were increased).

As described previously, RAS is an integral part of the definition of SLR. Changing the RAS rate will influence clarifier SLR. Increasing the RAS rate will draw down the sludge blanket and transfer more sludge to the aeration basin. Over time, this action will increase the MLSS concentration in the aeration basin and, therefore, increase the clarifier SLR. Lowering the RAS rate will have the converse effect; sludge will progressively accumulate in the bottom of the clarifier, allowing more time for thickening and increasing the sludge blanket solids concentration. The net effect is to transfer more solids to the clarifier and the MLSS concentration will slowly decline, reducing the clarifier SLR.

There is no optimum recommended RAS concentration or flowrate because the RAS rate is correlated with so many other parameters that are varied in the operation of an activated sludge system.

Another effect of reducing the RAS flowrate is an increase in the sludge blanket depth. This is discussed in Section 2.7.

2.7 Sludge Blanket Level

A sludge blanket will be established at the bottom of a secondary clarifier in response to the manner in which RAS flow is controlled, as described in Section 2.6. Typically, sludge blanket depths are maintained between 0.16 and 0.7 m (0.5 and 2 ft) from the bottom of the clarifier. An optimum depth is approximately 0.33 m (1 ft).

Greater blanket depths can lead to several different scenarios. For activated sludge systems that are not designed to control nutrients, fermentation may occur in the blanket, particularly at higher temperatures; fermentation gases will be released and will carry solids to the surface, where they could overflow the weir and decrease effluent quality. If the activated sludge system is designed to achieve biological phosphorus removal, similar fermentation reactions can occur that generate soluble BOD that will stimulate the anaerobic release of phosphorus. Phosphorus can elutriate from the sludge blanket and migrate to the clarifier overflow, thereby increasing the effluent phosphorus concentration beyond what was achieved in the aeration basin.

For nitrifying systems, soluble BOD generated by fermentation in the sludge blanket will induce denitrification of nitrate that is present in the mixed liquor, release nitrogen gas that will carry solids to the clarifier overflow and, potentially, deteriorate effluent quality. Apart from minimizing blanket depth, alternative approaches to preventing denitrification from occurring in sludge blankets include lowering the SRT to prevent nitrification in cases where it is not required, increasing the level of internal mixed liquor return in nitrogen removal facilities, and/or increasing the fraction of aeration tank volume that is devoted to anoxic denitrification. In cases that are carbon-limited, denitrification can be improved by carbon supplementation.

Also, significant hydraulic surges, such as that which can occur when taking clarifiers out of service, can disturb a deep sludge blanket and cause solids to carry over the weirs.

Based on the aforementioned scenarios, storing solids in a secondary clarifier by allowing deep sludge blankets to be established is not a recommended operational strategy because of the deleterious effects of such a practice. It also makes it more difficult to track sludge inventory.

Sludge blanket depths should be measured once a shift using a Sludge Judge® or equivalent sampling device to ensure that blanket accumulation is not occurring. This is a clear plastic tube marked in 0.33-m (1-ft) increments, approximately 2.5 cm (1 in.) in diameter, in lengths that thread together to allow the operator to reach from the clarifier bridge to the bottom of the clarifier. The bottom section of the tube has a check valve that allows the tube to fill as it is lowered into the clarifier, but seats when the operator draws the tube up. The operator then is able to actually see the depth of sludge in the tank. Based on the measured depth, the RAS rate may be increased or decreased according to the desired blanket depth.

2.8 Sludge Quality (Mixed Liquor Settleability), Sludge Volume Index

Because it affects many of the variables considered during design and operation of secondary clarifiers, solids settling is the most important of all of the factors considered in this section. Also, because control of settling characteristics takes place in the aeration basin, good secondary clarifier performance begins in the aeration basin (Wahlberg, 1996b). The success of an activated sludge facility depends heavily on the ability of the MLSS to settle quickly by gravity. Successful operation of the secondary clarifier depends on mixed liquor settleability, which, as discussed in Section 1.1, depends on the efficiency of flocculation that is achieved in the aeration basin and in the clarifier center well. Organisms must combine into properly sized particles that have sufficient density to allow them to sink to the floor of the secondary clarifier. The biomass must compact well enough that the sludge blanket does not occupy an excessive amount of space in the clarifier or solids may be lost in the effluent. Settling tests may be carried out to assess sludge quality in terms of how well mixed liquor settles. The test is simple, but when conducted daily, provides an ongoing measure of sludge quality: a poorly settling sludge is the first sign that the microbial environment in the aeration basin is not conducive to inducing the required flocculating properties necessary for rapid settling in the clarifier.

There are several ways of conducting such a test; it may be conducted in a stirred or unstirred measuring cylinder or in a large beaker referred to as a *settleometer*, which is typically a clear plastic beaker with a capacity of 2 L, such as a Mallory jar. The settleometer-based test is the most widely used of these. Graduations on the beaker range from 100 to 1000 mL of settled sludge per liter. A sample of mixed liquor should be obtained from the discharge end of the aeration tank, being careful not to include scum in the sample. The sample should be tested without delay. The mixed sample should fill the settleometer to the highest graduated mark. An interface will form between the settling mass of solids and the clear supernatant and gradually fall through the settleometer. The position of the interface should be measured every 5 minutes; these data may be used to plot a settling curve from which solids settling velocity can be computed. The settled volume of solids is noted after 30 minutes of settling and is a measure of the degree of compaction that can be achieved by settling the mixed liquor. These measurements may be used to compare performance on a day-to-day basis for a given WRRF but are not useful to compare performance against another facility that is operating with a different MLSS concentration. A mixed liquor with a higher MLSS concentration will contain a higher mass of solids per unit volume and inherently occupy more volume than a mixed liquor with a lower MLSS concentration; the settled volume after 30 minutes will not be the same for these two facilities. The SVI takes MLSS concentration differences into account.

The SVI is the volume in milliliters occupied by 1 g of a suspension after 30 minutes of settling. The SVI is calculated in the following manner: (1) allow a mixed liquor sample from the aeration basin to settle for

30 minutes as previously described, (2) determine the suspended solids concentration for a sample of the same mixed liquor, and (3) calculate SVI by dividing the measured (or observed) wet volume (mL/L) of the settled sludge by the dry weight concentration of MLSS (g/L) as follows:

$$\text{SVI, g/mL} = \frac{\text{Settled sludge volume (mL)}}{\text{Sample volume (mL)}} \times \frac{1000\,\dfrac{\text{mg}}{\text{g}}}{\text{MLSS}\,\dfrac{\text{mg}}{\text{L}}} \tag{5.3}$$

Clearly, the SVI determination takes longer than the settleometer test because of the need to measure MLSS concentration. However, there are microwave-based solids analyzers available that measure solids concentrations within minutes. The SVI procedure is described in detail in *Standard Methods for the Examination of Water and Wastewater* (American Public Health Association et al., 2022). For facilities operating with high MLSS concentrations (>4000 mg/L) or for poor settling sludges (30-minute settled sludge volume greater than 300 mL/L), it may be more appropriate to use the diluted SVI version of the test to obtain more accurate SVI measurements.

With respect to process control, SVI is used to monitor mixed liquor settling characteristics and MLSS' ability to settle and compact. The SVI data may be compared between facilities because they are normalized according to MLSS concentration. The lower the mixed liquor SVI, the better it is said to settle; conversely, when the mixed liquor SVI is high, the more likely it is to settle poorly. For MLSS concentrations in the range of 1200 to 3000 mg/L, a good settling sludge has an SVI in the range of 100 to 150 mL/g.

2.9 Clarifier Hydraulics

The intent of secondary clarifiers is to provide a quiescent environment that allows mixed liquor to flocculate and settle unimpeded to the bottom of the clarifier. In reality, a quiescent environment is difficult to maintain because of the influence of many ongoing effects such as wind over the surface of the clarifier, circular motion of the solids-removal device, the withdrawal of settled sludge, currents induced by mixed liquor passing through center well slots, and temperature gradients. Traditionally, the design of the clarifier allowed for these effects by increasing clarifier diameter and depth. Clarifier inlet design must include provisions for decreasing the momentum of influent flow, distributing flow equally throughout the clarifier, and maximizing the potential for flocculation. The rate at which effluent overflows clarifier weirs is controlled to avoid localized excessively high flow velocities that can bring up solids over the weir. Baffles are used to break up density currents in the clarifier. These are sometimes referred to as *Stamford Baffles*.

The weir loading rate is the volume of water passing over a unit length of the weir per day. Typical weir overflow rates range from 125 to 175 m^3/m·d (10 000 to 14 000 gpd/ft).

Today, computational fluid dynamics (CFD) models are able to simulate these hydraulic effects and counter them by installing improved energy-dissipating devices in the center of the clarifier, larger and deeper center wells to allow for improved flocculation, and modified solids-removal devices and overflow weir configurations.

2.10 Summary

The operator can change the number of clarifiers in service and the RAS rate but will have little opportunity to change a given clarifier configuration once it has been designed and built. What is important for the operator to focus on is ensuring that mixed liquor generated in the aeration basins can settle, compact, and flocculate well. Secondary effluent quality depends on these characteristics.

3.0 COMMON MISCONCEPTION ABOUT SECONDARY CLARIFIERS: CLARIFIERS CAN THICKEN AND STORE SOLIDS

The movement of solids between the aeration basin and secondary clarifier is a dynamic process. An overloaded condition can result when a net transfer of solids occurs from the aeration basin to the secondary clarifier as, for example, during wet weather flow conditions. If adequate storage capacity is available in the clarifier (clarifier depth is sufficient) to accept this transfer of solids, a solids blanket will form. If not, loss of solids will occur when the solids blanket rises to the surface and is discharged in the effluent. It is possible to size a clarifier to have sufficient capacity to prevent the accumulation of a blanket even under high flow conditions, but the cost of doing so is prohibitive. Designing a clarifier for a minimal sludge blanket by having adequate solids-removal capacity is good practice. The intent of design is not to provide clarifier volume for solids storage during stressed operating condition that occur during peak wet weather flow conditions. Not following this practice will lead to sludge blanket accumulation with the attendant operational difficulties discussed in Section 2.7.

A limited degree of sludge blanket accumulation is appropriate, as discussed in Section 2.7, to provide sludge thickening such that RAS concentrations can increase to allow reasonable RAS flowrates. Without this occurring, RAS would be diluted, leading to higher RAS flowrates (and the attendant operating cost) to accommodate solids transfer back to the aeration basins. If sludge blanket depths are allowed to increase with the intent of further sludge thickening, at some point thickening failure will occur and the blanket will rise to the surface of the clarifier and over the weirs. This is not a recommended operational strategy.

Solids stored in the secondary clarifier are like money in a savings account not collecting interest (Wahlberg, 1996b). Microorganisms in the sludge blanket are unable to function as intended at the bottom of a secondary clarifier; they are best removed quickly from the secondary clarifier and returned to the aeration basin, where they can continue to perform their microbiological conversion processes.

4.0 STATE POINT ANALYSIS

Probably the most important secondary clarifier process control tool available to operators is the state point analysis (SPA) tool (Wahlberg et al., 1995; Wahlberg, 1996a). A solids inventory control method, the state point concept takes into account the relationships between secondary clarifier effluent flowrate, RAS flowrate, clarifier surface area, MLSS concentration, and the settling characteristics of mixed liquor. This relationship is shown graphically in Figure 5.2. A state point analysis consists of a solids balance

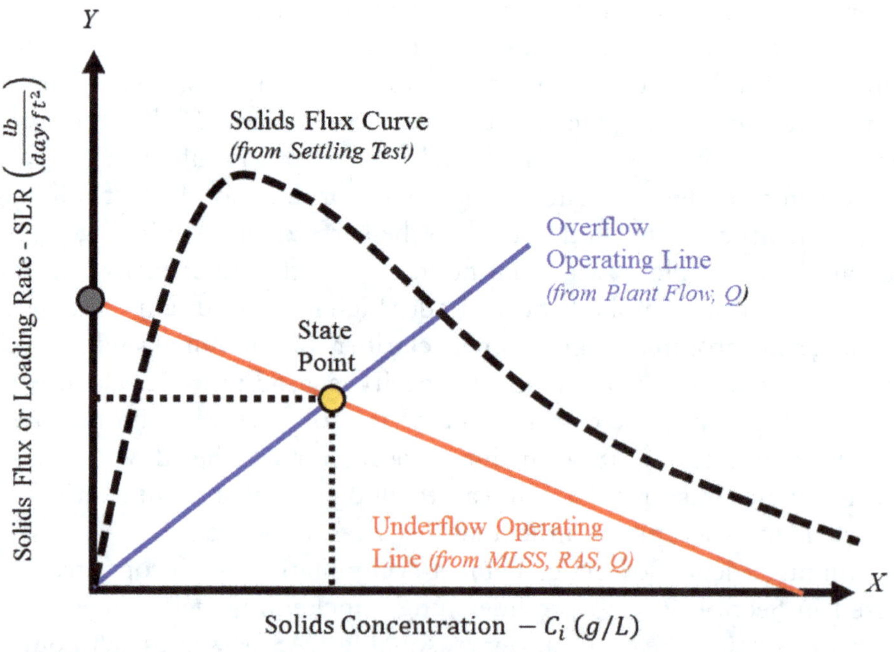

FIGURE 5.2 State Point Diagram Components (Adapted from WEF, 2023)

around the secondary clarifier overlaid with solids-settling characteristics. The method has been made more convenient by empirical models relating SVI to the parameters necessary to generate the settling flux curve used in a state point analysis. A state point analysis is useful for making process control decisions, troubleshooting secondary clarifier performance, and optimizing the activated sludge process.

The SPA diagram (Figure 5.2) is constructed in the following manner. Solids concentration is plotted on the x-axis of the graph and SLR is plotted on the y-axis. The SLR is also referred to as *solids flux*, that is, the rate of mass flow of solids through the cross-sectional area of the clarifier over a given time.

The overflow operating line (blue) is drawn through the origin of the graph (coordinates: 0, 0) with a slope equal to the SOR (Q/A). A dashed vertical line is drawn from the x-axis at the point of the MLSS concentration. Where these lines intersect represents the state operating point. This line represents a solids mass balance around the top of the clarifier.

The underflow operating line (red) is drawn through the point on the y-axis representing the clarifier SLR and with a slope of Q_{RAS}/A. This line represents a solids mass balance around the bottom of the clarifier. The representation of the clarifier SLR by the intersection of the line with the y-axis can be confusing because it is not the true measure of *solids flux* through the clarifier. The SLR is a measure of total solids loading to the clarifier that includes solids in the mixed liquor and RAS that is being returned to the clarifier. Under stable operating conditions, because the solids associated with the RAS both enter (at the center well) and leave the clarifier (at the bottom), the net mass of solids settling through the clarifier is the difference between total solids load and RAS solids; this is the true solids flux through the clarifier and is represented by the intersection of the dashed horizontal line from the state point to the y-axis.

Where the underflow line intersects the x-axis represents the RAS solids concentration when the clarifier is not overloaded. The point of intersection of the overflow and underflow operating lines represents the clarifier solids mass balance (orange circle) and is the state operating point of the clarifier for the clarifier prevailing loading conditions and sludge settleability.

A graph of mixed liquor settling performance is superimposed on the lines discussed above. This is a representation of the settling capability of mixed liquor at different mixed liquor solids concentrations. The shape of the curve is intuitive in that settling rate increases as the solids concentration increases and the size of flocs increase. At some point, the solids concentration reaches a maximum rate at which the solids are able to settle because there are too many solids to pass through the available cross-section area of the clarifier. As the solids continue to increase, the settling rate declines, as described by the descending arm of the right-hand side of the curve. If this line is a representation of the solids mass flow of settling solids, then any condition represented below it (as represented by the state point) is a region in which the clarifier may operate without failure. Therefore, the state point diagram provides a visual interpretation of the capability of the clarifier (when operating under the conditions used to construct the diagram) to achieve solids separation. Any area above the solids flux curve will, therefore, lead to a failed operating condition. Some examples of its application are given in subsequent figures. Correlations between SVI and the solids flux curve have been published, allowing operators to generate figures similar to those given below using their daily operational data (Wahlberg and Keinath, 1988). Another useful resource for SPA analysis is given by the New England Interstate Water Pollution Control Commission (2016).

The solids flux curve may be compared to observations made of the interface of mixed liquor as it settles in the settleometer. In this test, as sludge settles, the velocity of the interface level is measured. While this is happening, the concentration of solids is increasing, so the mass of solids moving through the liquid also increases. As the sludge flocculates and settles, both the concentration and velocity increase to a maximum, equivalent to the highest point on the flux curve. As compaction becomes the dominant type of settling, velocity decreases rapidly and the rate of increase in concentration slows considerably. The flux curve levels out because no more settling or compaction can occur.

Figure 5.3 shows an example of a clarifier when it is overloaded with respect to thickening (Keinath, 1985). The state point is below the solids flux curve, but the underflow line intersects with the solids flux curve. In this condition, solids are settling at a greater rate than they are being removed, and the sludge blanket accumulates over time; the blanket depth will increase until solids overflow the weirs. The response to this situation is to increase RAS flowrate to draw down the sludge blanket, as shown in Figure 5.4. Increasing the RAS rate increases the slope of the underflow line, which will shift it below the settling flux curve. This action will generate sufficient time for the operator to determine why the clarifier reached the thickening failure condition and evaluate alternative corrective actions such as reducing flow to the clarifier by bringing more clarifiers online.

Figure 5.5 shows an example of a clarifier that is hydraulically overloaded and failing with respect to clarification (Keinath, 1985). In this condition, the state point is located above the settling flux curve, where the upward velocity of flow through the clarifier is greater than the settling velocity of the solids. This results in the solids being lifted to the surface of the clarifier and over the weirs.

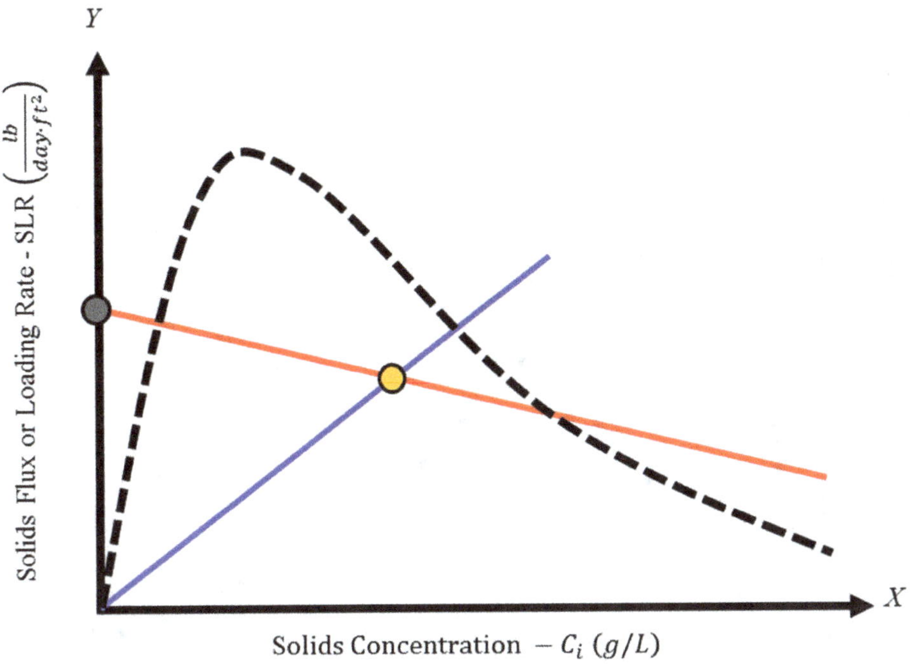

FIGURE 5.3 Simulation of Clarifier Overloaded With Respect to Solids Loading (Adapted from WEF, 2023)

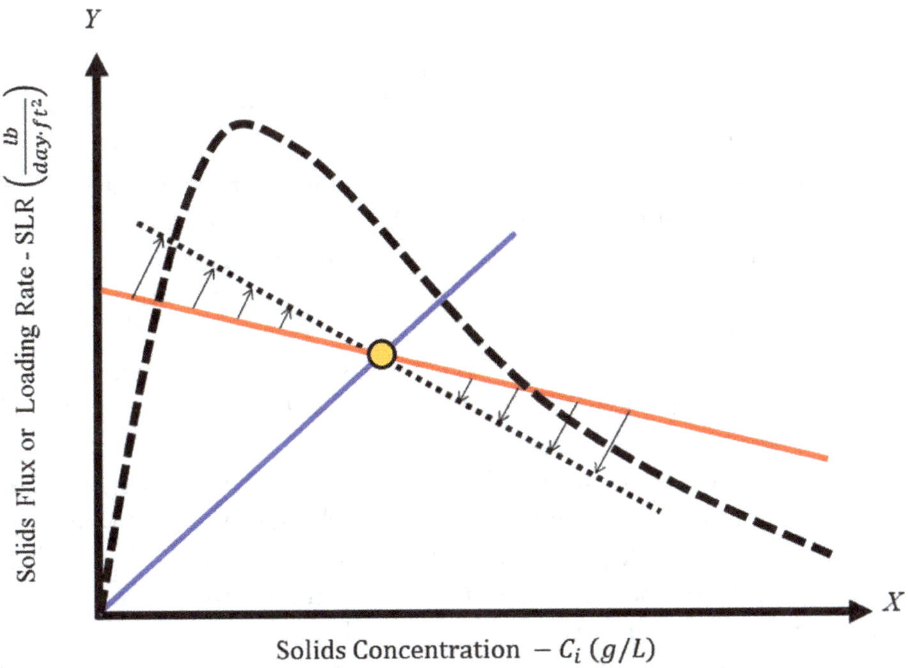

FIGURE 5.4 Corrective Action to Solids Overload (Adapted from WEF, 2023)

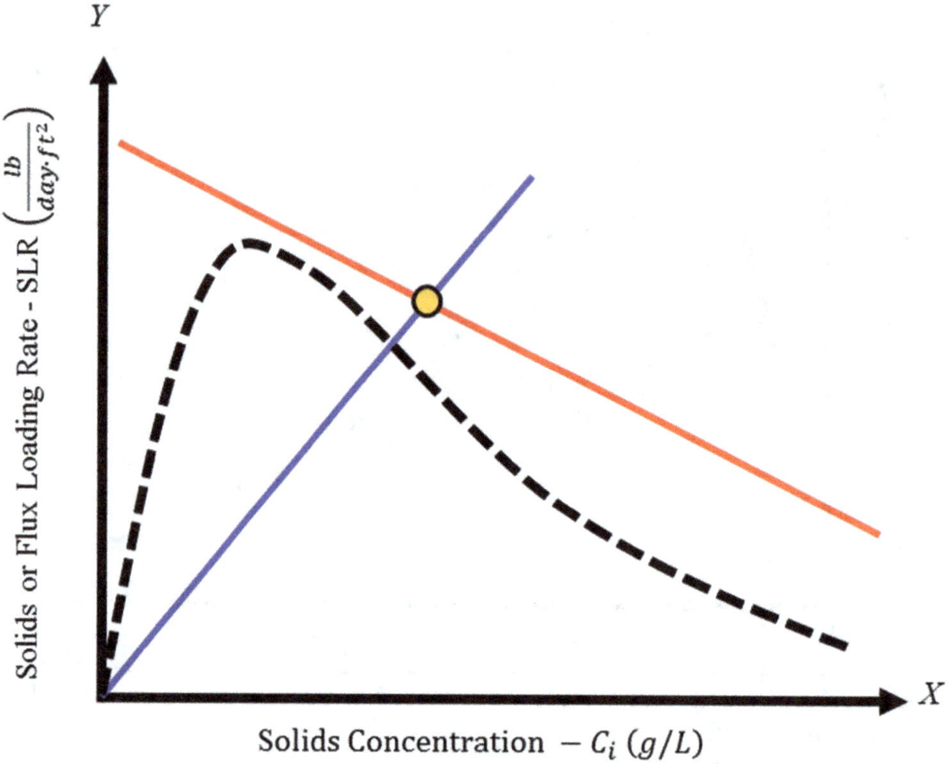

FIGURE 5.5 Simulation of Clarifier Overloaded With Respect to Hydraulic Loading (Adapted from WEF, 2023)

Apart from bringing additional clarification capacity online, a remedy for clarification failure is to reduce the MLSS concentration, which results in a downward shift of the underflow line, as shown in Figure 5.6. The MLSS concentration can be reduced by bringing additional aeration basin volume into service. Of note is that the SLR is also reduced by this action, as shown by the downward shift of the intersection of the underflow line with the y-axis. This makes sense because the MLSS concentration is reduced in the expression for SLR.

When mixed liquor settleability declines as a result of filamentous bulking, SVI values can escalate to values as high as 400 mL/g. The higher SVI sludge will settle poorly and generate a solids flux curve that is smaller than for better-settling mixed liquor. The area under the curve is significantly reduced and offers a visual interpretation of the fact that the operator is limited in operational flexibility. This is illustrated in Figure 5.7, which is a representation of the potential decline in performance from that portrayed in Figure 5.2 as a result of the increase in SVI.

Chlorination of RAS is a typical response that can be applied to eliminate filamentous organisms. The effect would be to restore the solids flux

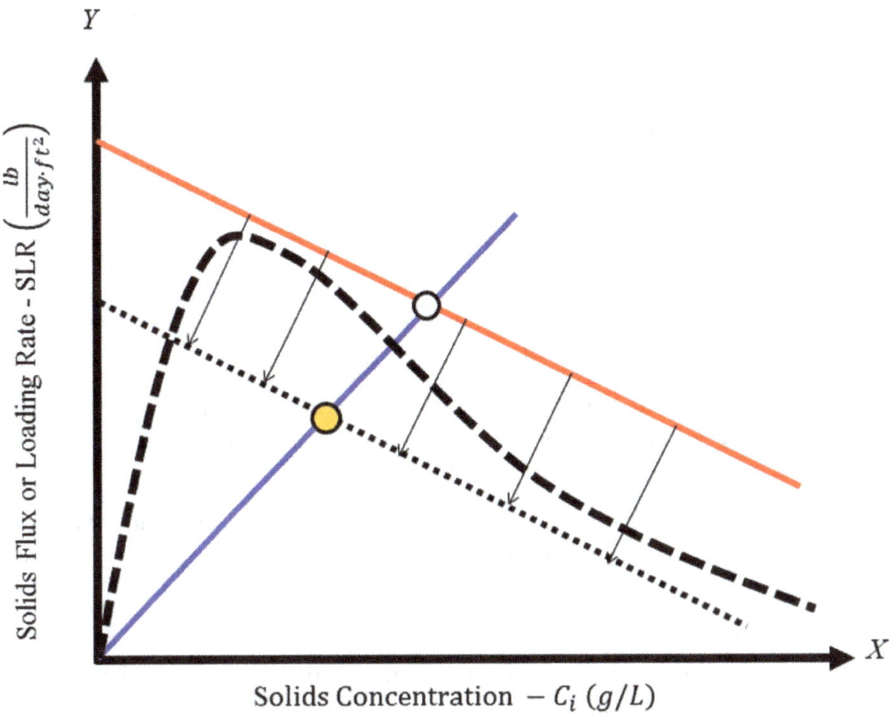

FIGURE 5.6 Corrective Action to Hydraulic Overload (Adapted from WEF, 2023)

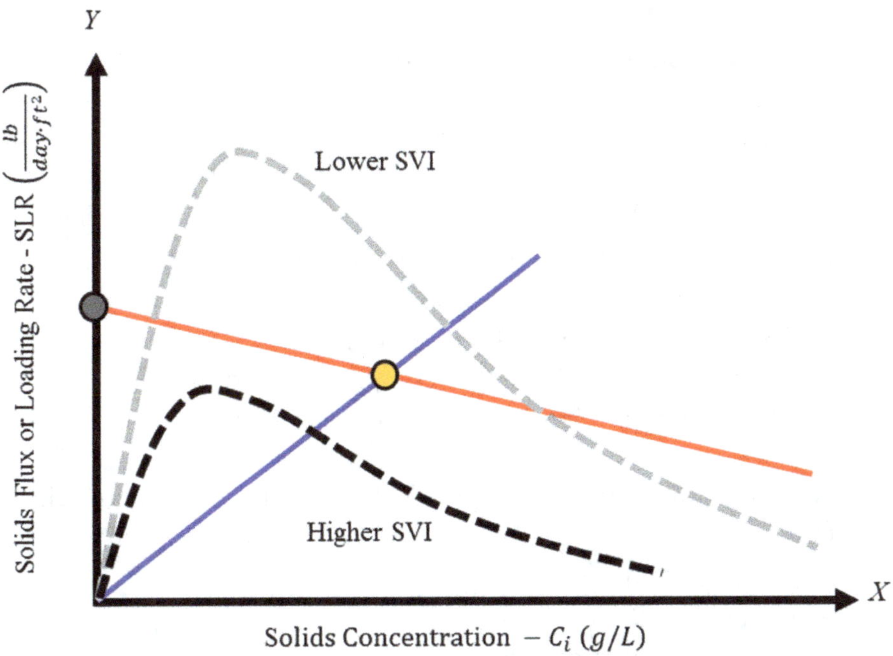

FIGURE 5.7 Effect of SVI on Settling Curve

curve to its original shape. Tools other than SPA are available to analyze clarifier operation. They are based on CFD analysis and are more sophisticated than SPA. These are not applicable to daily operational control but may be used to develop tools or to create site-specific SPA calibrations that may be used by operators.

5.0 PRACTICE QUESTIONS

1. Which of the following is Stokes' law:
   a. Settling velocity of particles is proportional to their density and the square of their diameters when they are settling under the influence of gravity
   b. An increase in the speed of a fluid occurs simultaneously with a decrease in pressure or a decrease in the fluid's potential energy
   c. Formula used to calculate the pressure drop, or head loss, of a fluid flowing through a pipe
   d. An application of continuity of flow and continuity of potential to iteratively solve for flows in a pipe network

2. Which type of settling is observed in secondary clarifiers with shallow blankets?
   a. Class I (discrete nonflocculent) settling
   b. Class II (discrete flocculent) settling
   c. Class III (zone) settling
   d. Class IV (compression) settling

3. In a graph of the velocity of the solids–supernatant interface over time, what does the curvature indicate?
   a. The constant velocity of zone settling
   b. The velocity of compression settling
   c. The transition from zone to compression settling
   d. The velocity of discrete nonflocculent settling

4. Which of the following typically has the least effect on clarifier failure?
   a. Solids overflow rate
   b. Inadequate RAS flow
   c. Poor inlet/outlet design
   d. Inadequate detention time for flocculation

5. All of the following will affect the SLR calculation EXCEPT
   a. MLSS concentration
   b. RAS flowrate
   c. Number of clarifiers
   d. Surface water depth of clarifier

6. The scenario in which SOR controls the sizing of the total clarifier surface area instead of SLR is
   a. Low RAS flowrate
   b. Low MLSS concentrations
   c. High influent flowrate
   d. High WAS flowrate

7. Which of the following is not a consequence of a too deep sludge blanket?
   a. Phosphorus release in systems designed for phosphorus removal
   b. Denitrification
   c. Improved SVI
   d. Solids overflow during hydraulic surges

8. What is the optimum RAS flowrate as a percentage of influent flow?
   a. 10% to 20%
   b. 25% to 50%
   c. 75% to 120%
   d. Depends on site-specific parameters

9. Which of these is not an advantage of the state point analysis?
   a. Estimate effluent suspended solids
   b. Optimize the activated sludge process
   c. Troubleshoot performance of secondary clarifiers
   d. Develop clarifier process control strategies

10. What chemical can be applied to eliminate filamentous organisms in RAS?
    a. Fluoride
    b. Chlorine
    c. Potassium permanganate
    d. Iron

Answer key:

1. a, 2. c, 3. c, 4. a, 5. d, 6. b, 7. b, 8. d, 9. a, 10. b

6.0 REFERENCES

American Public Health Association, American Water Works Association, & Water Environment Federation. (2022). *Standard methods for the examination of water and wastewater* (24th ed.). American Public Health Association.

Keinath, T. M. (1985). Operational dynamics of secondary clarifiers according to the thickening criterion. *Journal of the Water Pollution Control Federation*, *57*(7), 770–776.

New England Interstate Water Pollution Control Commission. (2016). *Guides for the design of wastewater treatment works* (TR-16). New England Interstate Water Pollution Control Commission.

Parker, D. S., Kinnear, D. J., & Wahlberg, E. J. (2001). Review of folklore in design and operation of secondary clarifiers. *Journal of Environmental Engineering*, *127*(6), 476–484.

Parker, D. S. (1983). Assessment of secondary clarification design concepts. *Journal of the Water Pollution Control Federation*, *55*(4), 349–359.

Wahlberg, E. J., & Keinath, T. M. (1988). Development of settling flux curves using SVI. *Journal of the Water Pollution Control Federation*, *60*(12), 2095–2100.

Wahlberg, E. J. (1996a). Activated sludge solids inventory control using the state point concept. In *Enhancing the design and operation of activated sludge plants*. Central States Water Environment Association.

Wahlberg, E. J. (1996b). Secondary clarifiers: The mechanisms behind the myths. *Operations Forum*, January, 12–18.

Wahlberg, E. J., Augustus, M., Chapman, D. T., Chen, C.-L., Esler, J. K., Keinath, T. M., Parker, D. S., Tekippe, R. J., & Wilson, T. E. (1995, October 21–25). Evaluating activated sludge secondary clarifier performance using the CRTC protocol: Four case studies. *Proceedings of the 68th Annual Water Environment Federation Technical Exposition and Conference*, Miami Beach, FL, United States. Water Environment Federation.

Water Environment Federation. (2023). *Standards for operation of water resource recovery facilities* (WEF 11).

7.0 SUGGESTED READINGS

Great Lakes-Upper Mississippi River Board of State and Provincial Public Health and Environmental Managers. (2014). *Recommended standards for wastewater facilities*. Health Education Services Division.

Keinath, T. M. (1990). Diagram for designing and operating secondary clarifiers. *Research Journal of the Water Pollution Control Federation*, 62(3), 254–258.

Water Environment Federation. (2005). *Clarifier design* (2nd ed., Manual of Practice FD-8).

Water Environment Federation. (2023) *Standards for design of water resource recovery facilities* (WEF 8).

6

Process Control Tests and Monitoring

1.0 INTRODUCTION

This chapter summarizes many of the process control tests that are critical for operating a successful, well-controlled activated sludge process. For complete, step-by-step laboratory procedures, refer to *Basic Laboratory Procedures for the Operator-Analyst* (Water Environment Federation, 2022). A few essential procedures for basic process control have been included here with descriptions of other tests. Recommended sampling frequencies are shown in Tables 6.1, 6.2, and 6.3.

2.0 VISUAL INSPECTION OF THE AERATION BASIN

The surface of the activated sludge basin should be free of debris, floatable material, and foam. For aerated zones or basins, the surface of the aeration

TABLE 6.1 Recommended Sampling and Analysis Frequencies for Influent (WEF, 2022)

	Analysis[a]	Frequency[b]	Sample type
Influent	BOD	Daily	Composite
	TSS	Daily	Composite
	pH	Daily	Grab
	TKN[c]	Daily	Composite
	Ammonia[c]	Daily	Composite
	Total phosphorus[c]	Daily	Composite

[a]May not show all analyses required.
[b]Sampling frequencies may need to be increased for process control or permit requirements.
[c]May not be required if disinfection is not required; check with discharge permit.

TABLE 6.2 Recommended Sampling and Analysis Frequencies for Primary Effluent (WEF, 2022)

Sampling location	Analysis[a]	Frequency[b]	Sample type
Primary influent	BOD	Weekly	Composite
	TSS	Weekly	Composite
	pH	Daily	Grab
	TKN	Daily	Grab
	Ammonia	Daily	Grab
	Total phosphorus	Daily	Grab
Primary effluent	BOD	Weekly	Composite
	TSS	Weekly	Composite
	pH	Weekly	Grab
	Dissolved oxygen	Weekly	Grab
	TKN[c]	Weekly	Composite
	Total phosphorus[c]	Weekly	Composite

[a]May not show all analyses required.
[b]Sampling frequencies may need to be increased for process control or permit requirements.
[c]May not be required if nutrients (ammonia, phosphorus, etc.) are not a concern.

TABLE 6.3 Recommended Sampling and Analysis Frequencies for Activated Sludge (WEF, 2022)

Sampling location	Analysis[a]	Frequency[b]	Sample type
Aeration basin influent	pH	Daily	Grab
	BOD	Weekly	Composite
	TSS	Weekly	Composite
	TKN[c]	Monthly	Grab
	Ammonia	Monthly	Grab
	Alkalinity	Monthly	Grab
Aeration basin	Dissolved oxygen	Daily (continuous)	In situ
	Temperature	Daily (continuous)	In situ
Aeration basin effluent	TSS (mixed liquor suspended solids)	Daily	Grab
	Settleability	Daily	Grab
	pH	Weekly	Grab
	Microscopic	Weekly	Grab
Return activated sludge	TSS	Daily	Grab
	Flow	Daily	Totalizer
Waste activated sludge	TSS	Daily	Grab
	Flow	Daily	Totalizer
Secondary clarifier	Sludge blanket	Daily	In situ
Secondary clarifier effluent	BOD	Weekly	Composite
	TSS	Weekly	Composite
	Ammonia[c]	Monthly	Composite
	Nitrate[c]	Monthly	Grab
	Nitrite[c]	Monthly	Grab
	Total phosphorus[c]	Monthly	Composite
	pH	Daily	Grab
Plant effluent	Turbidity	Daily	Grab
	Fecal coliform[d]	Daily	Grab
	Chlorine residual[d]	Daily	Grab

[a]May not show all analyses required; check with permit requirements.

[b]Sampling frequencies may need to be increased for process control or permit requirements.

[c]May not be required if nutrients (ammonia, phosphorus, etc.) are not a concern.

[d]May not be required if disinfection is not required; check with discharge permit.

basin should demonstrate even mixing throughout the basin, with a minimal amount of light, tan-to-brown foam that shifts easily, making a honeycomb type pattern. Foam coverage should be less than 5% of the total surface area. An example activated sludge basin is shown in Figure 6.1. A small amount of foam will always be present because of surfactant production by bacteria in the process; however, it should dissipate quickly. Larger amounts of foam or foam that is difficult to move or break apart is indicative of an abnormal condition.

2.1 Mixing Patterns

Daily walkthroughs should be conducted once per shift, when possible. Activated sludge basins should be inspected for evidence of even, complete mixing. A well-mixed, aerated basin will have a similar appearance to the basin in Figure 6.1, with an ever-changing pattern of bubbles and light foam across the surface. Typically, anoxic and anaerobic zones and basins will be mixed with submersible mixers or, more frequently in recent years, with compressed air mixers (also known as "big bubble mixers"). These basins will not exhibit the bubble pattern shown in Figure 6.1. In the

FIGURE 6.1 Surface of an Aerated Activated Sludge Basin in Clifton, Colorado (Reprinted with permission from Indigo Water Group)

case of submersible mixers, the basin surface should be smooth and clear throughout, without clear differences in color or solids density from one area of the basin to another. With a compressed air system, the basin should have a relatively smooth surface with minimal or no foam. Regularly timed pulses of air should be seen reaching the surface, reflecting the pattern of the accumulator plates installed below. In both cases, areas with noticeable foam buildup or clear water evident at the surface suggest a lack of mixing that should be investigated. Figure 6.2 shows areas of clear water within the aeration basin caused by turning off the diffused air to this portion of the basin. Without mixing, mixed liquor suspended solids (MLSS) rapidly begin to settle.

2.2 Evidence of Broken Headers and Diffusers

Broken headers or missing, torn, and damaged diffusers can often be identified by a boiling effect that is localized to one area of the aerated zone or basin. This may be accompanied by an inability to maintain dissolved oxygen levels in the aeration basin. In the case of multiple basins being fed from a single blower, basins without broken headers and/or diffusers

FIGURE 6.2 Clear Water Demonstrates Solids Separation and a Lack of Mixing (Reprinted with permission from Indigo Water Group)

may have lower dissolved oxygen concentrations because the air from the blower will flow predominantly to the broken header as this is the path of least resistance.

2.3 Foam Color Indicators

Activated sludge basins should be free of excessive foam buildup. The absence or presence of foam should be noted on daily walkthroughs, with a written record of foam color, thickness, and an estimation of the percentage of basin coverage. More detailed information on foaming and sludge settleability problems can be found in Chapter 7, Process Troubleshooting. In all cases, the presence of foam on an activated sludge basin should be investigated and the root cause determined, ideally before 30% of the basin is covered with foam.

Three filamentous organisms can cause activated sludge foaming: *Nocardia* sp., *Microthrix parvicella,* and Type 1863 (Jenkins et al., 2003). These organisms can produce a stable, viscous, brown foam on the surface of the activated sludge basins that can carry over to secondary clarifiers and final effluent. Foaming can range from a nuisance to a serious problem. In cold weather, this foam may freeze solid and have to be manually removed. In warm weather, it can generate odors. Unchecked, it can overflow tanks and create a slippery mess on walkways, as shown in Figure 6.3.

White foam may be present during system startup, under high organic loading rate conditions, or in response to a toxic shock load. Historically, certain household detergents have contributed to white foaming episodes in activated sludge processes. These detergents have been replaced by biodegradable detergents, which do not cause foaming issues (Jenkins et al., 2003). Grey, pumicelike foam can be caused by recycling fine solids from anaerobic processes or receiving large quantities of hauled septic wastes. Nutrient deficiency can cause a sticky, viscous foam.

3.0 DISSOLVED OXYGEN

The dissolved oxygen test can be conducted with either handheld or online dissolved oxygen meters. Both types consist of a process dissolved oxygen probe connected by a signal line to a readout device. Ideally, with either type of meter, multiple points in the aeration basin are monitored on a regular basis.

Dissolved oxygen probes are available as either traditional membrane probes or luminescent-type probes. Traditional membrane probes measure an electric current caused by oxygen diffusing through the membrane and

FIGURE 6.3 Activated Sludge Foaming (Reprinted with permission from Indigo Water Group)

interacting with an internal electrode. Membrane probes are not accurate at lower dissolved oxygen concentrations and require regular maintenance to replace the membrane as it fouls and the internal electrode solution. Luminescent probes work by emitting blue light, which excites luminescent material on the dissolved oxygen sensor. When the blue light is turned off, the material relaxes and red light is emitted and measured by a sensor. The time lag between when the blue light is turned off and the red light is emitted correlates to the dissolved oxygen concentration in the water. The higher the dissolved oxygen concentration, the shorter the time it takes for red light to be emitted. Luminescent probes are more accurate at low dissolved oxygen concentrations, are less susceptible to fouling, and do not require membrane and electrode solution replacement. Both types of probes and meters require regular calibration, which should be performed according to the manufacturer's recommendations. A routine cleaning program is needed for in-basin meters.

The use of a handheld dissolved oxygen meter requires an operator to monitor various set locations in the aeration basin. The important thing to remember is to test the dissolved oxygen in approximately the same location(s) and depth every time. This way, changes in the dissolved oxygen level can be attributed to an actual dissolved oxygen drop or rise rather than to a change in location. In-basin dissolved oxygen meter calibration should be verified using a laboratory-calibrated handheld meter on a regular basis. Be sure to compensate for barometric pressure (i.e., elevation above sea level) when calibrating.

The optimum level of dissolved oxygen depends on the design and actual operating conditions of the water resource recovery facility (WRRF) and can vary substantially from facility to facility. Conventional activated sludge systems typically operate in a dissolved oxygen range of 1 to 2 mg/L. Systems for biological nutrient removal can operate at lower dissolved oxygen ranges and have been observed to achieve nutrient removal with dissolved oxygen concentrations below 0.7 mg/L and as low as 0.3 mg/L. Pure oxygen systems can be operated as high as 8 to 15 mg/L. A sudden drop in dissolved oxygen may indicate a significant increase in organic loading or the start of nitrification. A sudden increase in dissolved oxygen may be the first sign of a toxic or inhibitory load passing through the WRRF that is having an adverse effect on microorganisms. For more information on dissolved oxygen setpoints, refer to Chapter 4.

4.0 TEMPERATURE

As discussed in previous chapters, monitoring in-basin water temperatures is critical for determining optimal solids retention time (SRT) targets for facilities that are required to remove ammonia. Water temperature in biological reactors should be monitored on a daily basis. A sudden increase in influent temperature may indicate an industrial discharge. Sudden decreases in influent temperature may indicate inflow and infiltration caused by a storm event or snowmelt. Biological activity doubles for every 10 °C (18 °F) increase in water temperature.

5.0 pH

pH is a numerical expression of the relative intensity of acidity or basicity. A pH of less than 7.0 Standard Units denotes acidity with the intensity of acidity increasing as the number decreases. pH values between 7.0 and 14.0 denote basicity with intensity increasing as the number increases. Technically, pH is the inverse log of the hydrogen ion concentration in a solution. The "p" in "pH" actually means negative log. Because pH is on a log scale, a substance with a 4 pH is 10 times more acidic than a substance with a 5 pH.

When analyzing samples for pH, it is important to use a pH meter that has been calibrated with a minimum of two buffer solutions and, preferably, three. Calibration should be done at least daily. The analyst should refer to the manufacturer's instructions for the particular meter. Sample pH can change quickly, especially in biological samples such as MLSS, and should be analyzed within 15 minutes of collection. Laboratory, handheld field, and in-basin meters are available for pH monitoring.

In the activated sludge process, the pH should be kept higher than 7.0. Optimum conditions for the growth of nitrifying bacteria occur between 7.5 and 8.0 pH. Nitrification is inhibited at lower pH, with rates dropping rapidly below a 6.8 pH (U.S. Environmental Protection Agency, 2010). Facilities that nitrify should monitor the activated sludge process pH at least three times per week and, preferably, daily.

6.0 ALKALINITY

Alkalinity and pH are related but do not measure the same thing. pH measures hydrogen ion concentration, whereas alkalinity measures buffering capacity. Buffering capacity or alkalinity is a measure of acid-neutralizing capacity. In other words, how much acid can water absorb before there is a substantial change in pH?

An alkalinity test is performed by collecting a sample and then gradually adding small amounts of acid to the sample while stirring and continuously monitoring pH. *Total alkalinity* is defined as the amount of acid required to lower the pH to an endpoint of 4.55. The endpoint can be monitored with a pH meter or by using special indicator dyes that change color at the desired pH. The phenolphthalein indicator changes from pink to clear at 8.3 pH. The bromocresol green methyl red indicator changes from blue to red at 4.5 pH. It is blue when the pH is higher than 4.5 and red when it is below.

Alkalinity may decrease or increase through the activated sludge process because of nitrification and denitrification. Every time 1 mg/L of ammonia-nitrogen is converted to nitrate, 7.14 mg/L of alkalinity as calcium carbonate ($CaCO_3$) is consumed. Likewise, every time 1 mg/L of nitrate-nitrogen is converted to nitrogen gas, 3.57 mg/L of alkalinity as $CaCO_3$ is recovered. Residual alkalinity of 50 to 100 mg/L as $CaCO_3$ is recommended for stable operation of nitrifying activated sludge facilities. Facilities that nitrify should monitor the alkalinity of secondary clarifier effluent several times a week to ensure that there is enough residual alkalinity to support nitrification and prevent an unacceptable drop in pH.

7.0 DEPTH OF BLANKET

7.1 General Description

As MLSS settles in the secondary clarifier, a layer of settled sludge called the *blanket* accumulates on the floor. Ideally, the blanket will be less than 1-m (2.57-ft) deep from the floor of the clarifier to the top of the blanket; however, the depth of blanket can vary depending on the time of day and the amount of flow coming into the WRRF. This is especially true in WRRFs that do not flow-pace return activated sludge (RAS) flow to influent flow. Blanket depth in clarifiers without flat floors will also vary depending on where it is measured.

7.2 Application

The depth of blanket test measures the depth of the blanket in the secondary clarifier and provides the following information:

- Enables the operator to determine the total mass of MLSS in the clarifier blanket. This information may be included in the mean cell residence time (MCRT) calculation for process control.
- Alerts the operator to an increase in the sludge blanket in the final clarifier, allowing the operator to take corrective action to prevent solids from washing out over the weirs.
- Enables the operator to determine the clarifier SRT. The depth of blanket test is also referred to as the *clarifier core sample* in some texts.

7.3 Apparatus and Materials

Typically, sludge blankets are measured with a core sampler, which is also known as a *Sludge Judge®*. Blanket depth may also be measured using a handheld total suspended solids (TSS) meter or an ultrasonic blanket detector; however, although these methods do provide information on blanket depth, they do not allow for collection of a core sample that is needed to determine the total mass of MLSS in the clarifier. Core samplers are made from clear, 2-cm (0.75-in.) plastic pipe, and typically consist of three sections that are 1.5 m (5 ft) long each. The sections screw together to form a single, 4.6-m (15-ft)-long sampler, as shown in Figure 6.4. Samplers are marked at 0.3-m (1-ft) intervals for easy determination of blanket depth and have a check valve at the bottom of the 4.6-m (15-ft) assembled sampler. The check valve allows the operator to lower the entire assembly into the

FIGURE 6.4 Sludge Judge® Demonstration (Reprinted with permission from Indigo Water Group)

clarifier, close the valve, and then withdraw the sampler with a core sample of the clarifier contents inside.

7.4 Sample Collection

Core samples should be taken at a point that is representative of clarifier contents. For rectangular clarifiers, the core sample should be collected at a point halfway down the length of the clarifier. For circular clarifiers with floors that slope downward toward the center, as shown in Figure 6.5, the

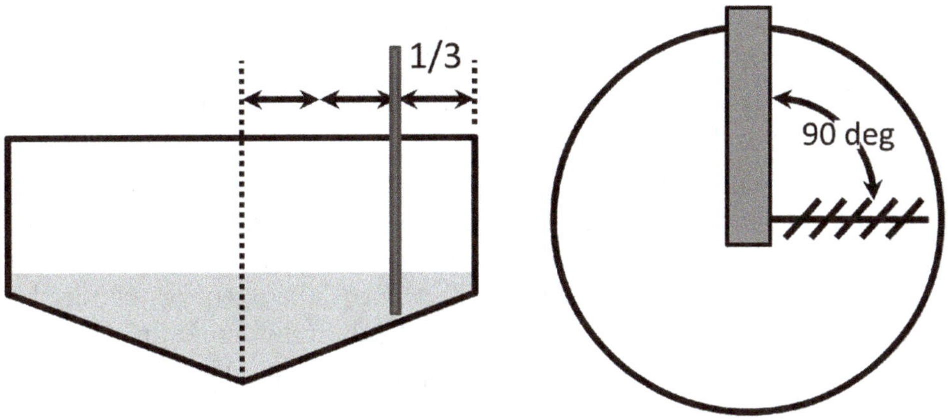

FIGURE 6.5 Ideal Sampling Location for Clarifier Core Sample (Reprinted with permission from Indigo Water Group)

core sample should be collected one-third of the distance from the outside wall of the clarifier to the center. Collecting the sample at this location compensates for the conical bottom of the clarifier. Recall that the formula for finding the volume of a cone uses the volume of a cylinder divided by three. If the sample is collected too close to the outer edge, the mass of MLSS in the clarifier will be underestimated. If the sample is collected too close to the center of the clarifier, the total mass will be overestimated. For flat-bottom clarifiers, the positioning of the core sampler is not as important because the blanket depth should be the same across the clarifier. For clarifiers that slope from a high point in the center down toward a collection ring on the outside of the clarifier, the core sampler should be positioned one-third of the way from the center of the clarifier to compensate for the cone created by the sloped floor.

For circular clarifiers, the core sample should be collected when the rake arm is at a 90° angle relative to the sample location, as shown in Figure 6.5. The motion of the rake arm disturbs the sludge blanket. Taking the sample when the rake arm is at a 90° angle ensures that the blanket has had an opportunity to resettle and is representative of the general conditions at the bottom of the clarifier.

7.5 Procedure

1. Position the core sampler at the clarifier railing approximately one-third of the way in from the outer edge of the clarifier for center slope circular clarifiers or one-half of the way across the length for rectangular clarifiers.

2. Slowly lower the core sampler to the bottom of the clarifier, taking note of the position of the rake arm.
3. Once on the bottom, give a quick jerk to close the check valve on the bottom of the sampler.
4. Slowly remove the sampler from the clarifier. Note the depth of the blanket and record it on the bench sheet. The core sampler will sometimes have a fuzzy layer where the interface between the settled sludge and clear supernatant is not well defined. The operator will need to use their best judgment on blanket depth in this situation.
5. Place the bottom end into a bucket or other sampling container that can hold the entire contents of the core sampler.
6. Release the entire sampler contents into the bucket.
7. Mix the contents of the sample container and collect a subsample for TSS analysis.
8. Repeat for each clarifier.

7.6 Special Variations or Applications

The depth of blanket test may be used as a troubleshooting tool as well as for routine process control. The blanket profile is used to assess the best location for routine measurements of average depth of blanket. Some WRRFs have poor flow splits to multiple final clarifiers and it becomes necessary to adjust flows based on blanket measurements because of the lack of flow-measurement devices. With multiple units and "equal" flow splits, blankets should be within approximately 0.3 m (1 ft) of each other. For a balanced system, any variation in depth of blanket between clarifiers deeper than 0.6 m (2 ft) is attributable to some type of hydraulic imbalance. The imbalance may be attributable to uneven loading to the clarifiers or uneven sludge withdrawal from clarifiers. For facilities with multiple clarifiers of different sizes, flow from aeration basins should be proportioned according to the sizes of clarifiers so that each clarifier has the same solids loading rate ($kg/m^2{\cdot}d$ [lb/sq ft·d]) and hydraulic loading rate ($m^3/m^2{\cdot}d$ [gpd/sq ft]).

8.0 SETTLEABILITY

8.1 General Description

The main purpose of the settleometer test is to give the operator an idea of how sludge is settling in the secondary clarifier. The settleometer test simulates how activated sludge settles in the secondary clarifier. The settleometer test should be performed at least daily at activated sludge facilities. For larger facilities, the settleometer test should be performed once per shift. Information that should be recorded for long-term trend tracking includes the settled sludge volume (SSV) at 5 minutes (SSV_5), SSV at 30 minutes (SSV_{30}), and the time if and when the blanket "popped." The SSV_5 indicates how well MLSS is settling, whereas the SSV_{30} measures how well MLSS is compacting. The analyst may wish to refer to Method 2710 C of *Standard Methods for the Examination of Water and Wastewater* (American Public Health Association et al., 2022).

8.2 Apparatus and Materials

Three pieces of equipment are necessary to run settleability tests: a settleometer, a paddle, and a timer. The equipment is described as follows:

- Mallory settleometers—three, 2-L Mallory settleometers are suggested, one Mallory settleometer for the regular test, the second for dilution, and the third for any special test that may need to be conducted, such as an extended-rise test or multiple dilutions. If only one Mallory settleometer is available, tests may be run in series. If Mallory settleometers are not available, 2-L beakers or similar devices may be substituted. A standard graduated cylinder should not be used because the narrow cylinder reduces settling velocity caused by excess friction with the walls. The base of the settleometer, where the wall meets the floor, should be square and not rounded.
- Paddle—a wide paddle is needed to gently stir, then quiet, mixed liquor in the settleometer at the beginning of the test. The paddle should be made of smooth acrylic or a similar material, with a width slightly less than the diameter of the settleometer.
- Timer—the timer should be capable of sounding an alarm at the end of 5-minute intervals.

8.3 Sample Collection, Preservation, and Holding Times

It is important that mixed liquor samples collected for the settleability test are representative of mixed liquor flowing out of the aeration basin and into the secondary clarifier. Typically, this sample is collected either in a weir overflow box or the secondary clarifier influent splitter structure or at a similar point before its entry to secondary clarifiers. If the mixed liquor comes from two or more aeration basins, the analyst should be sure to sample the mixed liquor after the sources have combined. However, it is also a good idea to periodically measure settleability in each individual aeration basin to check for differences in sludge condition between the basins. Finally, it is important to remember to collect mixed liquor samples and not scum and foam that may be present on the mixed liquor.

The necessity for collecting mixed liquor samples in wide-mouthed containers and starting the settleometer testing as soon as possible cannot be overemphasized. Indeed, testing should begin within 10 minutes of collection. In addition, shaking and agitation of the samples should be minimized, especially during transportation to the testing site.

8.4 Procedure

Test conditions and sampling techniques can strongly influence test results. As such, the test should be set up in a location that is free from vibration and direct sunlight. If possible, the temperature should not vary a great deal from one test to another. The following steps should be followed for the settleability procedure:

1. Mix the mixed liquor sample carefully, then pour it into the settleometer carefully and rapidly, with the least possible amount of additional aeration or turbulence. Stir the settleometer contents gently with a wide paddle to ensure a thorough mixing, then stop all movement of the mixed liquor with the paddle.
2. As the paddle is smoothly and carefully removed from the settleometer, start the timer that was previously set for 5-minute intervals. It is important to not leave the settleometer or the analyst will miss out on some of the most important information to be gained from the settleometer test.
3. During the initial 5 minutes, the analyst should observe visual characteristics of the settling sludge. A conscientious operator will critically observe how sludge particles agglomerate while forming the blanket. He/she will also observe whether sludge compacts slowly and uniformly while squeezing clear liquid from the sludge mass or whether tightly knotted sludge particles are simply falling down through the turbid effluent. The operator will observe how much and what type of straggler floc, if any, remains in the supernatant liquor above the main sludge mass.
4. The importance of conscientious, perceptive observation during the first 5 minutes cannot be overemphasized. During these important 5 minutes, the operator will acquire additional insight to sludge character and quality and will be in a much better position to evaluate what the settleometer test reveals.

5. At the end of the first 5 minutes, the analyst should read and record the volume of the settleometer occupied by settled sludge. This is the most important reading taken during the test.
6. The analyst should continue to read and record settled SSV at 5-minute intervals for the first 30 minutes.
7. The analyst should continue taking SSV readings until SSV changes become small from reading to reading. If a slow settling sludge (i.e., SSV greater than 900) is being tested, readings at 90, 120, 150, 180, and 240 minutes may be required.
8. After the last reading has been taken, the sample should be allowed to stand for several hours more. The analyst should record the time at which previously settled sludge starts to swell and/or rise to the surface of the settleometer. It is important to wash the settleometer and paddle with mild soap and thoroughly rinse and dry them after the test.

8.5 Special Variations or Applications

There are essentially three variations of the test, both of which have been alluded to previously. The first is the rise test, which indicates that denitrification might be occurring in the secondary clarifier. The second is the dilution test for determining sludge age. The third test is to compare the settleometer supernatant to the secondary clarifier effluent, which is useful for troubleshooting clarifier performance. This is the flocculated/dispersed suspended solids test described in Section 10.0 of this chapter.

8.5.1 Rise Test

After conducting the standard 30-minute settleometer test, allow the beaker with the sample to set for several hours and observe the time that the sludge rises. This is an indication of denitrification. If the sludge rises quickly (e.g., less than 90 minutes), observe the depth of sludge in the secondary clarifier and the SRT. The lower the RAS (underflow) flowrate, the higher the detention time. If this condition is occurring, pay close attention to the sludge blanket level in the core sampler. See if it is rising or expanding.

8.5.2 Dilution Test

The dilution test is conducted to determine whether there are too many (hindered settling) or too few solids in the aeration tank. It is important to rely on other tests, including visual observations and microbial tests, to make a judgment on necessary process control adjustments. The analyst should run three settleometer tests at the same time. Undiluted MLSS should be placed in the first beaker, 50% MLSS and 50% secondary effluent in the second, and 25% MLSS and 75% secondary effluent in the third. The analyst should conduct 60-minute settleabilities, prepare settling curves, and compare slopes. If the slopes are the same or parallel, this probably indicates settling problems caused by filamentous sludge. If an increase in settling rate occurs with the diluted samples (greater slope or rise over run), the MLSS is probably too high and wasting should be increased to lower the SRT. The increase should not be drastic, but incremental each day. The analyst should try gradually increasing the wasting over a 3-day period until the desired SRT and settling rate are achieved.

8.6 Calculations

1. Observation of settling for the first 5 minutes.
2. Settling rate (volume of settled sludge vs time).
3. Calculation of the 30-minute settled sludge concentration:

$$SSC_{30} = \left(\frac{\text{Initial settleometer volume, mL}}{SSV_{30}}\right) \times \text{MLSS, mg/L} \qquad (6.1)$$

Where

SSC_{30} = settled sludge concentration after 30 minutes and
SSV_{30} = settled sludge volume after 30 minutes.

The settled sludge concentration after 30 minutes (SSC_{30}) is the maximum achievable sludge concentration in the secondary clarifier, assuming that solids remain in the clarifier blanket for 30 minutes or fewer. In practice, the RAS concentration will often be significantly lower than the SSC_{30} because solids are removed from the clarifier in fewer than 30 minutes. Occasionally, the measured RAS concentration may be higher than the calculation SSC_{30}. This may be caused by denitrification in the settleometer "puffing up" the blanket or by leaving the solids in the clarifier blanket for longer than 30 minutes. The RAS concentration observed in the field is dependent on the RAS flow to influent flow ratio and can be calculated using the following equation:

$$X_{RAS} = \left(\frac{Q_{influent}}{Q_{RAS}} + 1\right) \times X_{MLSS} \qquad (6.2)$$

Where

X_{RAS} = RAS flow,
$Q_{influent}$ = influent flowrate,

Q_{RAS} = RAS flowrate, and
X_{MLSS} = MLSS concentration.

The operator–analyst should be mindful that turning the RAS down to exceedingly low levels to further thicken and compact solids in the clarifier blanket is counterproductive and can cause effluent quality to deteriorate. The clarifier cannot compact solids further than the concentration predicted by the settleometer.

4. Time to rise.
5. If the dilution test is being conducted, comparison of settling rates, which provides information to evaluate the age of the sludge.
6. If the flocculated/dispersed suspended solids test is being conducted (Section 10.0), comparison of supernatant TSS or turbidity to secondary clarifier effluent TSS or turbidity, which provides information to evaluate the cause of elevated effluent TSS or turbidity.

Figure 6.6 provides a sample bench sheet.

8.7 Interpretation of Results

The settleometer test is arguably one of the most useful process control tests for activated sludge. Much information can be gained from a properly conducted and interpreted test. Three data points should be tracked and plotted on control charts: the SSV_5, SSV_{30}, and the time, if any, when denitrification causes settled sludge to rise to the top of the settleometer. Figure 6.7 shows settleometer data from a typical activated sludge process with the SSV_5, SSV_{30}, and SSV_{60} plotted over 3.5 years.

The SSV_5 is an indicator of sludge-settling characteristics (e.g., how fast the particle is moving through the water column). For the first 5 minutes of the test, the solids are essentially in free fall, with few collisions occurring between particles. Primary variables controlling how fast particles settle in the first 5 minutes are density of particles relative to the density of the water and presence or absence of filamentous bacteria. For activated sludge facilities that do not practice enhanced biological phosphorus removal (EBPR), a typical floc particle is only slightly more dense than the surrounding water. Typically, EBPR floc is denser than non-EBPR floc. As water cools, its density increases. In Figure 6.7, the SSV_5 is shown in blue. For this facility, the summer-month in-basin water temperature may be as high as 20 °C (68 °F); however, in the winter months, the in-basin water temperature is typically 10°C (50°F). As a result, there is a strong seasonal pattern to the SSV_5, with settling rates decreasing in the winter months (higher SSV_5) and increasing

Location: ____________

Method: ____________

Date: ______________________
Sample ID: ______________________
Analyst: ______________________
Time of Test: ____________ ATC (%) ______

$$SSC = \frac{(ATC) \times (1000)}{SSV}$$

Time	SSV, mL/L	SSC, %
0	1000	
5		
10		
15		
20		
25		
30		
35		
40		
45		
50		
55		
60		

Observations in the first 5 minutes:

Floc
☐ Granular ☐ Compact
☐ Fluffy ☐ Feathery

Particle Size
☐ Large
☐ Medium
☐ Small

Bio-solids/ Supernate Interface
☐ Well Defined
☐ Ragged

Supernatant
☐ Clear
☐ Cloudy

Straggler Floc
☐ Yes
☐ No

Observation after 30 minutes:

☐ Crisp/Sharp Edges ☐ Fluffy/Feathery
☐ Sponge-like ☐ Homogeneous

Date: ______________________
Sample ID: ______________________
Analyst: ______________________
Time of Test: ____________ ATC (%) ______

$$SSC = \frac{(ATC) \times (1000)}{SSV}$$

Time	SSV, mL/L	SSC, %
0	1000	
5		
10		
15		
20		
25		
30		
35		
40		
45		
50		
55		
60		

Observations in the first 5 minutes:

Floc
☐ Granular ☐ Compact
☐ Fluffy ☐ Feathery

Particle Size
☐ Large
☐ Medium
☐ Small

Bio-solids/ Supernate Interface
☐ Well Defined
☐ Ragged

Supernatant
☐ Clear
☐ Cloudy

Straggler Floc
☐ Yes
☐ No

Observation after 30 minutes:

☐ Crisp/Sharp Edges ☐ Fluffy/Feathery
☐ Sponge-like ☐ Homogeneous

FIGURE 6.6 Data Collection Sheet for Settleometer Test (WEF, 2022)

in the summer (lower SSV_5), even when excess filaments are not present. Despite large seasonal fluctuations in SSV_5, SSV_{30} remained fairly constant. For this facility, settling characteristics changed seasonally, but compaction characteristics did not. The SSV_{30} is increasing slightly in the winter months because SRT has been increased to maintain stable nitrification. Higher SRT increases the total amount of MLSS in the system and MLSS concentration. As a result, MLSS is taking up more space in the settleometer blanket.

The presence of filamentous bacteria can also cause SSV_5 increase, especially when bridging between floc particle occurs. Filaments create drag, which causes sludge particles to settle more slowly. In Figure 6.7, there is a sudden increase in both SSV_5 and SSV_{30} in December 2007. This sudden increase is characteristic of an increase in the numbers of filamentous

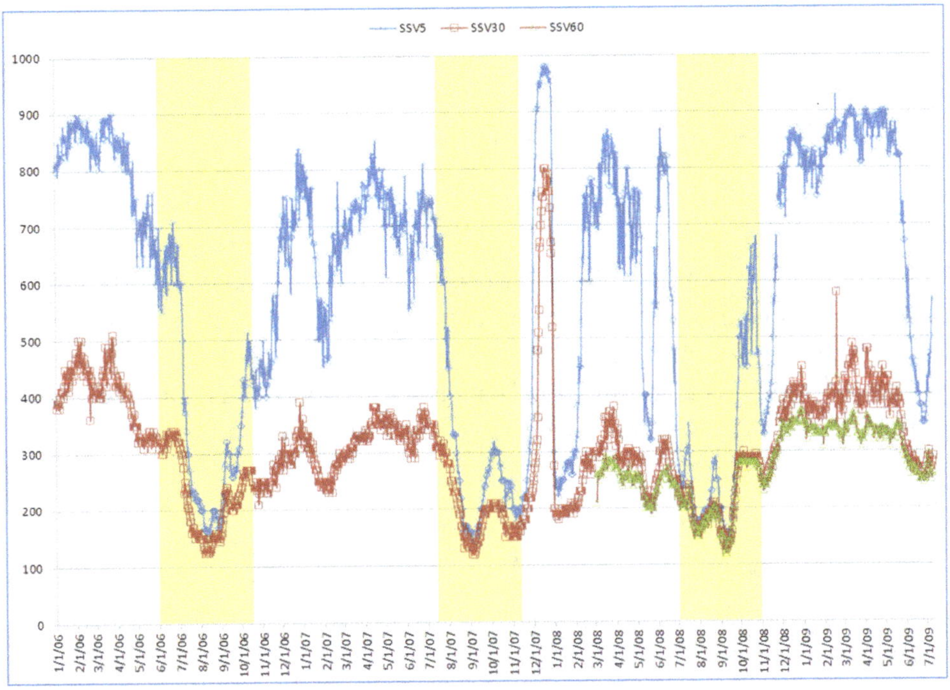

FIGURE 6.7 Typical SSV_5, SSV_{30}, and SSV_{60} Results (Reprinted with permission from Indigo Water Group)

bacteria. A microscopic examination should be made to confirm the presence or absence of filaments followed by staining to identify which filament is dominant.

The SSV_{30} is an indicator of sludge compaction. As sludge particles settle in the settleometer and in the clarifier, they pile up on one another and sludge concentration in the blanket increases. There is a limit to the amount of compaction achievable in both the settleometer and the clarifier. Compaction is limited, in part because floc particles are negatively charged. Floc particles repel one another in much the same way as two ends of magnets can. In Figure 6.7, there is some additional compaction after 30 minutes, as evidenced by the slight differences between the SSV_{30} and SSV_{60}; however, compaction slows as the settled sludge concentration increases. For slow-settling sludges, compaction may continue for longer than 60 minutes. Practically speaking, SSV_{30} represents the maximum achievable compaction in the secondary clarifier.

As an example, the MLSS concentration is 2800 mg/L and the settled sludge volume after 30 minutes is 240 mL. The RAS is flow-paced to the influent flow and is currently set at 50% of influent flow. Calculate the maximum achievable RAS concentration as predicted by the settleometer test. Then, calculate the actual RAS concentration from facility operating data.

$$\text{SSC}_{30} = \left(\frac{\text{Initial settleometer volume, mL}}{\text{SSV}_{30}}\right) \times \text{MLSS, mg/L}$$

$$\text{SSC}_{30} = \left(\frac{1000 \text{ mL}}{240 \text{ mL}}\right) \times 2800 \text{ mg/L}$$

$$\text{SSC}_{30} = 11\ 667 \text{ mg/L}$$

The concentration of the sludge in the settleometer blanket after 30 minutes of settling is 11 667 mg/L. Sludge quality says that if the solids are left in the clarifier for 30 minutes, then the RAS concentration will be 11 667 mg/L. Slight differences may occur in field measurements because of inaccuracies in flow measurement. What is actually being achieved in the field?

$$X_{\text{RAS}} = \left(\frac{Q_{\text{influent}}}{Q_{\text{RAS}}} + 1\right) \times X_{\text{MLSS}}$$

$$X_{\text{RAS}} = \left(\frac{1 \text{ m}^3}{0.5 \text{ m}^3} + 1\right) \times 2800 \text{ mg/L}$$

$$X_{\text{RAS}} = 8400 \text{ mg/L}$$

9.0 SLUDGE VOLUME INDEX

9.1 General Description

Sludge volume index (SVI) is defined as the volume of sludge in milliliters (cubic inches) occupied by 1 g (0.04 oz) of activated sludge after settling for 30 minutes. This index relates the 30-minute settled sludge volume from the settleometer process control test to the concentration of solids in the sample on which the settleometer test was performed. The SVI will help the operator evaluate settling characteristics of the activated sludge as the concentration of solids in the system change. All WRRFs have unique sludge characteristics based on their design, specific influent characteristics, and operational parameters. Therefore, an SVI that represents a good settling sludge at one facility is not comparable to SVIs at other facilities. This is especially true if the facility is treating industrial or other unusual wastes. The reader may wish to refer to Method 2710 D of *Standard Methods* (American Public Health Association et al., 2022).

9.2 Application

Operators often develop an SVI number from the settleometer test. The SVI is calculated using the 30-minute settled sludge volume from a 2000-mL

Mallory settleometer. The units for SVI are milliliters per gram. Sludge volume index is determined as follows:

$$\text{SVI}, \frac{\text{mL}}{\text{g}} = \frac{\text{SSV}_{30} \times 1000 \ \frac{\text{mg}}{\text{g}}}{\text{MLSS} \frac{\text{mg}}{\text{L}}} \tag{6.3}$$

Sludge volume indexes between 75 and 150 mL/g indicate a good settling sludge. The SVIs may be lower than this range for facilities that use alum or ferric, receive water treatment residuals, or perform biological phosphorus removal. Increasing SVI numbers mean that sludge settleability is deteriorating. When the SVI reaches 200 mL/g, the sludge (MLSS) is technically "bulking." Bulking sludge may be caused by a variety of things, including nutrient deficiency, filamentous bacteria, and solids concentrations that are too high.

Using the Mallory settleometer and the aforementioned test procedures, the settleometer test is run and the settled sludge volumes for 5, 10, 15, 20, 25, 30, 40, 50, and 60 minutes are recorded. For slow-settling sludge, additional readings may be required at the 90-, 120-, 150-, and 180-minute points, or for as long as sludge continues to settle. The corresponding settled sludge concentration (SSC) at each time is then calculated as follows:

$$\text{SSC} = \left(\frac{\text{MLSS} \times 1000}{\text{SSV}}\right) \tag{6.4}$$

As SSV gets smaller with time, SSC gets larger as sludge concentrates in the bottom of the settleometer. In an old, overoxidized sludge that settles quickly, maximum compaction occurs within approximately 30 minutes. A medium-settleability sludge will typically compact to maximum within 1 hour. A young, underoxidized or filamentous sludge may not compact maximally until 2 to 3 hours of settling time. As noted in Section 8.7, sludge is rarely left in the secondary clarifier blanket for longer than 30 minutes.

The settleometer test can be used to develop additional information about the system. If the settleometer is allowed to sit idly after the settling test has been completed, it can be determined whether significant nitrification, and the potential for denitrification, is occurring. As settled sludge sits in the settleometer, microorganisms run out of oxygen quickly; they begin to denitrify, converting nitrate to nitrogen gas. If the concentration of nitrate is high enough, sufficient nitrogen gas will be released to float the settled blanket or at least split the blanket and send some portion of it back to the top of the settleometer, as shown in Figure 6.8.

FIGURE 6.8 Denitrification in the Settleometer Test (Reprinted with permission from Indigo Water Group)

9.3 Interferences

It is important for the analyst to note that the SVI calculation is only valid for MLSS concentrations with a normal range of 1000 to 4000 mg/L for activated sludge. Outside of this range, the SVI calculation can give meaningless results. For example, a MLSS with a concentration of 8000 mg/L and a 30-minute SSV of 950 mL will have an SVI of 119 mL/g. If one only considered the SVI number, one would think that the sludge was settling well, which is not the case. Conversely, this sludge is hardly settling at all. When creating daily log sheets for the settleometer and SVI tests, operators are encouraged to record the MLSS concentration, SSV, and SVI. These numbers are all meaningless separate from one another.

9.4 Required Data

Required data for calculating the SVI include the following:

- Thirty-minute settling volume in milliliters per liter from the settleometer test and
- Mixed liquor suspended solids concentrations in milligrams per liter for the same sample on which the settleometer test was performed.

9.5 Calculation

The following equation should be used to calculate SVI:

$$\text{SVI}, \frac{\text{mL}}{\text{g}} = \frac{\text{SSV}_{30} \times 1000 \,\frac{\text{mg}}{\text{g}}}{\text{MLSS}\,\frac{\text{mg}}{\text{L}}} \tag{6.5}$$

10.0 FLOCCULATED/DISPERSED SUSPENDED SOLIDS TESTING FOR SECONDARY CLARIFIERS

10.1 General Description

Water Environment Research Foundation developed a protocol for evaluating secondary clarifier performance in activated sludge systems (Wahlberg, 2001). Components of the clarifier testing protocol are the dispersed suspended solids (DSS) and flocculated suspended solids (FSS) tests. Dispersed solids are fine solids that are too small and/or light to settle on their own within the allotted test time. Flocculated solids are agglomerations of smaller particles that are large enough and heavy enough to settle on their own. Flocculated solids are capable of reaching the bottom of the secondary clarifier. (The following description and testing protocol given in Sections 10.1 and 10.4 and the information in Table 6.4 have been reprinted with permission from WERF Project 00-CTS-1: *WERF/CRTC Protocols for Evaluating Secondary Clarifier Performance* [Wahlberg, 2001]).

The DSS concentration is operationally defined as the supernatant suspended solids concentration after 30 minutes of settling in a Kemmerer sampler (see Figure 6.9). This kind of sampler is used because it allows sample collection and settling to be done in the same container, thereby sparing the biological solids any aggregation or breakup effects. For this reason, the DSS can be thought of as a "snapshot" of the state of flocculation (or breakup) existing at the time the sample is collected.

TABLE 6.4 Interpretation of DSS/FSS Test Results (Wahlberg, 2001) (Reprinted with permission from WERF)

Result	Indication	Corrective action
FSS are >10 mg/L (applies to domestic WRRFs only).	Biological flocculation problem. Poor sludge quality.	Check to see if dispersed growth is occurring in the aeration basin because of a low MCRT. Check to see if some constituent in the influent is causing deflocculation to occur and initiate source control if identified.
Influent DSS > effluent DSS and effluent DSS = FSS and ESS = FSS.	Good flocculation is occurring in the clarifier and there are no hydraulic problems in the clarifier.	None.
Influent DSS = effluent DSS and effluent DSS = FSS and ESS = FSS.	Mixed liquor was well-flocculated before entering the clarifier and there are no hydraulic problems in the clarifier.	None.
Effluent DSS > FSS and DSS < ESS.	Both flocculation and hydraulic problems may be present.	Conduct additional testing to determine action needed.

The FSS concentration is operationally defined as the supernatant suspended solids concentration after 30 minutes of flocculation followed by 30 minutes of settling. The FSS test is conducted by placing 2 L of mixed liquor in a square flocculation jar or circular jar with stators. The mixed liquor sample is stirred at 50 rpm using a stirrer for 30 minutes. The choice of the stirring intensity and time is based on work referenced in Wahlberg (2001). The FSS concentration represented the secondary effluent suspended solids (ESS) possible if mixed liquor entering the settling zone in the secondary clarifier were optimally flocculated and flow through the secondary clarifier were hydraulically ideal (no currents with a horizontal component to hinder liquid–solids separation).

10.2 Application

The DSS/FSS testing is a method of determining whether elevated ESS are a result of poor sludge quality (biology) or poor flocculation in the aeration

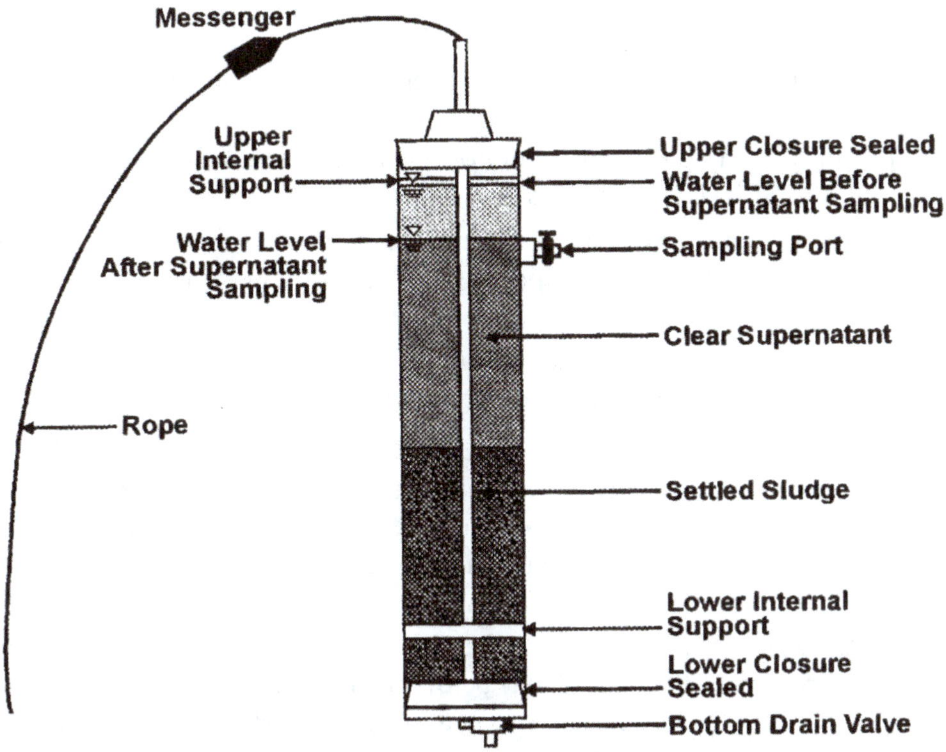

FIGURE 6.9 Kemmerer Sampler in DSS/FSS Test (Source: Wahlberg, 2001 © the Water Environment Research Foundation)

basin or clarifier wet well, are caused by a hydraulic problem within the secondary clarifier, or are attributable to some other problem existing within the secondary clarifier such as denitrification. Samples are collected at three different locations: the activated sludge basin, clarifier inlet, and clarifier outlet to determine whether supernatant quality and sludge-settling characteristics are changing as mixed liquor moves from the aeration basin to the clarifier. It is possible for the mixed liquor to experience shear in the pipeline between the activated sludge basin and clarifier or within the clarifier itself. By testing sludge performance at each location, it becomes possible to isolate the cause or causes of dispersed suspended solids.

10.3 Apparatus and Materials

The following equipment will be needed to conduct DSS/FSS testing:

- Two 4.2-L Kemmerer samplers,
- 2-L square flocculation jar or circular jar with stators, and
- Jar testing device capable of mixing at 50 rpm for 30 minutes.

10.4 Procedure

Determine the influent DSS (DSS_{inf}) concentration as follows:

1. Collect a sample of mixed liquor from the inlet zone of the test clarifier in a 4.2-L Kemmerer sampler.
2. Quickly lower the water surface in the sampler down to the midpoint of the upper internal support using the bottom drain valve. Discard this part of the sample.
3. Tie the Kemmerer sampler off on a handrail or other convenient location; prop top closure open with the messenger.
4. Start stopwatch.
5. After approximately 10 minutes or after the solids interface has passed below the bottom of the loops that secure the upper closure to the sampler, gently twist the top closure to dislodge any solids adhering to the loops.
6. After 30 minutes, collect the entire supernatant volume above approximately 15 cm (6 in.) below the midpoint of the top internal support, either by siphon or via a sampling port drilled into the side of the sampler. Leave approximately 1.3 cm (0.5 in.) of supernatant to avoid floating debris in the supernatant sample.
7. Analyze the supernatant sample for the TSS concentration. This is the influent DSS (DSS_{inf}) concentration.

Determine the ESS and effluent (DSS_{eff}) concentrations as follows:

1. While the secondary influent DSS sample is settling, collect a sample of the supernatant approximately 15 to 30 cm (6 to 12 in.) below the water surface at the effluent weir in another 4.2-L Kemmerer sampler. (If only one Kemmerer sampler is available, this test may be conducted at the completion of the other test.)
2. Quickly lower the water surface in the sampler down to the midpoint of the upper internal support using the bottom drain valve. Collect this sample and analyze for the TSS concentration. This corresponds to the ESS concentration.
3. Tie the Kemmerer sampler off on a handrail or other convenient location. Prop top closure open with the messenger.
4. Start the stopwatch.
5. After approximately 10 minutes or after the solids interface has passed below the bottom of the loops that secure the upper closure

to the sampler, gently twist the top closure to dislodge any solids adhering to the loops.

6. After 30 minutes, collect the entire supernatant volume above approximately 15 cm (6 in.) below the midpoint of the top internal support, either by siphon or via a sampling port drilled into the side of the sampler. Leave approximately 0.3 cm (0.5 in.) of supernatant to avoid floating debris in the supernatant sample.
7. Analyze this supernatant sample for TSS concentration. This is the DSS_{eff} concentration.

Determine the FSS concentration as follows:

1. While the secondary influent and effluent DSS samples are settling, collect a sample of mixed liquor from the aeration basin exit or anywhere in the mixed liquor transfer system between the aeration basins and secondary clarifier in a 2-L, square flocculation jar or circular jar with stators.
2. Flocculate this sample for 30 minutes at 50 rpm on a Phipps & Bird jar tester or comparable stirrer.
3. After flocculation, allow the flocculated sample to settle for 30 minutes.
4. After settling, carefully collect approximately 250 mL of supernatant sample by siphon or via a sampling port drilled into the side of the sampler. Leave approximately 0.3 cm (0.5 in.) of supernatant to avoid floating debris in the supernatant sample.
5. Analyze this supernatant sample for TSS concentration. This is the FSS concentration.
6. Repeat each component of the procedure at least 3 times to obtain average ESS, influent DSS, effluent DSS, and FSS concentrations.

10.5 Interpretation of Results

At the end of testing, there will be four results, as follows:

- Flocculated suspended solids—the suspended solids remaining in the supernatant after gentle stirring to combine smaller floc particles followed by settling in a settleometer. The settleometer represents ideal settling conditions compared to the secondary clarifier because it does not have any currents in it caused by surface overflow rate, rake arm movement, or wind action. The FSS represent solids that do not flocculate because of sludge quality. They are independent of the clarifier.

- Influent DSS—This sample is collected at the inlet to the clarifier. The DSS_{inf} are the solids that either did not flocculate in the aeration basin or were created by the breaking up of floc particles as mixed liquor was moved from the activated sludge basin to the clarifier. Taking the difference between the FSS and DSS_{inf} measures the amount of floc breakup.
- Effluent suspended solids—This sample is collected at the outlet of the clarifier. It is not flocculated or allowed additional settling time. It measures the TSS passing out of the clarifier to the final effluent. The ESS represent how well the clarifier is performing.
- Effluent DSS—This sample is also collected at the outlet of the clarifier. The sample is allowed to settle for an additional 30 minutes before the solids in the supernatant are analyzed. The DSS_{eff} represent the portion of solids in the final effluent that had the potential to settle to the bottom of the clarifier. These solids end up in the final effluent because of a hydraulic inefficiency in the clarifier. Hydraulic inefficiencies may be caused by currents in the clarifier, excessively high surface overflow rates, rapidly moving rake arms, and denitrification.

11.0 CENTRIFUGE METHOD FOR ESTIMATING SUSPENDED MATTER

11.1 General Description

In 1974, the U.S. Environmental Protection Agency developed the centrifuge method for estimating suspended solids in wastewater as a substitute for the gravimetric determination of TSS. This technique was intended to reduce the time and equipment required for process control of activated sludge systems and the skill level needed to operate a WRRF. The result of this analysis is given in percent solids by volume and is used to calculate solids inventory in "sludge units" by multiplying by tank volume. Sludge units then become a substitute for kilograms (pounds) of solids.

This method is empirical in nature, dependent on specific sludge characteristics, and subject to several limitations. Comparisons between facilities are made difficult by the highly specific nature of the test, which is dependent on sludge compactability. Therefore, the result of a centrifuge spin test depends on SVI. The same plant can give radically different centrifuge spin results for the same sludge concentration if sludge settleability has changed. Two different WRRFs in Colorado report correlations of plus or minus

50% when comparing uncorrected centrifuge spin results with laboratory gravimetric analysis.

Sludge solids inventory is used to determine critical core components of process control: SRT, aerobic SRT ($SRT_{aerobic}$) (or MCRT), wasting rates, return rates, and even aeration rates. It should follow that laboratory techniques used to calculate solids inventory should be as accurate as possible. The analyst may, therefore, want to consider daily TSS analyses in lieu of the estimation given by this procedure.

11.2 Application

The laboratory centrifuge provides a quick and easy way to estimate activated sludge mixed liquor or return sludge suspended solids concentrations when used with the gravimetric method. Solids concentration is used in process control to determine SRT; $SRT_{aerobic}$; (or MCRT); and, ultimately, wasting rates.

Given the dynamic nature of biological systems, the centrifuge factor as determined in Step 1 in Section 11.7 will be reliable only as long as the biological system remains more or less stable. For practical purposes, the centrifuge factor should be reestablished at least once per week.

11.3 Interferences

Individual sludge characteristics, including the presence of filaments and inorganic matter, affect the compactability and percent volume of sludge.

11.4 Apparatus and Materials

The following equipment will be needed to conduct the centrifuge spin test:

- All equipment needed for gravimetric determination of suspended matter,
- Laboratory centrifuge capable of 1200 to 2000 rpm,
- Tachometer for centrifuge,
- Timer, and
- Centrifuge tubes graduated to 0.1 mL.

11.5 Sample Collection, Preservation, and Holding Times

Samples should be collected in plastic bottles and analyzed immediately. No preservation is possible. It is important for the analyst to be aware that, as a sludge sample ages in the bottle, biological activity continues to take place and the nature of the sludge will change.

11.6 Quality Assurance and Quality Control

One of several significant drawbacks of this method is that a certified standard does not exist. The analyst may run duplicates by centrifuging two aliquots of the same sample.

11.7 Procedure

The following steps should be followed when conducting the centrifuge spin test:

Step 1—determination of the centrifuge factor.

- Determine either mixed liquor or return sludge suspended solids gravimetrically.
- Next, fill centrifuge tubes with a well-mixed portion of the same sample.
- Load centrifuge according to the manufacturer's instructions.
- Then, start the centrifuge and timer. Use a tachometer to adjust centrifuge revolutions per minute. Time and speed of centrifugation are not critical but must be the same for all determinations. Fifteen minutes at 1500 rpm is recommended.
- After the centrifuge has stopped turning, remove centrifuge tubes. Read and record the volume (to the nearest 0.1 mL) of packed solids in the bottom of the tube. Calculate the centrifuge factor as follows:

$$\text{Centrifuge factor} = \frac{\text{Total suspended solids, mg/L}}{\text{Volume of condensed solids, mL or \%}} \tag{6.6}$$

For example, given the following:

$$\text{Suspended solids} = 1800 \text{ mg/L}$$
$$\text{Volume of condensed solids} = 1.5 \text{ mL}$$

Step 2—estimation of suspended solids. Fill centrifuge tubes with a well-mixed sample. Then, load and run the centrifuge described in Step 1, but omitting the first two substeps (filling the centrifuge tubes and determining the sludge concentration gravimetrically).

11.8 Calculations

The centrifuge percent spin reading may be used to estimate the suspended solids concentration in milligrams per liter by multiplying percent spin by the centrifuge correction factor. The centrifuge factor must be re-created weekly or whenever sludge settling and compaction characteristics change to ensure accuracy.

For example, given the following:

Centrifuge tube reading = 1.8%
Centrifuge factor = 1200 mg/L/%

Suspended solids, mg/L = (Centrifuge tube reading, %) (Centrifuge factor, mg/L/mL) (6.7)

Suspended solids = (1.8%) (1200 mg/L/%) = 2160 mg/L

12.0 SETTLEABLE SOLIDS

12.1 General Description

This simple test can be performed to show quickly and qualitatively if primary and secondary clarifiers are functioning at the highest possible efficiency. An Imhoff cone, shown in Figure 6.10, is used to determine whether settleable material is present in the effluent of either a primary or secondary clarifier. Material that settles in the Imhoff cone is material that could have settled in the clarifier but did not. Surface overflow rate in clarifiers may push colloidal particles over clarifier weirs. An Imhoff cone does not include any current from surface overflow rate and represents perfect settling conditions. Some colloidal particles may settle in an Imhoff cone that cannot be captured in a clarifier.

12.2 Procedure

Follow these steps to conduct a settleometer test:

1. Fill the Imhoff cone to the 1-L mark with a fresh, well-mixed sample;
2. Settle for 45 minutes;
3. Gently stir the upper portion of the sample with a glass rod to dislodge solid material clinging to the tapered sides of the cone;
4. Settle for an additional 15 minutes; and
5. Read the volume of settleable material in milliliters per liter.

12.3 Interferences

Occasionally, settleable matter may exceed the 40-mL/L graduation of an Imhoff cone. In these instances, a 500-mL volume of the sample should be

FIGURE 6.10 Imhoff Cone (Reprinted with permission from Indigo Water Group)

used and transferred to an Imhoff cone. Follow the procedure in Section 12.2 and multiply the volume of settled material by a factor of 2 to express the final results in milliliters per liter.

If there is a separation of settleable and floatable materials, the floating material should not be included in the result.

13.0 TOTAL SUSPENDED SOLIDS

13.1 General Description

Method 2540 D-97 of *Standard Methods* (American Public Health Association et al., 2022) measures suspended solids, also referred to as *nonfilterable residue*. These parameters are determined by filtering a sample through a glass-fiber filter disk.

13.2 Application

The TSS test is one of the most important tests that an analyst can run. The results of the test can be used to estimate process loading to WRRF as a whole; calculate the efficiency of screening and sedimentation processes; calculate the SRT, $SRT_{aerobic}$, or MCRT; and determine the sludge wasting rate, loading to solids-handling processes, and removal and capture efficiency of solids-handling processes. The TSS test result is also needed to calculate SVI.

13.3 Interferences

If too many solids are collected on the filter paper, a water trapping crust may form. For this reason, it is important to limit the amount of material collected to between 2.5 and 200 mg total. If smaller-diameter filter papers are used, the amount of sample filtered may need to be decreased because there is less available filter area. As solids accumulate on the available filter area, it will become more difficult to filter additional sample. The analyst should be aware that collecting a large amount of material on a smaller filter area may result in a water trapping crust even when less than 200 mg of solids are collected.

Samples dried at 103 to 105 °C may contain chemically bound water (water of crystallization) and free water trapped under and within dried portions of the sample (mechanically occluded water). This extra water can bias results high. At the same time, some material can be lost during drying by being converted to carbon dioxide.

Nonrepresentative chunks of material may be removed from the sample and/or filter paper at the analyst's discretion. If it is determined that the offending material is not representative (i.e., only a few pieces in the entire 2-L sample), they may be removed.

Samples with visible oil and grease should be mixed well in a blender to disperse the oil and grease before taking a sample aliquot.

13.4 Apparatus and Materials

Several different filter apparatuses are acceptable for the TSS test. Figure 6.11 shows three different filter types: a magnetic filter funnel, Gooch crucible, and a Buchner funnel. The magnetic filter funnels consist of two easily separated parts, the base and the funnel. The filter paper is supported by a plastic or glass screen. The funnel snaps into place and eliminates the ability of the sample to get under the filter paper. These funnels may also

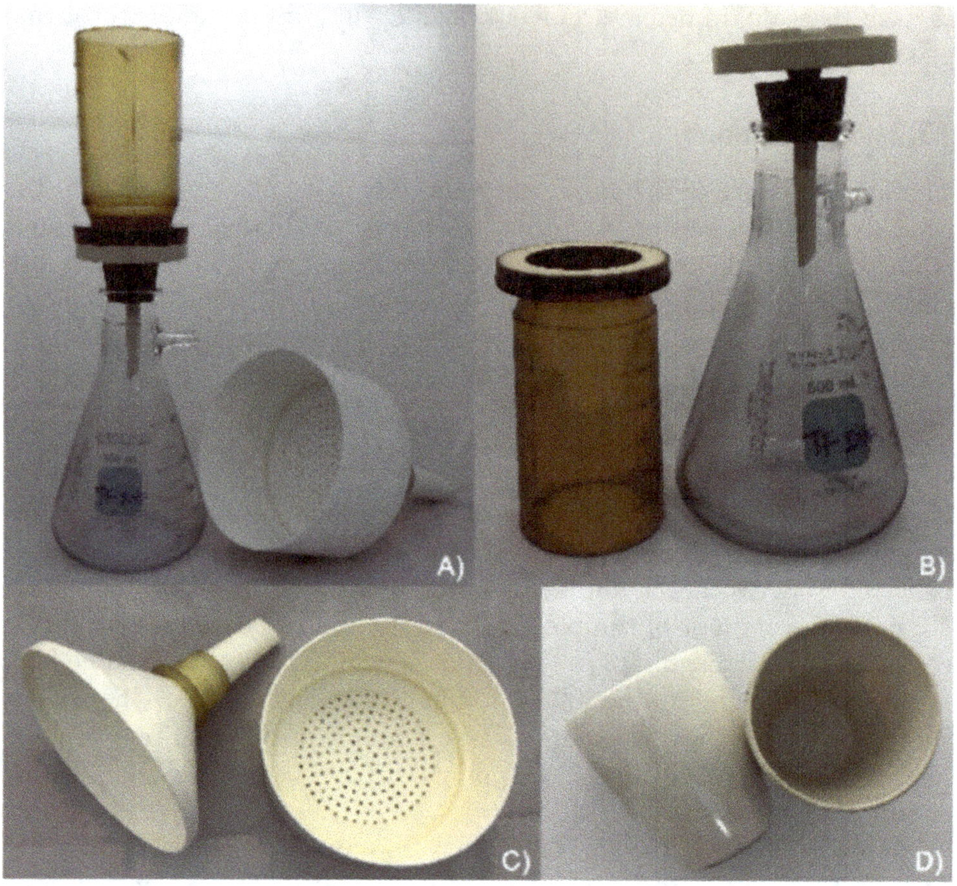

FIGURE 6.11 Funnel Types (Reprinted with permission from Indigo Water Group)

be purchased as a two-part glass funnel with a large clamp that holds the two pieces together. The magnetic or glass filter funnels are best suited to the TSS test because they take a larger-diameter filter paper (4.7 cm) than the older style Gooch crucibles. Larger-diameter filter papers can accommodate larger sample volumes without clogging and tend to filter faster than smaller-diameter filter papers.

Gooch crucibles may be purchased in a variety of sizes. For the TSS test, *Standard Methods* (American Public Health Association et al., 2022) calls for a 25- to 40-mL capacity Gooch crucible. A 25-mL Gooch crucible is 36 mm in diameter at the top and takes a 20-mm diameter filter paper. A 40-mL Gooch crucible is 40 mm in diameter at the top and takes a 24-mm diameter filter paper. Compared to the 47-mm filter paper commonly used with magnetic or glass filter funnels, the Gooch crucible filter papers provide about 25% of the surface area of larger filters. Consequently, it is necessary to reduce the total sample size when using Gooch crucibles. Reducing the sample size can result in nonrepresentative subsamples and magnify analytical error. Maintaining the same sample size requires the analyst to measure and add multiple aliquots of sample to the crucible. Each time a sample measurement is made, there is an opportunity for additional error. One advantage of Gooch crucibles is that they decrease the amount of filter handling by the analyst, which reduces the chances of contamination and/or filter damage.

The last filter shown is a Buchner funnel. This type of filter is not appropriate to use for the TSS test. Rather, it is used for rapidly filtering large volumes of sample. It does not work well for the TSS test because the filter paper does not completely cover the bottom of the funnel and some solids may not be captured. It is also difficult to remove filter papers from this type of filter funnel without damaging them. Buchner funnels, however, work well for the TSS test.

The following apparatuses are recommended:

- Filter funnel apparatus or Gooch crucible.
- Glass-fiber filters—Whatman-grade 934AH, Gelman-type A/E, Millipore-type AP40, E-D Scientific Specialties grade 161; Environmental Express Pro Weigh filters or equivalent are acceptable examples. Glass-fiber filters should have a pore size of 2.0 µm or less. Choose a filter size between 2.2 and 12.5 cm in diameter. Match the filter size to the filter funnel apparatus so the filter covers the entire bottom of the funnel.
- Vacuum flasks, 500 mL or larger.
- Vacuum pump or water aspirator.
- Drying oven, 103 to 105°C.
- Aluminum weighing pans.
- Porcelain evaporating dishes (for volatile suspended solids [VSS] only).
- Muffle furnace, controlled at 550 °C ± 50 °C.
- Analytical balance capable of measuring to the nearest 0.0001 g (0.1 mg).
- Dessicator with dessicant.
- Graduated cylinder.
- Wide-mouth pipet—a wide-mouth pipet is critical for accurate measurement of samples like activated sludge, WAS, and RAS. A narrow-mouth pipet will be prone to clogging and can shear up the solids in an activated sludge sample.

Figure 6.12 shows how two vacuum flasks can be connected in a series to form a water trap between the filter funnel and the vacuum pump. The second flask can be loosely filled with cotton or other absorbent material. The sole purpose of the water trap is to prevent moisture from entering the vacuum pump, where it can cause corrosion and eventual failure of the pump.

13.5 Reagents

No reagents are required for the TSS test other than deionized water.

13.6 Stock and Standards

For permit reporting, a certified standard is required. Standards may be purchased from a variety of sources.

13.7 Quality Assurance and Quality Control

When analyzing samples for process control, the analyst may not have the time or resources to run all of the recommended quality assurance samples.

FIGURE 6.12 Water Trap for TSS Test

Certified standards can be expensive. It is acceptable to decrease the analysis frequency for most or all quality assurance and quality control (QA/QC) samples when analyzing process control samples; however, QA/QC samples should still be analyzed on a regular basis to ensure data quality. At a minimum, a blank and a duplicate should be analyzed with each batch. The analyst must balance the benefit gained by running a particular QA/QC sample against available resources. Process control is only as good as the laboratory data it is based upon.

For permit reporting, *Standard Methods* (American Public Health Association et al., 2022) recommends the following minimum quality assurance samples:

- Analyze all samples in replicate using two filter papers per sample. Replicate results are averaged together to get a single reportable result for each sample.
- Blanks should always be lower than the method detection limit (MDL). When blanks are higher than this, it is an indication that some or all of the samples in the batch may have been contaminated. A blank should be run once every 20 samples or once per batch, whichever is more frequent. For the TSS test, the filter blank consists of 100 mL of Type III water that is run through a prewashed filter paper and treated as if it were a regular sample. The filter blank initial weight and final weight should not differ by more than 0.5 mg.
- The certified standard should agree within 10% of its certified value or within the limits specified on the manufacturer's certificate. A standard should be run once per 20 samples or once per batch, whichever is more frequent.
- At least 10% of all samples should be analyzed in duplicate. Duplicates should agree within 5% of their average weight.
- It is good laboratory practice to analyze at least one standard and one duplicate for each set or batch of samples, even if fewer than 20 samples are analyzed at a time.

13.8 Sample Collection, Preservation, and Holding Times

Samples should be collected in glass or plastic bottles. The analyst should ensure that solids in the sample do not stick to container walls. In addition, the sample should be refrigerated at ≤ 6 °C until ready for analysis; the samples should be brought to room temperature just before analysis. Ideally, samples should not be held for more than 24 hours before analysis. Samples should never be held longer than 7 days.

13.9 Procedure

Follow these steps to conduct the TSS test:

Step 1—Prewash the filter. The analyst should note the following about Step 1:

- The prewashing step removes loose filter material. This step is critical for obtaining accurate and precise results. If the filters are not prewashed, final results may be biased low. Always include a filter blank, even when analyzing process control samples.
- If pre-prepared glass-fiber filter discs are used, eliminate this step.
- Two glass-fiber filter papers should be washed for each sample to be analyzed.
- Place new, clean filter paper on the funnel stand. Use tweezers. Touching the filter paper may contaminate it with oils and dirt from your fingers. Magnetic filter funnels or Gooch crucibles may be used for this step.
- Add funnel top and secure.
- Add 20 mL of laboratory-grade water to the filter funnel to wash the filter.
- Repeat previous step two more times for a total of 60 mL of rinse water. Be sure to allow the filter to drain completely before adding the next aliquot.
- Remove the funnel.
- Inspect the filter paper visually to verify that it is as dry as possible before removing it. If the paper it too wet, it will stick to the bottom of the drying pan. This may cause part of the filter to tear off after drying.
- Use tweezers to carefully remove the washed filter paper and place it into a numbered aluminum pan.
- Angle the filter paper so that as little of the filter paper is touching the pan as possible.

Step 2—Dry filters until a constant weight is reached.

- Dry the filters in their numbered pans at 103 to 105 °C for at least 1 hour.
- Remove filters and pans from the drying oven.

- If total volatile suspended solids are needed, place the filter papers in a muffle furnace at 550 °C for at least 15 minutes to remove organic residue.
- Place filters and pans into the dessicator to cool.
- Ensure that filter papers are completely cool before proceeding to the next step. (It is important to note that many laboratories prewash filters the night before and leave them in the dessicator until morning. That way, they are always ready to use.)

Step 3—Record the initial filter weight to the nearest 0.0001 g (0.1 mg) and weigh using an analytical balance. Record the weight in grams on the data sheet and place the filter paper in the numbered aluminum pans. The analyst should make sure that the number on the data sheet corresponds to the number on the pan for each filter paper.

Step 4—Aliquot samples. The analyst should note the following about Step 4:

- The goal is to have between 2.5 and 200 mg of solids on the filter paper.
- Because excessive residue on the filter may form a water-entrapping crust, limit the sample size to that yielding no more than 200 mg of residue. If this requirement is met, samples should not take more than a few minutes to filter. For sample volumes of 25 mL or greater, a graduated cylinder should be used. Sample volumes less than 25 mL should be measured using a wide-mouth pipet.
- Suggested sample volumes are listed in Table 6.5. These volumes are only a starting point. Each WRRF is different and sample volumes may need to be adjusted up or down accordingly. It is important to remember to always use the largest volume of sample that will easily pass through the filter paper. Larger sample volumes are more likely to be representative than smaller volumes. Recommended volumes assume a 4.7-cm filter paper diameter in use.
  (1) Place each filter paper in consecutive order onto the vacuum filtration apparatus,
  (2) Place a magnetic filter funnel over each filter, and
  (3) Add the sample to the filter funnel.
    (a) Filter the blanks and the standard first. The next set of filters should be the final effluent samples. Filter all liquid samples before starting on thick sludges. Moving from clean to dirty samples will help prevent cross-contamination problems.

TABLE 6.5 Suggested Sample Volumes for TSS Test (WEF, 2022)

Sample type	Suggested volume
Facility influent	25 to 50 mL
Primary clarifier influent and effluent	25 to 100 mL
Trickling filter influent and effluent	25 to 100 mL
Aeration tank (MLSS)	5 to 10 mL
Return activated sludge	3 to 5 mL
Waste activated sludge	3 to 5 mL
Secondary clarifier effluent	200 to 1000 mL
Tertiary unit processes	200 to 1000 mL
Facility final effluent	250 to 1000 mL

(b) Thoroughly mix or shake each sample immediately before taking an aliquot. Pour sample volumes according to the data sheet through the corresponding filter paper and apply suction. If the sample looks as if it has substantially more or less particulates in it than normal, adjust the aliquot size accordingly and note the new aliquot size on the data sheet.

Step 5—If a graduated cylinder was used to measure the sample, rinse the inside of the graduated cylinder with Type III water. Add the rinsate to the filter funnel with the sample.

Step 6—If a pipet is used to aliquot the sample, care should be taken to avoid touching the pipet to the sides of the filter funnel. Be sure to completely cover the filter paper with sample. If dry spots are left on the filter paper, there will not be enough suction to completely dry the filter paper. Air will simply pass through the drier portions of the paper.

Step 7—Once the sample has completely passed through the filter, rinse the filter paper with three portions of 10 mL of Type III water. Wait between each rinse until the water has completely drained. It is important to note that, for high solids samples that were aliquoted with a pipet, it may not be desirable to wash the funnel and filter paper. For these samples, it is likely that the water will pond on top of the filter paper instead of passing through. Some samples in which washing the filter paper may cause difficulties include MLSS, RAS, and WAS.

Step 8—Carefully remove the top of the filter funnel. Visually inspect the filter paper to ensure that it is dry. It may be necessary to wait for several minutes to get the filter as dry as possible.

Step 9—Remove the filter paper with tweezers and return it to the appropriately numbered aluminum pan.

Step 10—Repeat the aforementioned steps for all remaining samples.

Step 11—Dry filters until a constant weight is reached:

- Dry the filters in their numbered pans at 103 to 105 °C for at least 1 hour,
- Remove filters and pans from the drying oven,
- Place filters and pans into the dessicator to cool, and
- Ensure that filter papers are completely cool before proceeding on to the next step.

Step 12—Record the final filter weight to the nearest 0.0001 g (0.1 mg) and weigh using an analytical balance. Record the weight in grams on the data sheet and place the filter paper in the numbered aluminum pans. Make sure that the number on the data sheet corresponds with the number on the pan for each filter paper. *Standard Methods* (American Public Health Association et al., 2022) requires that Steps 11 and 12 be repeated until the weight no longer changes. This is defined as no more than a 4% or 0.5-mg difference between weighings. In practice, the analyst will quickly learn how long it takes for samples to dry to a consistent weight. For process control samples, consistency is more important than following the precise details of the method. If the same drying time is used for process control samples each time they are analyzed, results from one day can easily be compared to the next.

Step 13—Calculate TSS in milligrams per liter.

Step 14—If TVSS are needed, place the dish from the previous step in a muffle furnace at 550 °C until the sample is completely burned. The length of time for complete burning (ashing) depends on the size of the sample. For most samples, 15 minutes is sufficient to completely ash the sample. *Standard Methods* (American Public Health Association et al., 2022) recommends repeated ashing, cooling, and weighing cycles until a constant weight is achieved. This is not necessary for process control samples.

Step 15—Remove the dish from muffle furnace. Wear heavy, heat-resistant gloves and use tongs to transfer the dishes from the muffle furnace to a heat-resistant surface. A few ceramic tiles on the bench top will protect the laboratory counter from heat.

Step 16—Allow the dish to partially cool in air until most of the heat has dissipated.

Step 17—Transfer the dish to a dessicator and allow it to cool to room temperature.

Step 18—Weigh the dish and record the weight to the nearest 0.0001 g (0.1 mg). If unsure that constant mass has been obtained, record the weight, then reheat and weigh again until the weight no longer changes.

Step 19—Calculate TVSS in milligrams per liter.

13.10 Calculation

Calculate TSS concentration from the initial and final filter weights according to the following formula:

$$\text{Total suspended solids, mg/L} = \frac{(A - B)^{*}1000}{\text{Sample volume, mL}} \tag{6.8}$$

Where

A = weight of filter plus dried residue (mg) and
B = weight of filter (mg).

Calculate the TVSS concentration from the initial and final filter weights according to the following formula:

$$\text{Total volatile suspended solids, } \frac{\text{mg}}{\text{L}} = \frac{(A - C)^{*}1000}{\text{Sample volume, mL}} \tag{6.9}$$

Where

A = weight of filter plus dried residue (mg) before ashing and
B = weight of filter plus dried residue (mg) after ashing.

13.11 Reporting

Sample results are reported as milligrams per liter with three significant figures. The calculated result of 2432 mg/L is reported as 2430 mg/L and the calculated result of 18.34 mg/L is reported as 18.3 mg/L.

13.12 Troubleshooting and Tips

Table 6.6 contains troubleshooting tips for the TSS test.

14.0 OXYGEN UPTAKE RATE AND SPECIFIC OXYGEN UPTAKE RATE

14.1 General Description

The oxygen uptake rate (OUR) test and resulting respiration rate estimate how active microorganisms are in the activated sludge process. The activity of microorganisms is related to the amount of oxygen the organisms

TABLE 6.6 Troubleshooting Chart for the TSS Test (WEF, 2022)

Problem	Probable cause
Standard result is lower than certified value.	1. Standard not shaken or mixed enough prior to aliquoting. 2. Standard poured too slowly or allowed to sit on bench after mixing. 3. Hole in filter disk.
Standard result is higher than certified value.	1. Filter disk not dried long enough. 2. Dirty glassware. 3. Standard may have been compromised. If previous uses were out of range low, the standard may have been concentrated in the bottle.
Filter blank changed by more than 0.5 mg between initial weighing.	1. Filter disk not prewashed. 2. Increasing weight may indicate (a) inadequate filter drying time and final or (b) dirty glassware contaminated filter. 3. Decreasing weight may mean that the filters were not prewashed enough. Additional filter material was lost during analysis. 4. Loss of filter material—possibly stuck to drying pan. 5. Balance not calibrated at initial or final weighing or both.
Sample replicates do not agree within 5% of their average weight.	1. Chunky samples. Look for obvious visual differences between filter disks.

consume. The OUR test measures how much oxygen a sample of activated sludge consumes over a specific time period. The rate of oxygen uptake can vary because of either microorganism activity or the number of microorganisms. A larger number of microorganisms will use more oxygen than a smaller number of microorganisms, even when they are respiring (breathing) at the same rate. The specific oxygen uptake rate (SOUR) test takes the OUR result and divides it by the concentration of MLSS in the activated sludge process. This normalizes the result to the amount of oxygen consumed by 1 g of mixed liquor volatile suspended solids (MLVSS) per hour and allows operators to compare results from one day to the next even though MLVSS concentration may be changing. Using MLVSS in the calculation relates oxygen consumption to the mass of active microorganisms in the basin instead of total MLSS, which includes inert, nonrespiring material. The SOUR test is also referred to as the *respiration rate*.

14.2 Application

The OUR test can be used to determine aerobic biological activity wherever the sample is taken (e.g., tail end of aeration tank [most common] or along the aeration tank to track treatment progression) to test for toxic effects of influent wastewater and demonstrate that digested biosolids are stabilized. It can also be used to determine the required time through the aeration tank for microorganisms to stabilize influent biochemical oxygen demand (BOD) and ammonia loads and become "activated" (i.e., reach endogenous respiration rate). The endogenous respiration rate is the rate of bioactivity when microorganisms have stabilized all stored food, and the rate of growth is less than the rate of cellular degradation (i.e., the cells are starving).

14.3 Apparatus and Materials

The following equipment will be needed for the OUR test:

- Calibrated dissolved oxygen meter.
- Two BOD bottles or two 1-L plastic bottles.
- A 7.6-cm (3-in.) section of 22-mm (0.875-in.) outer diameter plastic pipe that has been shaped on both ends to fit into the mouth of a BOD bottle. The pipe will be used to transfer MLSS between BOD bottles and should fit snuggly. Length is approximate.
- A self-stirring BOD bottle probe and/or separate magnetic stirring device (magnetic device recommended in addition to the self-stirring BOD bottle probe because fast-settling sludge may still settle with only the bottle probe mixer).
- Stopwatch or other timing device.

Other data required for the SOUR calculation include VSS in grams per liter (ounces per gallon) of the activated sludge sample from which the OUR test was performed. For SOUR calculations on aerobic digester sludge, the total solids concentration will be needed. For operational purposes, a TSS meter can be used for the MLSS determination and last week's volatility percentage can be used for estimating the actual MLVSS value. Standards for the Use and Disposal Of Sewage Sludge, 40 CFR Part 503 (1993) require a SOUR calculation for biosolids treated in aerobic digesters that will be beneficially reused through land application. For aerobically digested biosolids to meet the vector attraction reduction requirement in the these regulations, digested biosolids must have a SOUR of less than 1.5 mg/L dissolved oxygen per hour per gram of total solids. The SOUR test used for activated sludge process control is per gram of VSS. The procedure outlined in Section 14.4 is not exactly the same as U.S. Environmental Protection Agency (U.S. EPA) Method 1683 (2001) and should not be used for regulatory compliance.

14.4 Procedure

Follow these steps to conduct the OUR test.

Step 1—Collect a fresh sample of activated sludge either from the tail end of the aeration tank or from other locations within the process to measure changes through the process. (Note: perform the test within 15 minutes of collecting the sample for accurate results. Discard the sample and obtain a fresh sample if holding for more than 15 minutes.)

Step 2—Fill a 300-mL BOD bottle with the activated sludge sample and insert a 7.6-cm (3-in.) section of shaped plastic pipe.

Step 3—Insert the other end of the pipe into an empty BOD bottle. Pour the sample back and forth between the bottles until the dissolved oxygen is approximately 5 to 6 mg/L. Altitude may affect the maximum dissolved oxygen that is achievable. Do not shake the sample to aerate it because that can cause floc shear, which may expose new floc surfaces, making them available for oxygen uptake. Increasing floc surface area can cause SOUR results to differ from actual conditions in the aeration basin. (Note: if BOD bottles and/or the fitted plastic pipe are not available, two 1-L bottles may be used instead.)

Step 4—Remove the transfer pipe and fill one of the BOD bottles to overflowing. Some bubbles will gather at the top. Tilt the bottle and/or tap the sides of the bottle to dislodge air bubbles, then place the bottle into an insulated sleeve or other material to help maintain constant temperature, as shown in Figure 6.13.

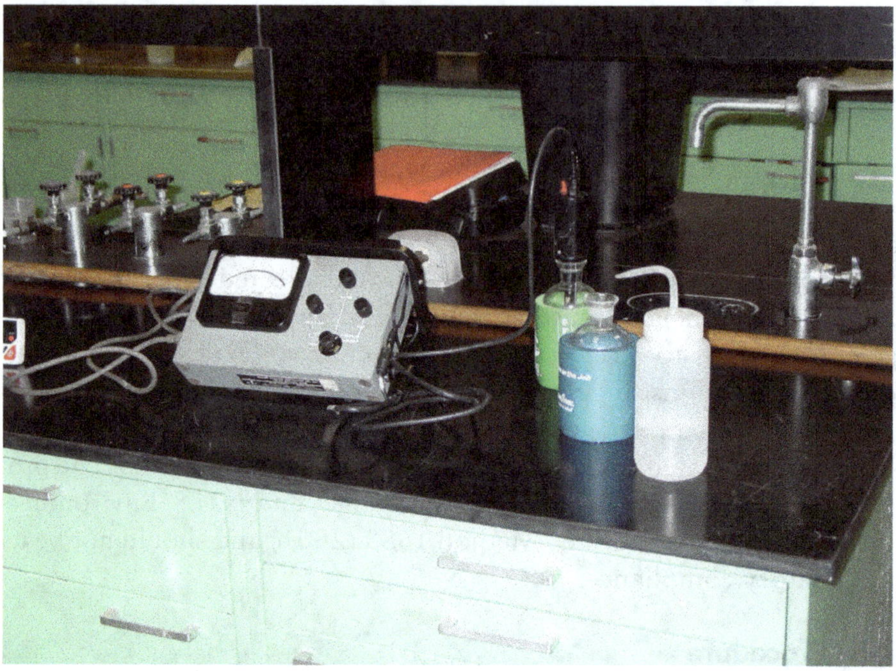

FIGURE 6.13 Insulated BOD Bottles for OUR/SOUR Test (Courtesy of Dave Flowers)

Step 5—Use a spatula to work the bubbles out of the sample. It is important to note that the activated sludge sample that remains after filling the BOD bottle can be used to perform the VSS test in Step 12 if a meter is not being used. Alternatively, excess MLSS not used in the bottle may be analyzed for TSS.

Step 6—Insert the dissolved oxygen meter probe in the BOD bottle and begin stirring. Turn the dissolved oxygen meter to a 0 to 10 scale.

Step 7—Wait approximately 30 to 60 seconds for the dissolved oxygen meter reading to stabilize. It is important to note that the indicator needle or readout should be constantly dropping during this procedure.

Step 8—Record the dissolved oxygen concentration and temperature of the sample in 30- or 60-second intervals. Stop the test before the dissolved oxygen concentration reaches 1.0 mg/L. If the temperature varies by more than 0.1 °C from the start to the end of the test, the test should be conducted again with a fresh sample. Results are not valid when temperature is changing because OUR is affected greatly by temperature, and changes in OUR may be attributable primarily to temperature changes and not bioactivity.

Step 9—Graph the results by plotting dissolved oxygen in milligrams per liter on the vertical axis and time in minutes on the horizontal axis.

Step 10—Draw a straight line connecting the majority of the points. Extend the line so that it crosses the horizontal and vertical axes. It is important to note that the operator may discover that the line is not straight at the beginning and end of the test. The straight line is drawn to negate the effects of these curves. Just the straight-line section is used. At the beginning, false values caused by undissolved bubbles can cause interference. The end of the curve flattens because of the limitation of the dissolved oxygen meter.

Step 11—Determine the slope of the line, in milligrams per liter of oxygen per minute. The easiest way to do this is to use the points where the line crosses the axes. Divide the milligrams per liter of oxygen by time in minutes.

Step 12—If not using a TSS meter and the latest volatility percentage, determine the VSS content of the activated sludge sample. Express VSS concentration in grams per liter. This is done by dividing milligrams per liter of VSS by 1000.

Step 13—Enter values in the equation and calculate the respiration rate. The OUR is expressed as milligrams of oxygen per hour per gram of VSS. Highly organically loaded industrial treatment systems may have such high OURs that the test sample has to be diluted with secondary clarifier effluent to slow the OUR enough to get readable values that can be graphed. At least three points in a straight line are required to establish the line.

14.5 Calculations

There are two methods commonly used for calculating OUR. Both methods should arrive at the same numerical results. In both instances, the respiration rate worksheet and graph will be used.

The first method requires drawing a straight line through the greatest number of points graphed. The line will cross both the horizontal (time) and vertical (milligrams per liter) axes of the graph. The values at these two points are used in the OUR calculation. The milligrams per liter of oxygen value is divided by minutes value and multiplied by 60 minutes per hour to derive milligrams per liter per hour.

The second method is applicable when using the second or right-hand side of the graph to calculate OUR for an aerobic digester or extended aeration system. These typically flat lines may not conveniently cross the horizontal axis. The straight line through the majority of points should still be drawn. The analyst should pick a point on the line as time 0 and enter the milligrams per liter of oxygen value in the calculation box in the upper part of the graph. The analyst should also pick a second point on the line as the end time and enter it. The points should be at least 5 minutes apart. Then, divide by the time span between the two points and multiply by 60 minutes per hour to derive the milligrams per liter per hour OUR, as follows:

$$\text{OUR (mg/L/h)} = \frac{\left(\text{DO}\frac{\text{mg}}{\text{L}}\text{ at time 2} - \text{DO}\frac{\text{mg}}{\text{L}}\text{ at time 1}\right)}{\text{Time at point 0} - \text{time at point 2}} * 60\text{ min/hr} \quad (6.10)$$

$$\text{SOUR (mg/h/g)} = \frac{\text{OUR (mg/L/h)}}{\text{MLVSS (g/L)}} \quad (6.11)$$

14.6 Oxygen Uptake Rate Test Variation—Stabilization Test

To this point, OUR testing has been limited to the end of the aeration tank; however, a different approach is called the "fed" test, where the correct amount of influent is mixed with return sludge to simulate the OUR at the head of the aeration tank (Wooley, 1981). This test is best conducted in a respirometer but also can be conducted by mixing the correct amount of RAS with the correct amount of influent, as follows:

$$\text{RAS, mL} = \text{MLSS, mg/L} * \frac{\text{Volume of test material}}{\text{RAS, mg/L}} \quad (6.12)$$

Here, MLSS and RAS concentrations are determined with a calibrated TSS meter. The volume of the container is 1 L. Place the calculated volume of RAS in a 1-L graduated cylinder and fill to 1 L with influent and then mix and fill a BOD bottle for aeration and determination of OUR. This OUR will be much higher than that taken at the tail end of the aeration tank and shows the effect of feeding. A ratio of fed MLSS to unfed MLSS less than 1.0 indicates inhibition/toxicity and a ratio between 1.0 and 2.0 suggests low soluble BOD load; a ratio between 2 and 5 is typical for municipal wastewater. A ratio greater than 5.0 is high for municipal wastewater, but industrial wastewater may be much higher depending on the industry.

Fed testing can be used in a different approach, that is, the stabilization test. This test is typically run only 3 to 4 times a year because of the high time requirement, but its value is great. The stabilization test is a fed test conducted on a larger scale, where approximately 7.6 to 15.1 L (2 to 4 gal) of RAS and influent are mixed in a 19-L (5-gal) bucket according to the fed test RAS volume calculation. For instance, if a total volume of 7.6 L (2 gal) is used and the MLSS concentration is 2000 mg/L and the RAS concentration is 8000 mg/L, then the volume of RAS would be 1.9 L (0.5 gal) mixed with 5.7 L (1.5 gal) of influent.

Follow these steps to conduct the stabilization test:

Step 1—Pour the RAS into the bucket and start aerating with two to four aquarium aerators with diffusers or a compressed air line with diffusers.

Step 2—While the RAS is aerating, collect a grab sample of 3.8 to 15.1 L (1 to 4 gal) of aeration tank influent and mix well with the aerated RAS.

Step 3—Immediately fill a 300-mL BOD bottle with the mixture from the bucket and aerate until the dissolved oxygen concentration is at least 5 mg/L. Facilities located at high elevation may have difficulty getting to 5 mg/L of dissolved oxygen.

Step 4—Run the OUR test.

Step 5—When the test is complete, empty the 300 mL of RAS/influent mixture from the BOD bottle back into the bucket and continue aerating for an hour, trying to keep dissolved oxygen approximately the same as that in the aeration tank.

Step 6—After an hour, sample again and run another OUR. Do this every hour until the OUR no longer changes. In a high SRT system, after the first hour, testing may be done every 2 hours for 4 hours, and then every 4 hours until endogenous respiration is reached. The time it takes to reach endogenous respiration is the stabilization time. Figure 6.14 shows the OUR curve for a typical stabilization test.

14.7 Oxygen Uptake Rate Test Variation—Aeration Basin Profile

In a plug flow type aeration basin, a series of OUR tests may be conducted along the length of the basin. Typical sampling locations may be at the inlet, one-quarter down the length of the basin, in the middle of the basin, three-quarters down the length of the basin, and at the outlet. There are five sampling locations total in this example; however, an operator may add additional sampling points as needed to fine-tune the aeration basin. Results are then graphed to produce a diagram similar to the one shown in Figure 6.15. In this example, endogenous respiration is reached a point midway down the aeration basin. The operator has an opportunity to reduce aeration in the second one-half of the basin to reduce energy costs. Take care to

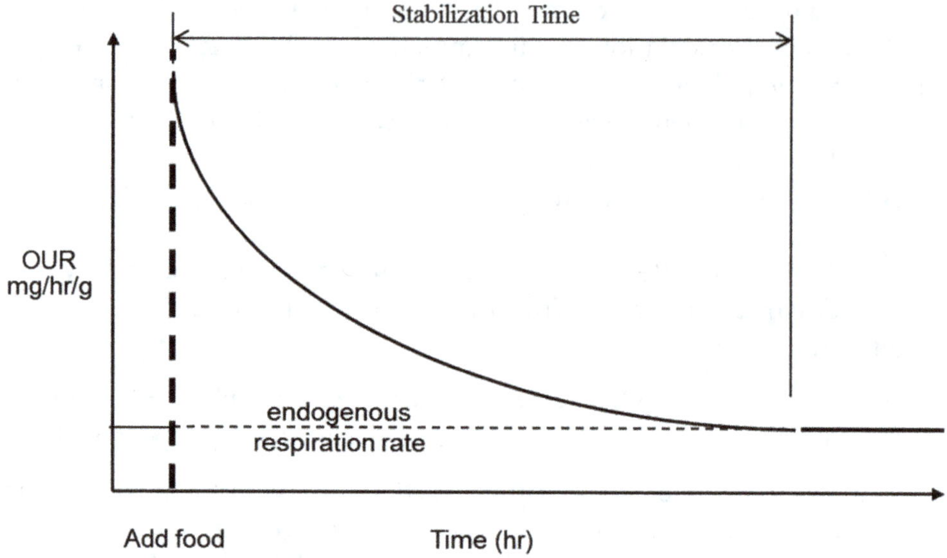

FIGURE 6.14 Typical Stabilization Time Plot (Courtesy of Ronald Schuyler)

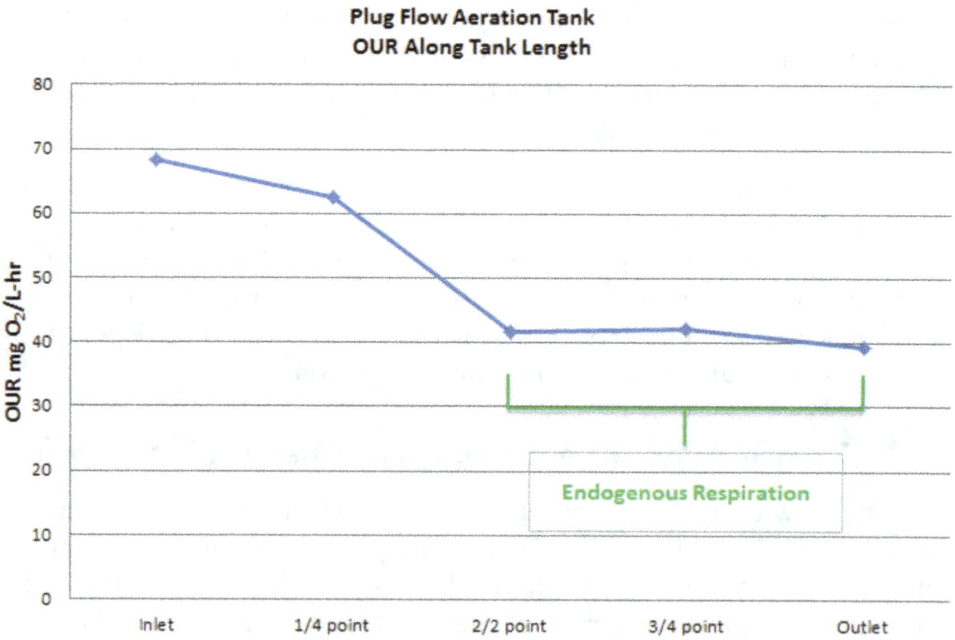

FIGURE 6.15 Oxygen Uptake Rate Profile of an Aeration Basin (Courtesy of Dave Flowers)

keep dissolved oxygen concentrations high enough to prevent the growth of some types of filamentous bacteria.

14.8 Significance of Respiration Rate or Specific Oxygen Uptake Rate and Stabilization Time

The SOUR can be related to typical values for different activated sludge processes. For instance, a conventional activated sludge process or biological phosphorus removal process using an SRT ranging from approximately 3 to 6 days might expect a SOUR between 12 and 20 mg/L/h/g, whereas a process loaded more in the range of extended aeration or nitrifying activated sludge process using an SRT from approximately 8 to 15 days might expect a SOUR between 3 and 12 mg/L/h/g. An aerobic digester will typically see SOUR values below 2 mg/L/h/g. Highly organic-loaded industrial processes may run well above 20 mg/L/h/g. These are typical ranges, but you may find that the process runs well at values outside these ranges. Higher SOUR values indicate a less-stabilized wastewater with a higher organic content.

15.0 BIOCHEMICAL OXYGEN DEMAND

15.1 General Description

For a complete discussion of the BOD testing protocol, refer to *Standard Methods* (American Public Health Association et al., 2022) or *Basic Laboratory Procedures for the Operator-Analyst* (Water Environment Federation, 2022). This test determines the amount of biodegradable organic material present in wastewater by measuring the amount of oxygen consumed by microorganisms as they degrade or "eat" organic constituents of the waste. Nitrogen may also be converted to nitrate if nitrifying bacteria are present.

The test consists of measuring dissolved oxygen concentrations before and after a 5-day incubation period of a sample at 20 °C (68 °F) to determine the amount of oxygen used for a known amount of sample. Because, typically, the test is conducted over a 5-day period, the shorthand 5-day BOD (BOD_5) is used in laboratory reports and other writing. In some cases, a 7-day incubation (BOD_7) or longer may be used. Biochemical oxygen demand can be broken down into soluble and particulate fractions by filtering a sample through a 1.2-µm glass-fiber filter paper. Biochemical oxygen demand may also be broken down into carbonaceous BOD (CBOD) and nitrogenous oxygen demand (NOD) by adding a compound to the sample that is poisonous to nitrifying bacteria to prevent the conversion of ammonia to nitrate. Adding the inhibitor allows the measurement of CBOD. Running the test without the inhibitor measures BOD. The difference between the two measurements, theoretically, is the NOD. In practice, the inhibitor is somewhat toxic to heterotrophic bacteria and will lower the BOD result slightly even when no nitrification is taking place in the test. For raw influent, domestic wastewater samples, CBOD is typically approximately 85% of total BOD.

Samples are diluted with standard dilution water or seeded dilution water. Dilution is required so that microorganisms do not run out of oxygen. Seed consists of live bacteria that are added to the sample to ensure that there are enough bacteria in the test to consume available organic material within the testing period. Industrial samples, samples with high or low pH, and samples that were previously disinfected should all be seeded. At the end of the incubation period, BOD is calculated using the following formula:

$$\text{BOD}, \frac{\text{mg}}{\text{L}} = \frac{[(\text{DO}_i - \text{DO}_f) - (\text{S}_i - \text{S}_f)]}{\text{Sample Volume}, \dfrac{\text{mL}}{300\ \text{mL}}} \tag{6.13}$$

Where

DO_i = Initial dissolved oxygen concentration in the sample,
DO_f = Final dissolved oxygen concentration in the sample,
S_i = Initial seed blank dissolved oxygen concentration, and
S_f = Final seed blank dissolved oxygen concentration.

When seed is added to samples, BOD is also added from the endogenous decay of those bacteria. For seeded samples only, the extra BOD contribution is subtracted from the sample result. The reader is highly encouraged to read the step-by-step BOD procedure in *Basic Laboratory Procedures for the Operator-Analyst* (Water Environment Federation, 2022).

15.2 Applicability

Results from the BOD or chemical oxygen demand (COD) test are used for permit compliance, to assess the loading to the WRRF as a percentage of permitted capacity, to calculate the food-to-microorganism ratio in an activated sludge process, and to estimate the amount of carbon available for denitrification and/or biological phosphorus removal. The BOD test has minimal practical value for activated sludge process control because the test takes 5 days to complete but is still used to calculate loading to the treatment process and for permit compliance. One caveat to note with respect to BOD testing is the effect that sampling procedure can have on results. Depending on permit requirements for a facility, sampling for BOD testing can be either grab sampling or composite sampling. Composite samples are taken using automatic samplers. Inspection and maintenance of sampling equipment, in particular, sample tubing, must be followed diligently. Over time of use, there is the possibility that residuals will accumulate in the sample tubing and become mingled with samples for testing and potentially skew results. As an example, in some facilities operating a high-rate process, a nitrifying biofilm can form on the inside of sample tubing if it has been in place too long. The nitrifying film can exert a BOD demand but not a CBOD demand, which if noted is an indication of the need to replace the sample tubing. For both grab sampling and composite sampling, periodic inspection, cleaning, and replacement of sampling equipment is fundamental.

16.0 CHEMICAL OXYGEN DEMAND

For a complete discussion of the COD testing protocol, refer to *Standard Methods* (American Public Health Association et al., 2022). Not all of the organic material entering the treatment process is biodegradable and can be measured in the 5- or 7-day BOD test. Some of this organic material cannot be broken down because bacteria lack the necessary enzymes. Some of this organic material can be broken down but, because it is so complex, bacteria need more than the 5 or 7 days given in the BOD test to consume it. Compounds are still organic and can be chemically burned. Once in the environment, they can exert an oxygen demand. For domestic wastewater without a large industrial component, the ratio of COD to 5-day CBOD is typically within the range of 1.8 to 2.2. A facility may collect COD and CBOD data for their influent and effluent over time to better refine the correlation for its wastewater.

The test consists of heating a small amount of sample at 150 °C for 2 hours in the presence of a catalyst, potassium dichromate ($K_2Cr_2O_7$) and concentrated sulfuric acid. The combination of the strong oxidizer and acid will chemically burn nearly all organic compounds in a sample. The COD test does not convert ammonia to nitrate. As organic compounds are oxidized, the dichromate is converted to trivalent chromium (Cr^{3+}).

17.0 NUTRIENT MONITORING

Nutrient monitoring in WRRFs is critical for ensuring that effluent discharged to natural water bodies does not cause environmental harm. Excessive nutrients, primarily nitrogen and phosphorus, can lead to water quality issues such as eutrophication, which results in algal blooms, oxygen depletion, and loss of biodiversity.

Effective nutrient monitoring in WRRFs is essential for regulatory compliance, environmental protection, and optimal performance. As nutrient pollution continues to be a global concern, advancements in monitoring technologies will play a key role in minimizing the environmental effects of effluent discharge.

Online sensors installed in the WRRF or at discharge points continuously monitor parameters such as ammonia, nitrate, and orthophosphate levels in real time. These sensors are valuable for immediate detection of process inefficiencies or violations of permitted discharge limits.

Calibration and maintenance of probe sensors is important to ensure accurate readings. Probes are also sensitive to fouling from organic materials and particles in the wastewater, so routine cleaning and maintenance are required.

Operators should select appropriate online monitoring probes/sensors based on the WRRF's needs, including factors such as sample type, frequency of measurement, and available budget.

17.1 Ammonia-Nitrogen

Nitrogen is a key constituent of wastewater, originating from various sources such as food processing, human waste, and industrial waste. Among nitrogenous compounds, ammonia (NH_3) is particularly significant because of its toxicity to aquatic life and its potential to cause eutrophication in receiving waters. To mitigate these effects, ammonia must be removed or transformed into less harmful compounds, such as nitrate or nitrogen gas.

Ammonia probes are commonly used in WRRFs for real-time monitoring. These devices can be installed at various points within the treatment process, including raw influent before the headworks to measure incoming ammonia concentrations, within treatment basins to monitor the completeness of nitrification, or the final effluent line to verify compliance with regulatory standards. Real-time monitoring provides operators with critical data to understand influent concentrations, mitigate high surges, troubleshoot process issues, and promptly detect any potential system upsets. For instance, high surges in ammonia concentrations can disrupt the treatment process, impair the nitrification process because of associated pH drops, and result in elevated total nitrogen and ammonia levels in the effluent flow.

It is important to note that ammonia probes have historically not been intended to directly control equipment, but with the advent and development in recent years of ammonia-based aeration control (ABAC), they are more frequently being made part of process control systems and put to use in managing dissolved oxygen levels and blower operations. A growing number of demonstrations and implementations in WRRFs indicate potential energy savings using ABAC, allowing operation at lower dissolved oxygen concentrations and more effective use of aeration basin volumes. As with any new development in wastewater treatment, the usefulness of ABAC will vary from facility to facility and should be investigated fully to determine applicability at a specific location.

Ammonia probes designed for wastewater application often feature robust sensor materials, resistance to biofouling, and automated cleaning mechanisms, allowing them to maintain accuracy and durability in environments with high solids content and variable chemical compositions. Many probes use ion-selective electrode (ISE) technology, which offers reliable performance and high sensitivity for monitoring ammonia concentrations in wastewater environments.

17.2 Nitrate/Nitrite-Nitrogen

Nitrate (NO_3^-) and nitrite (NO_2^-) are key nutrients in the nitrogen cycle, and their concentrations must be closely monitored and controlled during the wastewater treatment process, particularly when final effluent quality must meet specific total nitrogen limits. Excessive discharge of these compounds to the environment can contribute to eutrophication and degrade water quality.

Nitrate and nitrite probes are used in WRRFs for real-time monitoring. Typically, these devices are installed at the final effluent line. It is important to note that these probes are not intended to directly control equipment, such as blowers or managing dissolved oxygen levels in aeration basins.

Instead, real-time monitoring provides operators with critical data to understand influent concentrations, mitigate high surges, troubleshoot process issues, and promptly detect any potential system upsets. For example, elevated nitrate or nitrite levels in the effluent may indicate an impaired denitrification process. If left unaddressed, this can further impair the nitrification process because of insufficient recovery of alkalinity in the mixed liquor, leading to lower pH levels.

Nitrate and nitrite probes typically use ISE technology to detect specific ions in wastewater. These probes operate by interacting with nitrate or nitrite ions in the sample, generating an electrical signal that is directly proportional to their concentration.

Nitrate probes feature electrodes with a specialized membrane that is selective to nitrate ions. The probe measures the electrical potential created by the interaction between nitrate ions in the sample and the electrode, enabling accurate, real-time measurements.

Nitrite probes function similarly to nitrate probes but are specifically designed to detect nitrite ions (NO_2^-). Because, often, nitrite is present in smaller quantities than nitrate in wastewater, nitrite probes are highly sensitive and capable of detecting low concentrations with precision.

17.3 Phosphorus

Phosphorus is another important constituent in wastewater. Primarily, it is the result of human activity such as the use of detergents and disposal of food waste and human excreta. Excessive discharge of phosphorus to water bodies encourages eutrophication and can result in toxic algal blooms and general degradation of the aquatic environment. In wastewater, typically, it is present in dissolved form consisting primarily of orthophosphate, along with polyphosphates and organic phosphates. Typically, removal involves chemical precipitation via the addition of metal salts or biological uptake in systems using anaerobic, anoxic, and aerobic basins.

Monitoring influent and effluent phosphorus can provide operators with valuable indicators of process conditions and help track effluent compliance. Real-time results can be used to determine the need for process adjustments such as changes in chemical dosage or biological treatment process parameters. Phosphorus analyzers are ideal for monitoring process conditions and effluent quality to help maintain optimal wastewater treatment operations and ensuring that phosphorus concentrations remain within regulatory limits, ultimately helping to reduce eutrophication in water bodies.

Online phosphorus analyzers are available for measuring phosphates and orthophosphates. Analyzers typically use various detection methods such as colorimetric analysis, ISE, and advanced optical sensors.

17.4 Oxidation–Reduction Potential

Oxidation–reduction potential (ORP) probes are valuable tools in WRRFs, offering real-time insights to the redox state of the liquid stream in the facility, which is essential for optimizing biological, chemical, and physical treatment processes. These probes measure overall electron activity in the liquid stream, helping to understand the conditions under which biological and chemical reactions are occurring. Oxidation–reduction potential probes can be inserted directly to a treatment basin to monitor conditions in real time. These probes can be used to monitor and control process conditions in biological treatment basins, as well as control nuisance conditions such as denitrification in clarifiers and odor generation. With respect to biological treatment, by ensuring that conditions are optimal for nitrification, denitrification, and other treatment processes, ORP probes help improve operational efficiency, reduce costs, and ensure compliance with environmental standards.

In biological treatment, ORP indicates the presence or absence of dissolved oxygen and nitrate, which are related to conditions required for specific biochemical reactions to occur. Typical biochemical reactions seen in biological wastewater treatment and corresponding ranges of ORP readings are illustrated in Figure 6.16.

In the vocabulary of wastewater treatment, the conditions of a biological reactor can be divided into three types as follows:

Oxic conditions occur when dissolved oxygen is nonlimiting, allowing organisms that need it to access it. Typically, ORP is +50 mV or higher in oxic conditions. Biochemical reactions include

- Oxidation and removal of organic carbon (BOD).
- Nitrification, where ammonia is oxidized to nitrate (NO_3), with BOD removal preceding nitrification. Oxidation-reduction potential peaks after ammonia depletion, indicating completed nitrification.
- Phosphate uptake by phosphorus-accumulating organisms (PAOs), which involves redox reactions but no change in phosphorus oxidation state. Phosphorus-accumulating organisms use energy from oxidizing stored carbon during anaerobic stages to take up phosphate.

Anoxic conditions occur when dissolved oxygen is low, but nitrate (NO_3) and nitrite (NO_2) are available for respiration. Typically, ORP is +50 mV or lower, with facultative bacteria using nitrate for respiration. Biochemical reactions include

- Biochemical oxygen demand removal, where facultative bacteria consume BOD using nitrate, converting it to carbon dioxide (CO_2) and nitrogen gas.

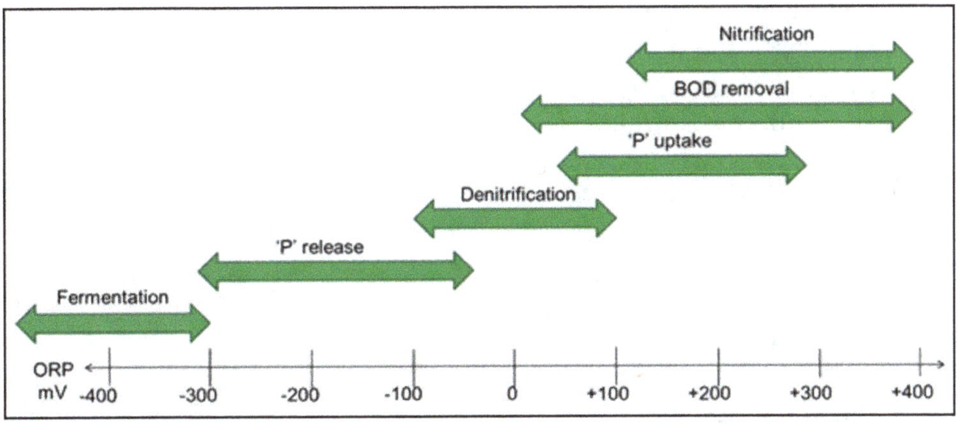

FIGURE 6.16 Approximate ORP Ranges for Biochemical Reactions

- Nitrogen removal through heterotrophic denitrification.
- Phosphate uptake, which occurs under both anoxic and oxic conditions and is not a redox reaction.

Anaerobic conditions occur when neither dissolved oxygen nor nitrate is present, and respiration does not occur. Typically, ORP is −50 mV or lower. Biochemical reactions include

- Phosphate release, where PAOs release phosphorus while sequestering BOD.
- Fermentation, breaking down organic molecules for PAOs and denitrifying microbes, with ORP below −300 mV needed for deep anaerobic conditions.
- Anaerobic digestion.
- Nitrogen removal via autotrophic ammonia oxidation (anammox).

18.0 PRACTICE QUESTIONS

1. An operator notices that one area of the aeration basin is particularly turbulent. The most likely cause is
   a. Damaged or missing diffuser
   b. Clogged air header
   c. Blower output is too high
   d. Water level is too low

2. An activated sludge basin is covered with large amounts of crisp, white foam. This indicates
   a. Presence of *Nocardia*
   b. Start-up conditions
   c. Low organic loading rate
   d. Nutrient deficiency

3. pH is a measure of
   a. Buffering capacity
   b. Hydrogen potential
   c. Hydrogen ion concentration
   d. Acidity

4. Alkalinity is a measure of
   a. Buffering capacity
   b. Hydrogen potential

c. Hydrogen ion concentration
d. Acidity

5. *Total alkalinity* is defined as the amount of acid required to lower the pH of the sample to
   a. 4.55 pH
   b. 7.1 pH
   c. 8.3 pH
   d. 9.3 pH

6. Clarifier blanket samples are typically taken one-third of the way in from the edge of the secondary clarifier for the following reason:
   a. Historic convention—an artifact of early clarifier design
   b. Sludge Judges® are too short to reach the bottom if he samples are taken closer to the center
   c. It is easier to avoid the rake arm at this location
   d. To compensate for the volume of the sloped clarifier bottom

7. When collecting a clarifier core sample, the sample is collected when the rake arm is at a 90° angle to the sampling location for the following reason:
   a. Avoids tangling the Sludge Judge® and clarifier mechanism
   b. Allows the sludge blanket to resettle after being disturbed by the rake arm
   c. Ensures that the lowest possible blanket depth is sampled
   d. Prevents the sludge blanket from getting too thick to sample

8. An activated sludge process has three identical secondary clarifiers. Sludge blanket depths in each are 1 m (3.3 ft), 1.2 m (3.9 ft), and 3 m (9.8 ft). What is the most likely cause for the differences?
   a. Sludge is being removed faster from the third clarifier
   b. The RAS pump on the first clarifier has failed
   c. Uneven flow distribution feeding the clarifiers
   d. Waste activated sludge is only being removed from the second clarifier

9. In the settleometer test, the settled sludge volume at 5 minutes is an indicator of
   a. Sludge-settling velocity
   b. Sludge compaction
   c. Sludge concentration
   d. Sludge toxicity

10. In the settleometer test, the settled sludge volume at 30 minutes is an indicator of
    a. Sludge-settling velocity
    b. Sludge compaction
    c. Sludge concentration
    d. Sludge toxicity

11. Which type of container should be used for the settleometer test?
    a. Graduated cylinder
    b. 1-L beaker
    c. 5-gal bucket
    d. Mallory settleometer

12. When running a settleometer test, MLSS samples should be collected at which of the following locations?
    a. At the end of the basins just before the clarifiers
    b. In the middle of the aerated zone of the activated sludge basin
    c. From the preanoxic zone
    d. After the primary clarifier and just before the activated sludge basin

13. When collecting samples for the settleometer test,
    a. Shake vigorously to ensure that they are well mixed
    b. Minimize shaking and agitation
    c. Place samples on ice and cool to 4 °C
    d. Add 2 mL of nitric acid as a preservative

14. Settled sludge may rise back to the top of the settleometer test for what reason?
    a. Breakdown of solids and production of methane gas
    b. Warming during the test releases dissolved gases
    c. Denitrification and production of nitrogen gas
    d. Hydraulic currents in the settleometer

15. Two settleometers tests are run side by side. The first settleometer contains undiluted MLSS at 4000 mg/L. The second settleometer contains 50% MLSS and 50% secondary clarifier effluent, for a concentration of 2000 mg/L. Both settleometer tests are started at the same time. After 5 minutes, the settled sludge volume in both settleometers is the same at 950 mL. What must be true?

a. Filaments are causing slow settleability in both tests
b. Sludge compaction is unaffected by the MLSS concentration
c. The final effluent will have high turbidity
d. Sludge settleability is unaffected by MLSS concentration

16. Two settleometer tests are run side by side. The first settleometer contains undiluted MLSS at 4000 mg/L. The second settleometer contains 50% MLSS and 50% secondary clarifier effluent, for a concentration of 2000 mg/L. Both settleometer tests are started at the same time. After 5 minutes, the settled sludge volume in the first settleometer is 900 mL but is 750 mL in the second settleometer. What must be true?

 a. Filaments are causing poor settleability in the first test
 b. Hindered settling is taking place in the first settleometer
 c. Wasting should be decreased
 d. The RAS should be increased

17. The 30-minute settled volume is 320 mL and the MLSS concentration is 2950 mg/L. Find the SVI in mL/g:

 a. 108 mL/g
 b. 222 mL/g
 c. 320 mL/g
 d. 512 mL/g

18. The MLSS concentration at the start of the settleometer test was 2500 mg/L. After settling for 30 minutes, the settled sludge volume is 230 mL. What is the settled sludge concentration?

 a. 2500 mg/L
 b. 7813 mg/L
 c. 10 870 mg/L
 d. 22 609 mL/g

19. The MLSS concentration in the aeration basin is 3200 mg/L. The influent flow is 7.57 ML/d (2 mgd) and the RAS is flow-paced at 60% of the influent flow. What is the RAS concentration?

 a. 4160 mg/L
 b. 5120 mg/L
 c. 13 870 mg/L
 d. 8530 mg/L

20. The SVI calculation
    a. Is not affected by the presence of filaments
    b. Can be estimated from the centrifuge spin results
    c. Is how much space 1 g of settled sludge occupies
    d. Uses the settled sludge volume at 60 minutes

21. The 5-minute settled sludge volume is dependent on
    a. Water temperature
    b. Presence of filaments
    c. Mixed liquor suspended solids concentration
    d. All of the above

22. The supernatant from a settleometer test and the secondary clarifier effluent are both analyzed for TSS. The settleometer supernatant contains 12 mg/L of TSS and the clarifier effluent contains 28 mg/L of TSS. What is most likely causing the difference?
    a. Presence of filamentous bacteria
    b. Nothing, the settleometer always performs better
    c. Hydraulic underloading of the clarifier
    d. Denitrification in the clarifier blanket

23. Dispersed suspended solids/FSS testing was performed on a secondary clarifier. At the end of the test, the FSS concentration is 25 mg/L. This suggests
    a. Poor sludge quality because of a bioflocculation problem
    b. Floc shear is occurring in the pipe between the aeration basin and clarifier
    c. Process performance cannot be improved
    d. Effluent ammonia should be below 2 mg/L

24. A clarifier has both flocculation and hydraulic problems. Results of DSS/FSS testing should show
    a. Effluent suspended solids equal to FSS
    b. Flocculated suspended solids greater than 10 mg/L
    c. Dispersed suspended solids greater than FSS but less than ESS
    d. Influent DSS greater than ESS and ESS equal to FSS

25. An MLSS sample was analyzed by the centrifuge spin method, with a result of 1.2%. The centrifuge factor for this week is 900 mg/L/%. Calculate the equivalent MLSS concentration:

a. 1080 mg/L
b. 1200 mg/L
c. 1330 mg/L
d. 2160 mg/L

26. Results from a centrifuge spin test may be affected by
    a. Water temperature
    b. Presence of filamentous bacteria
    c. Number of centrifuge tubes used
    d. Placement of baffles

27. An Imhoff cone is filled with secondary clarifier effluent. After 30 minutes, the operator notices approximately 2.5 cm (1 in.) of settled material at the bottom. Which of the following conditions should concern the operator?
    a. Nitrification in the secondary clarifier blanket
    b. Hydraulic underloading of the secondary clarifier
    c. Clarifier blanket depth
    d. Clarifier rake arm speed

28. Solids are classified by
    a. Carbon content
    b. Size and volatility
    c. Color
    d. Settling velocity

29. Solids that pass through a 1.2-µm glass-fiber filter are considered
    a. Suspended
    b. Dissolved
    c. Fixed
    d. Volatile

30. At what temperature are solids dried when determining TSS?
    a. 103 °C
    b. 180 °C
    c. 550 °C
    d. 720 °C

31. The total solids test has been completed. Dried solids are placed into a muffle furnace at 550 °C for 1 hour. At the end of the test, what type of material will remain in the dish?

a. Calcium carbonate
b. Organic or volatile
c. Starch
d. Inert or nonvolatile

32. For the TSS test, the amount of material allowed to accumulate on the filter paper should not exceed 200 mg because
    a. Too many solids may overload the balance
    b. Filter pores may become clogged
    c. Difficulty distinguishing between nonrepresentative particles
    d. A water-trapping crust may form

33. An operator filters 500 mL of final effluent for the TSS test. At the end of filtration, a large piece of algae is observed sitting on the filter. When the operator looks into the sample bottle, they see only a few small pieces of algae. The operator should
    a. Start the test over again with a more representative sample
    b. Carefully remove the algae with tweezers
    c. Nothing—continue with the procedure
    d. Dry the filters for longer than usual to desiccate the algae

34. If filter papers for the TSS test are not prewashed and dried before use, what may happen?
    a. Drying time will need to be increased
    b. Filters may curl up inside filtration apparatuses
    c. Results may be biased low
    d. Filter pores will not open properly and may clog

35. For the TSS test, how often should a duplicate be analyzed assuming two replicates are analyzed for every sample?
    a. One for every 10 samples or once per batch, whichever is more frequent
    b. Once a week when the instrument is calibrated
    c. Annually, as part of the discharge monitoring report quality assurance blind sample
    d. All samples should be analyzed in duplicate for permit reporting

36. The purpose of analyzing a blank is to
    a. Check the precision of the test
    b. Check the accuracy of the test
    c. Check the quality of reagents
    d. Check for contamination

37. The purpose of analyzing a laboratory duplicate is to
    a. Check the precision of the test
    b. Check the accuracy of the test
    c. Check the quality of reagents
    d. Check for contamination

38. The purpose of analyzing a certified standard is to
    a. Check the precision of the test
    b. Check the accuracy of the test
    c. Check the expiration date of reagents
    d. Check for contamination

39. The initial filter weight was 0.1234 g and the final filter weight was 0.1350 g. A sample volume of 50 mL was used. What was the total suspended solids concentration in the sample?
    a. 0.23 mg/L
    b. 23.2 mg/L
    c. 232 mg/L
    d. 2320 mg/L

40. The filter blank from a TSS test had an initial weight of 0.1230 g and a final weight of 0.1210 g. What is the most likely cause for the difference?
    a. Contamination from a dirty filtration apparatus
    b. Analyst did not adequately mix the sample
    c. Change in barometric pressure
    d. Filter was not prewashed before use

41. The primary difference between an OUR and a SOUR test is
    a. Oxygen uptake rate is conducted at 25 °C and SOUR is conducted at 22 °C
    b. Specific oxygen uptake rate corrects for solids concentration
    c. Oxygen uptake rate testing is used with anaerobic digesters
    d. Specific oxygen uptake rate is a shorter test

42. The SOUR test measures
    a. Blower and diffuser efficiency
    b. Dissolved oxygen probe fouling
    c. How fast microorganisms are consuming oxygen
    d. Effluent soluble BOD_5

43. An operator collects samples every 3 m (10 ft) down the length of an aerated activated sludge basin. As soon as each sample is collected, the operator conducts an OUR test on each. What is the operator trying to figure out?

a. Whether contents of the aeration basin are well mixed
b. The point where nitrification begins to take place
c. If short-circuiting is occurring through the basin
d. How much of the aeration basin is needed to treat the wastewater

44. Mixed liquor volatile suspended solids is used for OUR and SOUR testing rather than MLSS because

a. The MLSS analysis takes longer
b. The MLVSS test is more accurate
c. The MLVSS test measures live microorganisms
d. The MLSS test does not include inert solids

45. Yesterday, the SOUR in the aeration basin was 8 mg O_2/L·h. Between 12:00 and 2:00 p.m. today, the SOUR dropped to 1.5 mg O_2/L·h. What is the most likely cause?

a. Operator decreased blower output too low
b. Toxic or inhibitory substance entered the aeration basin
c. The SRT increased enough to support nitrification
d. Sudden increase in the total BOD_5 load

46. Typically, the SOUR for a particular WRRF is between 5 and 8 mg O_2/L·h. When the operator tested it at 3:00 p.m. today, the SOUR had increased to 12 mg O_2/L·h. What is the most likely cause for the increase?

a. Toxic or inhibitory substance entered the aeration basin
b. Solids retention time dropped below the minimum required to support nitrification
c. Increased BOD_5 loading to the aeration basin
d. Number of filamentous bacteria doubled

47. Typically, the BOD test is conducted over a 5-day period for what reason?

a. After 5 days, all organic material has been consumed
b. It only takes the Thames River 5 days to reach the ocean
c. It is convenient for the analyst to set up on Monday and take down on Friday
d. Test bottles run out of oxygen after 5 days

48. While the BOD test takes 5 days to complete, the COD test only takes
    a. 2 hours plus setup time
    b. 8 hours
    c. 12 hours
    d. 24 hours

49. Which of the following statements is true?
    a. Chemical oxygen demand analysis takes longer than BOD analysis
    b. Biochemical oxygen demand testing is more accurate than COD testing
    c. Biochemical oxygen demand test results include nonbiodegradable organic compounds
    d. Chemical oxygen demand must always be equal to or greater than BOD

50. Which of the following methods is most accurate for measuring MLSS?
    a. Centrifuge spins
    b. Handheld or in-basin TSS meter
    c. Gravimetric testing by filtration
    d. Calibrated eyeball or taste test

51. Which of the following is a potential consequence of not addressing elevated nitrate or nitrite levels in effluent?
    a. Washout of nitrifying organisms
    b. Insufficient recovery of alkalinity in mixed liquor
    c. Increasing pH in the treatment basin
    d. Deterioration of the nitrate/nitrite probe membrane

52. The ammonia probe at the end of the aeration basin is reading near zero. What is this an indication of?
    a. Dissolved oxygen has exceeded saturation levels
    b. The probe membrane has been damaged
    c. Nearly complete nitrification is occurring
    d. There has been a surge in influent ammonia

53. The ORP probe in the treatment basin is reading between +10 and +30 mv under normal operation. What is the most likely biochemical process occurring in the basin?

a. Anaerobic digestion
b. Nitrification
c. Phosphorylation
d. Denitrification

Answer key:

1. a, 2. b, 3. c, 4. a, 5. a, 6. d, 7. b, 8. c, 9. a, 10. b, 11. d, 12. a, 13. b, 14. c, 15. d, 16. b, 17. a, 18. c, 19. d, 20. c, 21. d, 22. d, 23. a, 24. c, 25. a, 26. b, 27. c, 28. b, 29. b, 30. a, 31. d, 32. d, 33. b, 34. c, 35. a, 36. d, 37. a, 38. b, 39. c, 40. d, 41. b, 42. c, 43. d, 44. c, 45. b, 46. c, 47. b, 48. a, 49. d, 50. c, 51. b, 52. c, 53. d

19.0 REFERENCES

American Public Health Association, American Water Works Association, & Water Environment Federation. (2022). *Standard methods for the examination of water and wastewater* (24th ed.). American Public Health Association.

Jenkins, D., Richard, M. G., & Daigger, G. T. (2003). *Manual on the causes and control of activated sludge bulking, foaming, and other solids separation problems* (3rd ed.). Lewis Publishers.

Standards for the use and disposal of sewage sludge, 40 CFR Part 503 (1993). https://www.epa.gov/biosolids/sewage-sludge-laws-and-regulations

U.S. Environmental Protection Agency. (2001). *Method 1683 specific oxygen uptake rate in biosolids*. EPA-821/R-01-014(.

U.S. Environmental Protection Agency. (2010). *Nutrient control design manual*. EPA-600/R-10-100. Office of Research and Development, National Risk Management Research Laboratory, Water Supply and Water Resources Division.

Wahlberg, E. J. (2001). *WERF/CRTC protocols for evaluating secondary clarifier performance*. Project 00-CTS-1. Water Environment Research Foundation.

Water Environment Federation. (2022). *Basic laboratory procedures for the operator-analyst* (6th ed.).

Wooley, J. F. (1981). *Oxygen uptake: Operational control tests for wastewater treatment facilities: Instructor's manual and student workbook*. Linn-Benton Community College. http://files.eric.ed.gov/fulltext/ED221396.pdf

7

Process Troubleshooting

1.0 INTRODUCTION

This chapter presents troubleshooting procedures for solving common operating problems. Operators are encouraged to exercise consistent process control using solids retention time (SRT) as the primary control mechanism, as described in previous chapters. Process control can be differentiated from troubleshooting in terms of where the attention is focused. Process control sets future goals and defines allowable tolerances and deviations from those goals. For example, a facility may set a process control goal of a 10-day SRT, with an allowable deviation of ±1.5 days. Process control is proactive. Troubleshooting identifies a current, undesirable condition and looks backward to find the cause. It is reactive. Having a good, consistent process control program in place should minimize situations from developing for which troubleshooting becomes necessary. Sudden changes in influent characteristics such as loading, flow, temperature, or the presence of inhibitory compounds can still cause process upsets even when a facility has an established process control program. Troubleshooting tables are included that list problems, observations, probable causes, necessary checks to determine the cause, and suggested remedies.

To effectively troubleshoot a problem, operators must first break the problem down into three significant categories: process problem, mechanical problem, or hydraulic problem. Once the correct category is identified, operators determine the probable cause of the problem and select one or more corrective measures to restore the process to full efficiency. Select the correct measure(s) with the least adverse effect on final effluent quality and at the lowest cost. To do this, a thorough knowledge of the facility's activated sludge process and the way it fits into overall water resource recovery facility (WRRF) operations is needed. When troubleshooting an activated sludge process, it is always recommended that one corrective action be implemented at a time and that the operator allow the process to assimilate. Typically, it requires 2 to 3 SRTs for the corrective action to be noticed. It is important as an operator that you do not implement too many corrective measures at once because it could make the problem worse. On the other hand, if the problem is resolved, then an operator will not know which corrective action worked. In addition, familiarity with influent wastewater characteristics, facility flowrates and patterns, design and actual loading parameters, performance of the overall facility and individual processes, and current maintenance procedures are also needed. Chapter 2, "Activated Sludge Process Fundamentals"; Chapter 4, "Process Control for the Activated Sludge Basin"; and Chapter 5, "Process Control Methods for the Secondary Clarifier," are recommended reading before using the troubleshooting guides. Many of these guides are based on the U.S. Environmental

Protection Agency's (U.S. EPA's) manual, *Process Control Manual for Aerobic Biological Wastewater Treatment Facilities* (1977).

When a problem or situation arises, it is important to first evaluate the problem and determine if it fits into one or a combination of the following areas:

- Process,
- Mechanical, and
- Hydraulic.

Hydraulic problems could be the result of several conditions. The first and foremost cause is a rainstorm. This is especially true if the WRRF is served from combined sewers. Another hydraulic problem that could occur in the activated sludge process is if, for example, there are not enough tanks in service because they are undergoing maintenance or repairs or there is an imbalance of flow to in-service process tanks. Smaller facilities experience this more than larger facilities that have multiple tanks. Typically, a hydraulic condition is easy to detect because a solids washout can be seen at the secondary clarifiers or if there is a difference in mixed liquid concentration in various aeration tanks.

Poor flow balancing can also result in hydraulic problems. It is common engineering practice to design redundant process units to be hydraulically equivalent, although exceptions do occur, particularly at retrofitted facilities with limited space available for expansion. Operators should strive to maintain equitable flow distribution to redundant, equally sized process units whenever possible. Typically, maintaining equitable flow distribution involves appropriately calibrating hydraulic control gates and weirs. Hydraulic imbalance issues can also occur if internal flows, such as return activated sludge (RAS) to the aeration tanks, are not evenly dispersed.

The second category is mechanical problems. A mechanical problem with a clarifier collector mechanism, RAS pump, low-pressure air blower, or mechanical aerator could result in an effluent violation if it is not noticed and corrected immediately. For example, if an RAS pump fails and this goes unnoticed, solids would not be removed from a secondary clarifier. This, in turn, could result in the clarifier filling up with solids until, eventually, solids flow over the weirs.

The third category the operator must be aware of is process problems. Typically, process problems are the most difficult problem to identify and correct. Typically, a facility that maintains good process control records and control charts can see warning signs that will alert the operator to an impending problem. A change in influent characteristics could result in a change in settling characteristics. In-facility sidestream changes, differences in dissolved oxygen concentrations, and other parameters are important aspects that an operator must know to effectively troubleshoot a process problem. Again, good facility records will enable an operator to determine and correct the problem. Operators should use their supervisory control and data acquisition (SCADA) system to produce trend charts so they can visually observe process conditions to determine if the process is functioning properly or if there is an issue. Also, an information management system can produce process dashboards on the facility SCADA system, which will allow facility staff to quickly see facility data and trends and process adjustments can be made efficiently. Process dashboards could show energy use, laboratory test results, process parameters, and other important data. Another excellent process troubleshooting mechanism is microscopic examination of activated sludge. A microscopic examination of a properly prepared and stained sample of mixed liquor may help operators determine the presence of, and identify the type of, filamentous organisms in the facility, which, in turn, will help determine the best course of action to correct a process problem.

Troubleshooting requires several skills, including investigative thinking and logical use of all tools available to help solve a problem. As mentioned previously, when troubleshooting, it is important to not perform numerous changes at once because that may result in further problems. It is also important to realize that the activated sludge system is a biological process and, as such, most responses will not be noticed immediately. It may take days or weeks before a process improvement can be seen.

2.0 KEYS TO TROUBLESHOOTING

As stated previously, the first step is to determine if the problem is related to hydraulics, mechanics, processes, or a combination of these factors. When a problem or situation is about to occur or if it has already surfaced, it is important to take logical and methodical steps to improve the process. The following seven areas are important to troubleshooting; they are critical for narrowing the search for solutions to most problems:

1. Thoroughly know the process being evaluated;
2. Know all facility flow patterns, including forward flow and sidestreams;
3. Know the facility's design parameters and how actual loadings compare to design values;
4. Know all maintenance procedures, including equipment maintenance considerations and staff responsibilities;

5. Know how to recognize an abnormal condition;
6. Know what alternatives are available when trouble develops; and
7. Know the amounts and characteristics of any industrial waste that may be discharged to the facility.

Because of stricter effluent requirements, more WRRFs are required to limit the amount of nutrients leaving the facility and entering receiving streams. Therefore, more WRRFs are being constructed or modified to operate as biological nutrient removal (BNR) facilities. These facilities may be designed to only reduce phosphorus, to nitrify, or to reduce phosphorus and nitrogen. Many variations of BNR processes are used to reduce nutrients; therefore, not every process can be discussed in this section with regard to troubleshooting needs. Additional BNR troubleshooting can be found in Chapter 8 of this manual. Instead, general activated sludge troubleshooting is provided in this chapter. Because BNR processes are primarily biological, some standard activated sludge troubleshooting guidelines can apply to BNR problems.

This section and the troubleshooting guides that follow include the following lists of solutions:

- Increasing or decreasing aeration,
- Balancing influent and return sludge flows,
- Adjusting return rates,
- Adjusting wasting rates,
- Adding chemical settling aids,
- Adding supplemental nutrients,
- Chlorinating return sludge or mixed liquor,
- Reducing or controlling recycle flows, and
- Rerouting recycled flows.

Some solutions produce quick results; some produce changes only after many days and some can be drastic, with potentially adverse side effects. Typically, the troubleshooting information presented here is oriented to solutions that minimize the potential for overcorrection and adverse side effects. The stronger the recommended solution, the more judgment and experience will be required in applying it.

In most cases, aeration adjustments, flow balancing, and RAS flow-rate adjustments will produce quick, mild results. Adjusting RAS rates is an immediate fix but, typically, adjusting RAS rates is only a temporary solution. For example, increasing the RAS flowrate to reduce the secondary clarifier sludge blanket is only a temporary solution because solids that were removed from the secondary clarifier will be quickly returned to the secondary clarifier. The unintended consequence is increased solids loading to the clarifiers, which can cause more serious problems if there are no standby clarifiers and/or if a wet weather event is imminent. Typically, the effect of altering sludge wasting takes longer and is less reversible, that is, it is more prone to overcorrecting. Chlorinating mixed liquor or return sludge produces quick results at moderate or greater application rates but poses significant risks if done improperly. It is considered by some a solution of last resort, while others consider it a standard solution if applied properly. The best approach when adding chlorine to the process stream is to apply the chlorine while maintaining a target sludge volume index (SVI) value. Typically, applying polymers or metal salts (such as alum or ferric chloride) is effective quickly but can be expensive and may increase sludge-wasting requirements. Also, adding chemicals could affect other processes such as solids-handling facilities. Correcting nutrient deficiencies can produce fast, positive results, but requires purchasing chemicals and chemical feed equipment. Determining what is causing nutrient deficiencies is also an important form of troubleshooting. For example, is it an industrial or commercial user discharging a waste that is causing the nutrient deficiency issue and, if so, then how can pretreatment measures help the situation? Reducing or controlling internal recycle sidestream flows produces quick, positive effects but may require changes in the operation of solids-treatment processes or the addition of offline storage or treatment facilities.

It is important to make only one corrective action at a time to avoid confusing the effects of multiple changes. As troubleshooting expertise is gained, it may be possible to apply solutions in a more direct manner. The beginner, however, should be cautious. Significant changes may result in overcorrecting, which can create additional problems just as severe as the initial problems they were intended to correct.

If operating problems cannot be solved in-house, help should be obtained from others such as operators of other facilities, state operators' associations, state and federal regulatory agencies, operations consultants, equipment manufacturers, and local colleges. Some problems are complex and difficult to solve because there may be several contributing factors. In these cases, only qualified individuals with firsthand knowledge and experience should be considered. Second guessing by well-intentioned people can waste time and money, divert attention from the correct solution, and possibly make matters worse.

Before making a change, collect as much data as possible and experiment with possible solutions in the laboratory. If a mistake is made, the results can be easily discarded and another solution tried. If a significant mistake is made within the facility, it is not as easily corrected.

When attempting to solve a problem, avoid treating only the symptoms. For example, chlorinating return sludge for a filamentous bulking problem will temporarily solve the problem, but the underlying cause (such as a nutrient deficiency or too low dissolved oxygen) must be corrected or the problem will return. However, in some cases, the underlying cause cannot be readily corrected because of design deficiencies or budgetary constraints and it may be necessary to settle for treating the symptom (e.g., chlorinating return sludge) until the underlying problem can be permanently solved.

3.0 AERATION SYSTEMS FOR BIOLOGICAL REACTORS

Mixed liquor must be aerated so that aerobic microorganisms will get enough oxygen to remain active and healthy. In addition, biological reactor contents must be mixed to bring microorganisms into contact with all of the organic matter in the wastewater being treated.

Typically, mixing in the biological reactor can be checked by observing turbulence on the reactor's surface. Surface turbulence should be reasonably uniform throughout the reactor; violent turbulence should not be present because it wastes energy and shears floc.

3.1 Diffused Air Systems

Typically, dead spots and nonuniform mixing patterns indicate a clogged diffuser or that diffuser header valves need adjustment to balance air distribution in the reactor. Typically, violent turbulence indicates a broken pipe or diffuser. Illustrations of both uniform and violent turbulence in a biological reactor are shown in Figure 7.1. For biological reactors with air diffusers, a dissolved oxygen profile should be made every 1 to 6 months or whenever the flow pattern is changed. Based on conventional design principles, most air distribution systems are designed to maintain a dissolved oxygen concentration of no less than 0.5 mg/L throughout the biological reactor. Setpoints for dissolved oxygen concentrations in aerated bioreactors will depend on treatment goals and organic loading rate. In conventional BNR processes, typically, conversion of ammonia to nitrate is achieved by operating with higher dissolved oxygen concentrations in the 1.5- to 3.0-mg/L range to maximum nitrification rates. However, several facilities have demonstrated that the biology responsible for nutrient removal can be acclimated to achieve highly efficient nutrient removal at dissolved oxygen concentrations at 0.7 mg/L or less.

(a)

(b)

FIGURE 7.1 Biological Reactor Surface Turbulence: (a) Violent Turbulence and (b) Good Turbulence

Some probable causes of nonuniform aeration include the following:

- Air rates too high or low for proper operation of the diffuser.
- Valves needing adjustment or balancing of air distribution.
- Diffusers needing cleaning or repair. A large quantity of air rising over the location of the diffuser. Blowoff legs, if so equipped, indicate that diffusers need cleaning or replacing or that there could be a line break. A plugged diffuser or broken membrane diffuser will result in uneven air distribution in the aeration tank. A broken membrane will cause large bulking bubbles at the tank surface rather than fine evenly distributed bubbles on the tank surface. Cleaning of fine-bubble diffusers is more prevalent than the cleaning of coarse-bubble diffusers.
- Mechanical equipment limitations.
- Diffusers not at the same elevation.

The following measures can be tried, when applicable, to correct aeration problems:

- Verify dissolved oxygen probe cleanliness and ensure that probes have been cleaned recently. (Troubleshooting low dissolved oxygen levels without verifying probe cleanliness may lead to misinterpretation of data if readings are not accurate.)
- Confirm that all valves are in the proper position by following process lines and ensuring that all water purge valves on air diffuser grids are closed. If an operator forgets to shut a purge valve, air loss will occur, affecting dissolved oxygen levels;
- Adjust diffuser header valves to balance air distribution and eliminate dead spots.
- Evaluate mixed liquor levels. If the WRRF has been decreasing sludge wasting to obtain a higher SRT (e.g., to obtain nitrification in winter months), check that mixed liquor concentrations are not too high, which will deplete dissolved oxygen levels in the tank. The facility may need to bring an additional aeration tank on line to lower the mixed liquor concentration so that target dissolved oxygen setpoints can be reached.
- Adjust airflow rate to maintain dissolved oxygen in the desired concentration range. Typically, mixing requirements of a biological reactor are 0.5 to 0.67 L/m^3·s (30 to 40 scfm/1000 cu ft). Typically, fine-bubble diffusers require 0.36 to 0.61/m^2·s (0.07 to 0.12 scfm/sq ft) of

reactor surface. However, oxygen-transfer requirements must also be met in bioreactors for either coarse- or fine-bubble diffuser systems.

- Check blower preventive maintenance requirements, service record, design capacity, and present output, if possible. The blower may need repair. Check blower intake filters.
- Verify blower header pressure. If the pressure is too low, it should be adjusted to appropriate levels to maintain optimal aeration. If a most-open valve control strategy is used, then the control system will make automatic adjustments to the header pressure to compensate for pressure loss.
- Regularly check and clean diffusers. Typically, diffusers should be cleaned every 6 months to 1 year to maintain good aeration performance.
- Inspect air diffuser grids for any accumulated water that needs to be purged as water in the diffuser grids can impair air distribution, affecting dissolved oxygen levels.
- If fine-bubble diffusers are extremely troublesome, investigate various air-cleaning devices, the airflow rate to each diffuser, and the type of clogging—internal because of dirty air, external calcium hardness buildup on the diffusers, or external biological fouling. Testing has shown that most fouling is external, and acid gas systems for ceramic diffusers have been developed to correct this problem. Because fine-bubble diffusers are more energy-efficient than coarse-bubble diffusers, municipalities currently favor them over coarse-bubble diffusers to save energy costs. There are different types of fine-bubble diffusers, such as ceramic and membrane, that should be investigated to determine which type of diffuser is best for the facility. Only if all else fails, replacing them with a coarse-bubble type of diffuser should be considered. Outside help should then be obtained to evaluate the necessary quantity of air to satisfy carbonaceous and nitrogenous removal.

3.2 Mechanical Aeration Systems

As with diffused air systems, inadequate mixing or aeration problems can occur with mechanical aeration systems. Mechanical aeration systems can either be surface splash-type aerators or submerged mixers with or without draft tubes, as shown in Figure 7.2. Typically, oxidation ditches use horizontal brush or disc-type aerators and some oxidation ditch manufacturers use single-splash-type mechanical aerators. Additionally, mechanical aerators can be subjected to surging and impeller fouling problems.

(a)

(b)

FIGURE 7.2 Mechanical Aerators: (a) Surface Aerator and (b) Mechanical Aerator With Draft Tube (Reprinted with permission from Kenneth Schnaars)

3.2.1 Inadequate Mixing and Aeration

A dissolved oxygen profile of the biological reactor can be made to check the adequacy of mixing and aeration. The following are some probable causes of inadequate mixing and aeration:

- Aerator speed is too slow,
- Surface aerator impeller submergence is inadequate,
- Impeller is fouled with rags or ice,
- Aerator is undersized, and
- Mechanical aerator is flooding on submerged mechanical aerator with draft tubes and air sparges.

The following measures can be taken, when applicable, to correct for inadequate mixing or aeration:

- Increase the aerator speed if the aerator has a two-speed motor that has been operating at slow speed; or, if the aerator is connected to a variable-frequency drive unit, increase the output to increase the speed of the aerator motor.
- Raise the tank weir setting, which will raise the tank liquid level and submerge the surface aerator blade. Increasing the blade depth in the water will result in reaching the maximum allowable impeller submergence and increase oxygen transfer to the water (do not exceed maximum allowable submergence because it could overload the motor).
- Remove rags that have been caught on the blades (if rags are a problem, check the headworks for screening efficiency).
- Remove ice (typically, this is a problem with surface aerators).
- Consider replacing the aerator with a larger or better designed aerator.
- For a submerged mechanical mixer with an air sparger, reduce airflow rate or turn the aerator on high speed.
- Install a submersible mixer in the basin to improve basin mixing.

3.2.2 Hydraulic Surging and Flooding

Mechanical aerators can be subjected to hydraulic surging. Surging occurs when impeller submergence is below the manufacturer's recommendation and a wave pattern is established in the biological reactor that causes the impeller to be alternately exposed to critical over- and under-submergence. The aerator motor will be intermittently overloaded and may shut down.

Surging is more likely to occur during low-flow conditions such as when an entirely new facility or additional biological reactor is put on line.

The following measures can be tried for surface aerators, when applicable, to correct a surging problem:

- If surging occurs during no-flow conditions, check the biological reactor effluent weir for leaks, especially if it is adjustable;
- Raise or lower the biological reactor effluent weir and/or lower the impeller to reach proper submergence (do not exceed maximum allowable submergence);
- Reduce the number of biological reactors on line, if possible, to increase flow and, therefore, submergence in reactors;
- Investigate the use of baffles, inlet draft tubes, and flow-straightening vanes to eliminate wave action; and
- IAs a last resort, increase the return rate.

Mechanical mixers with air sparges and draft tubes use a downward flow pattern. If that flow pattern reverses and travels upward, the condition is known as *flooding*. Flooding occurs when the flow of the column of water that travels downward in the aerator draft tube reverses and starts flowing upward. Typically, this condition occurs if the buoyancy of entrained air is greater than the aerator can force downward. Also, this can occur if the mechanical aerator is at slow speed and the airflow through the sparger is at a high rate. When an aerator is operating in a flooded condition, it reduces oxygen-transfer efficiency and could damage the aerator. A flooded condition should be corrected as soon as possible. Care must be taken when making the correction, however, because equipment damage could occur if the column of water is forced to reverse direction too quickly. Typically, the equipment manufacturer recommends that, during a severe case of flooding, the aerator be stopped and the air turned off. After several minutes, when the flow of the water through the draft tube stops, the aerator can be restarted and a correct flow pattern can be established. The air can then be restarted and regulated to the correct flowrate. Sometimes, if the flooding condition is not severe, operators could either reduce the airflow rate or increase the mixer speed without stopping the air or mixer. Operator experience will determine if adjustments can be made without a complete shutdown of the air and/or mixer.

3.2.3 Impeller Fouling

Mechanical surface aerators can be subjected to impeller fouling problems caused by rags and ice. Rags and ice interfere with aeration and mixing, may damage the gear box, and may cause a motor overload. Rags are most likely to accumulate in facilities without primary clarifiers. Icing on surface aerators may result from heat losses in upstream tanks (such as aerated grit chambers and primary clarifiers), quantity of water dispersed (spray) into the atmosphere, impeller and shroud design, and climate.

Rags and ice must be removed manually. Ice must be completely removed or it may cause a significant imbalance. It may be controlled by installing metal shrouds above the aerators to keep spray in the reactor.

3.3 Jet Aeration Systems

Jet aeration systems combine air and mechanical pumping. The jet aeration pump discharges water and air through nozzles located along the bottom of the aeration tank (biological reactor). The pumping system, typically a submersible pump, circulates mixed liquor through diffusers (nozzles) that are arranged either in a circular pattern or a straight header pattern, as shown in Figure 7.3. Air is combined with mixed liquor and dispersed through nozzles (diffusers). The air is provided either by a blower or it can be aspirated into jet aeration. Typically, the water recirculation pump is a constant-speed pump and the aerator capacity is adjusted by modulating airflow rate.

FIGURE 7.3 Jet Aeration System (Reprinted with permission from Fluidyne Corp.)

4.0 SLUDGE QUALITY PROBLEMS

4.1 Biological Reactor Foaming

The presence of some foam (or froth) on the biological reactor is normal for the activated sludge process. Typically, in a well-operated process, a small percentage of the reactor surface will be covered with a layer of light-tan foam in a loose pattern that moves and changes with aeration. A typical biological reactor is shown in Figure 7.4.

Under certain operating conditions, foam can become excessive and affect operations. Four general types of problem foam are often seen:

- Stiff, white foam;
- Greasy, dark-tan foam;
- Thick, scummy, dark-brown foam; and
- Very dark or black foam.

If foam is allowed to build up excessively—especially white foam—it can be blown by the wind onto walkways and facility structures and create

FIGURE 7.4 Well-Operated Activated Sludge System (Photo by Indigo Water Group at Arapahoe County Water and Wastewater Authority)

hazardous working conditions. It can also create an unsightly appearance, cause odors, and carry pathogenic microorganisms. If a greasy or thick, scummy foam builds up and is carried over with flow to secondary clarifiers, it can build up behind influent baffles and create additional cleaning problems. In addition, it can plug the scum-removal system.

4.1.1 Stiff, White Foam

Stiff, white, billowing foam can result from several conditions, but most of all, white foam problems result from low solids in the aeration tank. White billowing foam could indicate a young sludge (low SRT), typically found in new or recently started facilities, as shown in Figure 7.5.

When SRT is low, mixed liquor suspended solids (MLSS) concentration may also be low (low activated sludge solids inventory) and the food-to-microorganism ratio (F:M) may be higher than the range defined for conventional activated sludge, which is 0.2 to 0.5 kg of 5-day biochemical oxygen demand (BOD_5)/kg (0.2 to 0.5 lb BOD_5/lb) of mixed liquor volatile suspended solids (MLVSS). Foam may consist of undegraded detergents or proteins that could not be broken down at that SRT. Typically, this type of foam goes away once the startup window has passed and SRT exceeds 3 to 4 days. A sudden increase in organic loading that creates conditions similar to startup conditions may also result in white foam. Foaming may

FIGURE 7.5 Stiff, White Foam (Reprinted with permission from Kenneth Schnaars)

also be caused by some types of poorly biodegradable household or industrial detergents; however, modern detergents have eliminated this issue for the most part (Jenkins et al., 2003). In addition, white, billowing foam may appear if there was an industrial spill that was toxic to microorganisms. This would result in poor settling and the washing out of solids from the system. Another reason for white, billowing foam could be the overwasting or the reduction of return sludge entering the aeration tanks.

Some probable causes of stiff, white, billowing foam include the following:

- Low SRT resulting from process startup.
- Low MLSS for current organic loading caused by excessive sludge wasting.
- Low SRT resulting from unintentional loss of activated sludge biomass caused by
  - Excessive or shock hydraulic loads;
  - High sludge blanket in the secondary clarifier, resulting in solids washout;
  - Activated sludge not being returned to the biological reactor;
  - Poor sludge settleability;
  - Mechanical deficiencies in the secondary clarifier, such as leaking seals, open drain valves, or collector mechanism failure (check to see if the torque alarm switches have tripped and/or if the shear pin is broken); and
  - Improper distribution of flows or solids loadings to multiple clarifiers.
- Improper distribution of wastewater and/or RAS flows to multiple biological reactors.
- Industrial loading that is toxic to the biomass.

The following measures can be tried, when applicable, to correct the foaming problem:

- Verify that return sludge is flowing to the biological reactor. Maintain sufficient return rates to keep the sludge blanket in the lower quarter of the clarifier, preferably between 0.3 and 0.9 m (1 and 3 ft) from the bottom.
- Increase SRT by decreasing or stopping sludge wasting for a few days (MLSS will increase and F:M will decrease concurrently).

- Control airflow rates or mechanical aerator submergence to maintain dissolved oxygen concentrations of 1.5 to 3 mg/L in the biological reactor.
- Consider seeding the process with activated sludge from a well-operating facility.
- Actively enforce sewer use ordinances to avoid process upsets and deterioration of facility effluent by controlling the type of industrial wastewater entering the WRRF.
- Modify piping or structures as necessary to maintain proper distribution (proportional to reactor volumes) of flows to multiple biological reactors and secondary clarifiers. The construction of some type of flow-distribution structure may be required.
- If flowmeters are not provided, visually check return sludge flow and compare sludge blanket levels in each clarifier, the return sludge suspended solids concentration from each clarifier, and the MLSS concentration in each biological reactor. Corresponding measurements should be nearly equal if the wastewater, biological reactor effluent, and return flows are being properly distributed and MLSS and RAS from different treatment trains are mixed.
- If the secondary clarifier mechanism failed or if the gasket around the center column has failed, then these will need to repaired.
- If the biological reactors have water sprays for froth control, operate the water sprays when there is danger of foam being blown onto walkways or other structures. White, billowing foam works better with a fine spray.
- If there are no water sprays, consider using defoaming agents or improvising a froth-control system using effluent water.

4.1.2 Excessive Brown Foam

Excessive brown foam is associated with facilities operating at longer SRTs and/or low organic loading ranges. Facilities designed to nitrify and that are operating in nitrifying mode may have low-to-moderate amounts of rich, chocolate-brown foam caused by *Microthrix parvicella* (Figure 7.6), especially during the winter months when longer sludge ages are necessary to maintain nitrification. Facilities with the filamentous organism *Nocardia* (Figure 7.7) will have a strong, greasy, dark-tan foam that may carry over onto the clarifier surface. Foam containing filamentous organisms should be wasted from the system and pumped to solids-handling facilities rather than be returned to biological reactors. However, pumping foam containing an abundance of filamentous organisms can result in excessive foaming

FIGURE 7.6 *Microthrix parvicella* Foam (Reprinted with permission from Kenneth Schnaars)

FIGURE 7.7 *Nocardia* Foam (Reprinted with permission from Kenneth Schnaars)

issues in solids-handling facilities, such as aerobic and anaerobic digesters. If possible, some facilities use a vacuum truck to remove foam and deliver it to a landfill for disposal; however, it would be better to determine what is causing the problem and correct it rather than using the vacuum truck. Successful studies were conducted to help WRRF operators control foam in their facilities by establishing a wasting station referred to as a *classifying selector*. Classifier selectors are used to control the population of foam-causing organisms in activated sludge facilities to prevent the development of nuisance foams (Parker et al., 2001). Throughout the years, classifying selectors were refined and designed into full-scale WRRFs. It is recommended that facility personnel read an informative paper about classifying selectors entitled *A Critical Review of Nuisance Foam Formation and Biological Methods for Foam Management or Elimination in Nutrient Removal Facilities* (Parker et al., 2014).

Thick, scummy, dark-brown foam indicates an old sludge (long SRT) and can result in additional problems in the clarifier by building up behind influent baffles and creating a scum disposal problem, as shown in Figure 7.8.

The following are some probable causes of this type of foaming problem:

- Biological reactor being operated at an excessively long SRT.
- Buildup of high MLSS concentration as a result of insufficient sludge wasting. This condition could unintentionally occur when seasonal

FIGURE 7.8 Secondary Clarifier Scum Issues

(winter to summer) wastewater temperature change results in greater microorganism activity and, consequently, greater sludge production.

- Operating in sludge reaeration mode. A heavy, dark-brown, greasy foam is normal in the sludge reaeration tank, particularly if the process is being operated at a low F:M.
- Improper wasting control program.
- Degasification because of deep tank aeration.

The following measures can be tried, when applicable, to correct foaming problems:

- Gradually increase the wasting rate to decrease the SRT
  - Facilities with effluent ammonia concentration limits will be limited in how far SRT may be safely reduced. Typically, an aerobic SRT of 4 to 6 days is needed to maintain full nitrification when the water temperature in the reactor is 20 °C (68 °F) and 10 to 15 days when the water temperature is 10 °C (50 °F).
  - Facilities with effluent ammonia concentration limits should monitor effluent ammonia concentrations while decreasing SRT. If effluent ammonia begins to increase, SRT must be increased. If the dissolved oxygen concentration is below 2 mg/L, increasing the dissolved oxygen concentration can increase nitrification efficiency, which may allow for a lower SRT.
- If the foam is not returned to the biological reactors, include solids removed in the foam in waste sludge calculations. During normal operation, the amount of solids removed with foam is too small to matter. However, during heavy foaming, as much as 10% of waste activated sludge (WAS) solids may be removed with foam. Note that if foaming is caused by the presence of filamentous bacteria, then complete removal of foam is advised. Placing foam in an aerobic or anaerobic digester or other process could result in operational issues in those processes. Sometimes, this foam may be recycled back to the bioreactors (aeration tanks), but that situation must be avoided.
- If filaments appear, try to identify the type and cause of the filament (refer to Section 4.2.1 Filamentous Microorganisms Present for additional information).
- Implement a better program for controlling wasting of activated sludge.
- If foam contains filaments, remove the foam from the surface of the water and send it to solids-disposal facilities. Ensure that the foam is not recycled back to the WRRF.

4.1.3 Very Dark or Black Foam

The presence of a very dark or black foam indicates either insufficient aeration, which results in anaerobic conditions, or that industrial wastes such as dyes and inks are present. The following measures can be tried, when applicable, to correct this foaming problem:

- Increase the aeration rate in the aeration tanks and
- Investigate the source of industrial wastes to determine if appearance is resulting from dyes or inks.

4.2 Bulking Sludge

The presence of clear supernatant above poorly settling sludge (see Figure 7.9) indicates that settling is being hindered by the presence of filamentous microorganisms. The presence of filamentous microorganisms is corrected by improving the treatment environment by adjusting SRT; correcting dissolved oxygen concentration in the biological reactor; correcting pH; eliminating excessive organic acids; or adding nutrients such as nitrogen, phosphorus, and/or iron. The presence of a cloudy effluent above a poorly

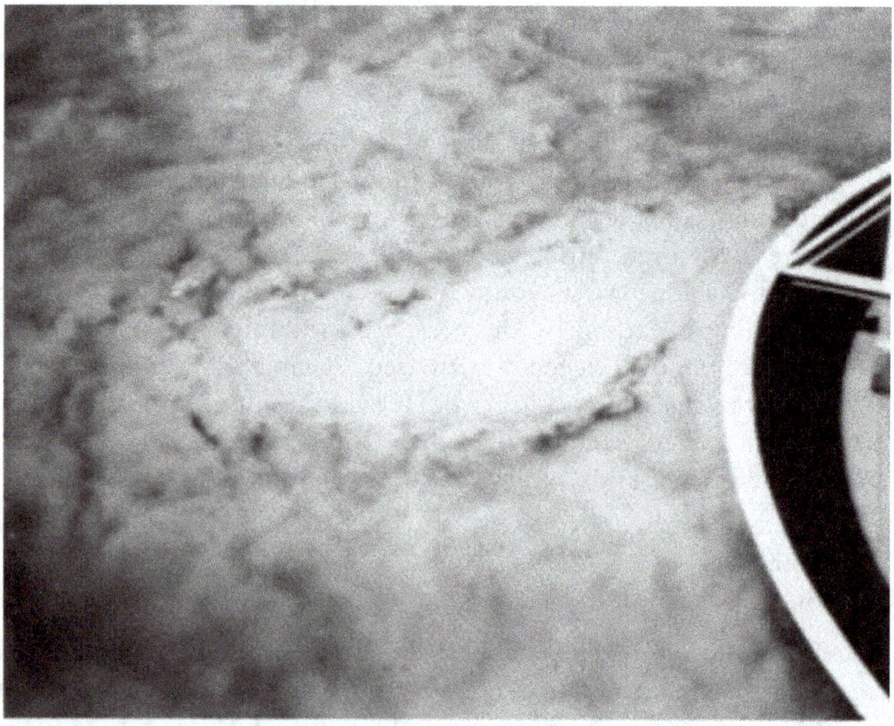

FIGURE 7.9 Sludge Bulking

settling sludge indicates dispersed-growth bulking and either improper organic loading, overaeration, or toxics.

The following are some probable causes of sludge bulking:

- Filamentous microorganisms are present (filamentous bulking):
    - Excessively long SRT.
    - Low-dissolved oxygen concentrations in biological reactors.
    - Insufficient nutrients.
    - Improper pH—either too low or widely varying.
    - Warm wastewater temperature.
    - Widely varying organic loading.
    - Industrial wastes with high BOD and low nutrients (nitrogen and phosphorus), such as simple sugars or carbohydrates (e.g., whey waste).
    - High influent sulfide concentrations that cause the filamentous microorganism, *Thiothrix*, to grow and produce filamentous bulking. (Note that if high sulfides are present in the wastewater, nodules in the *Thiothrix* filament will glow after the sample is stained. If nodules in the *Thiothrix* filament are not glowing after staining, it could be an indication of nutrient deficiency.)
    - Large amounts of filaments present in influent wastewater or recycle streams.
    - Insufficient soluble BOD_5 gradient (as found in completely mixed biological reactors).
- Few or no filamentous microorganisms are present (dispersed-growth bulking):
    - Excessively high or low SRT,
    - Improper organic loading—either too high or too low F:M,
    - Overaeration,
    - Nutrient deficiency, and
    - Toxic material.

The first step in analyzing this condition is to examine mixed liquor under a microscope.

4.2.1 Filamentous Microorganisms Present

If a moderate to large number of filamentous microorganisms are present, the following effective actions should be performed:

- Identify the type of filament. If the facility staff does not have the equipment or expertise, outside help should be used. Proper identification requires Gram and/or Neisser staining and a good microscope. Use filament identification charts to determine which filament or filaments are present and the most likely cause for their proliferation.
- Many filaments proliferate at higher SRTs. Verify current operating conditions and confirm that SRT is within the desired range. Optimal SRTs depend on treatment goals and the water temperature within the aeration basin. For a facility that is required to remove ammonia or summer operation when the water in the aeration basin is 20 °C (68 °F) and nitrifiers are more active, the aerobic SRT target may be 4 to 6 days. For the same facility, winter operation when water in the aeration basin is 10 °C (50 °F) and nitrifiers are less active, the aerobic SRT target must be longer to maintain nitrification (in the 10- to 15-day range). Longer SRTs encourage the proliferation of some types of filaments, such as *M. parvicella*. Lowering the SRT may help minimize *M. parvicella* issues; however, lowering SRT can affect the nitrification process, so selecting the correct troubleshooting alternative is important to solving the problem without creating a different problem. Another effect that has worked to reduce *M. parvicella* is the addition of polyaluminum chloride (PAX) to the system. It should be noted that not all forms of PAX will work, but some facilities have had success using PAX-14. Using a chemical to solve a problem is expensive. It is recommended that process changes, such as adjusting SRT, be the first troubleshooting options an operator uses to try to correct the issue.
- Measure the dissolved oxygen concentration at various locations throughout the biological reactor. Many WRRFs have multiple zones within biological reactors or separate reactors to provide anaerobic, anoxic, and aerated conditions. This section applies to aerated zones and basins only. For sequencing batch reactors, this section applies to the aerated phase of the cycle.
  - If the typical dissolved oxygen concentration throughout the reactor is less than 0.5 mg/L, there may be insufficient dissolved oxygen in the biological reactor. Increase aeration until dissolved oxygen increases to 1.5 to 3 mg/L throughout the tank. Some facilities do operate at dissolved oxygen concentrations of 0.5 mg/L at the effluent end of the bioreactor after satisfying nitrification requirements.
  - If dissolved oxygen concentrations are nearly zero in some parts of the reactor but are higher in other locations, the air distribution system on a diffused air system may be off balance or diffusers in an area of the reactor may need to be cleaned. Balance the air

system and clean diffusers. If a mechanically aerated system is used, increase aerator speed, raise overflow weirs, or increase the air if the mechanical aerator is also provided with an air sparger.

- If dissolved oxygen concentration is low only at the head of the tank, which is being operated in the plug flow pattern, consider changing to step-feed or complete-mix flow patterns or use tapered aeration, if possible. However, a low dissolved oxygen concentration only at the head of the reactor may not be a problem as long as adequate reaction time is available in the rest of the bioreactor.

- Measure the pH within the biological reactors. If pH in the biological reactor is less than 6.5 or is widely varying, mixed liquor settleability may be affected because bacteria that settle readily are inhibited. Chronic low pH operation can also encourage the growth of fungi in MLSS, which can affect settleability. Reactor pH may be affected by influent wastewater characteristics and biological processes taking place within the reactor.
  - If raw wastewater pH is less than 6.5 and the water supply pH is higher, the low pH problem may be a result of industrial waste, and a survey should be conducted to identify the industry discharging low pH water. The best procedure, if possible, is to prevent the problem by controlling it at the source. A good industrial waste monitoring and enforcement program will avoid many difficulties in this area.
  - Nitrification, the conversion of ammonia to nitrate, destroys alkalinity and can reduce biological reactor pH. The nitrification process can be inhibited at levels below 7.0 pH and could cease if the pH goes below 6.5. If this is the cause of the problem, raise the MLSS pH by adding magnesium hydroxide ($Mg(OH)_2$), sodium bicarbonate ($NaHCO_3$), caustic soda (NaOH), or lime ($Ca(OH)_2$) at the head of the biological reactor. If nitrification is required, the MLSS pH should be kept higher than 7.0 and held as constant as possible to encourage an acceptable rate of nitrification. Be sure that the system is not shocked by high pH levels or overdosed to pH levels greater than 9.
  - When determining the amount of $Mg(OH)_2$, $NaHCO_3$, NaOH, or $Ca(OH)_2$ needed for pH adjustment, obtain and weigh (or measure) a small amount (1 to 2 g) of the actual chemical to be used in the treatment process. While stirring a measured sample, add small increments of the chemical until the pH is approximately 7.2. Then, weigh (or measure) the portion not used to determine the amount used in the titration.

  - Adding $Mg(OH)_2$, $NaHCO_3$, NaOH, or $Ca(OH)_2$ to raise pH is expensive; therefore, chemical use must be carefully monitored so that excessive chemical is not fed into the process. Mixed liquor settleability should be closely monitored to observe changes to ensure that the chemical added is effective. If no improvement occurs within 2 to 4 weeks and nothing else in the process has changed in the meantime, chemical addition should be stopped.
- A uniform, low concentration of soluble BOD, as found in completely mixed biological reactors, is considered by some to encourage the growth of some types of filamentous microorganisms that compete with floc formers to cause low F:M bulking. One technique suggested to control F:M filaments is the use of a nonaerated (anoxic or anaerobic selector) mixing zone for wastewater and RAS. In this mixing zone, a high-growth-rate environment is provided to promote the growth of floc-forming organisms at the expense of filamentous organisms.
- If nutrient limitations are suspected, calculate the ratios of BOD_5 to nitrogen (use total Kjeldahl nitrogen [TKN] expressed as N), BOD_5 to phosphorus (expressed as P), and BOD_5 to iron (expressed as Fe). Typically, a proper nutrient balance in an activated sludge system is 100:5:1 (BOD:N:P). Anhydrous ammonia is used to add nitrogen, trisodium phosphate is used to add phosphorus, and ferric chloride is used to add iron. The ratio of BOD_5 to required nutrients changes with SRT. For example, high mean cell residence times (MCRTs) produce less sludge and result in lower nitrogen and phosphorus requirements. Typically, 3 to 5 mg/L of nitrogen (3 mg/L at high SRTs and 5 mg/L at low SRTs) and 1 mg/L of phosphorus are needed for every 100 mg/L of BOD_5.
- Typically, a nutrient-deficient waste will not occur with domestic wastewater. However, if a facility receives industrial discharge, such as from a canning company, a nutrient-deficient waste could occur.

If there are insufficient nutrients in wastewater, three things can occur:

1. Filamentous bacteria will predominate or take over MLSS;
2. Organic materials will only be partially converted to end products; that is, there will be inefficient BOD_5 or chemical oxygen demand (COD) removal; and
3. Sludge bulking may occur.

If excess nutrients are added to wastewater to overcompensate for a nutrient deficiency, a large fraction of these nutrients may not be incorporated to MLSS and may, therefore, pass into the effluent. Nutrients, if required, should be added at the head of the biological reactor. Mixed liquor settleability should be carefully observed to see if it is improving. If settleability improves, the nutrient dose can be reduced by 5% per week until settleability decreases. Then, the dose should be increased by 5% and the settleability should be observed.

Nutrients are expensive and should be carefully applied. For example, adding excess ammonia may create a nitrification demand. Nutrient dose should be increased as BOD concentrations increase, which takes into account the effects of additional microorganism growth that will occur. If settleability does not readily improve, nutrient dosing should be continued until the actual problem is identified and solved because the problems that are causing poor settleability may be interrelated.

4.2.2 Filamentous Foaming

Three filamentous organisms can cause activated sludge foaming: *Nocardia* sp. (common), *M. parvicella* (common), and Type 1863 (rare) (Jenkins et al., 2003; Richard, 1989). These three types of filamentous organisms can be distinguished from one another through microscopic examination and staining. *Nocardia* foaming seems to be the most common and severe. Little is known yet about foaming by the other filaments (Richard, 1989).

Filamentous organisms can produce a stable, viscous, brown foam on the biological reactor surface that can shift over onto the clarifier surface and may even escape in the effluent, violating permit limits. Foaming can range from being a nuisance to being a serious problem. In cold weather, foam may freeze and need to be removed manually with a pick and shovel. In warm weather, it often becomes odorous. Because several species of *Nocardia* and *Mycobacterium* have been isolated from activated sludge foams that can be opportunistic human pathogens, activated sludge foaming should not be taken lightly because of possible harmful health effects. Samples of the foam and mixed liquor should be microscopically examined to determine if the foaming is, in fact, attributable to filamentous growths (Jenkins et al., 2003; Richard, 1989).

4.2.2.1 Nocardia *Foaming*

The analysis of *Nocardia* foaming involves two factors:

- The conditions that allow *Nocardia* to grow in activated sludge and
- The conditions that cause foaming (Jenkins et al., 2003; Richard, 1989).

Typically, *Nocardia* growth is associated with warmer temperatures; fats, oil, and grease present in treated wastes; and longer SRTs (typically more than 9 days), although it has been frequently encountered at SRTs of 2 days. Foam, however, has a longer SRT than underlying MLSS and calculations of retention time may be incorrect (Richard, 1989). Washout SRT experiments at the Sacramento Regional Wastewater Treatment Plant saw *Nocardia* populations decrease to nondetectable levels when SRT was 2.2 days at 16 °C (Cha et al., 1992).

Nocardia foaming seems to involve the hydrophobic (water-repellent), waxy nature of the *Nocardia* cell wall, which causes flotation under aeration (Richard, 1989). *Nocardia* cells in mixed liquor concentrate in foam and continue to float even after they die.

Facilities prone to *Nocardia* foaming often receive oil and grease wastes (e.g., from restaurants without grease traps), have poor or no primary scum removal, recycle scum rather than remove it from the facility, and have biological reactor configurations such as submerged wall cutouts or effluent gates that trap foam.

The best way to deal with *Nocardia* foaming is to prevent conditions that encourage *Nocardia* growth. Once established, *Nocardia* foaming can be extremely difficult to eliminate because of the following:

- Foam is difficult to break with water sprays, especially with fan-type sprays.
- Typically, foam does not respond to chemical antifoaming agents.
- Chlorinating RAS and/or surface foam, although often helpful, does not eliminate *Nocardia* because most of it is in floc and not exposed to chlorine.
- Increased wasting has its limitations because
    - Foam is not wasted with WAS
    - Even if foam and scum are removed from the process, they can cause problems in downstream units such as digesters and can be recycled with decant or supernatant to the activated sludge process.
    - Reducing the SRT to fewer than 9 days (the classic cure for *Nocardia* growth and foaming) may be inadequate. Many facilities must reduce SRT to fewer than 2 days, which may conflict with other process goals such as nitrification or sludge-handling capabilities. Numerous *Nocardia* species are involved in foaming. Of the two most dominant, one is slow growing and the other is fast growing. The SRT needed to control *Nocardia* foaming may depend on which species is involved (Jenkins et al., 2003; Richard, 1989).

- If *Nocardia* is pumped to aerobic or anaerobic digesters, they could cause foaming in the digesters.

One reactive approach to dealing with *Nocardia* is to provide surface-wasting facilities to remove foam and scum from the surface of biological reactors or a convenient channel. Then, foam and scum should be taken to a dissolved air flotation (DAF) thickening process (with WAS) and removed from the system. Another approach is to provide a fine contained spray of a highly concentrated chlorine solution (0.5% to 1.0%) directly to the surface of the biological reactor where foam is located.

A positive way to treat severe *Nocardia* foaming is to correct the cause (control greases, fats, and SRT) while also physically removing (e.g., with a vacuum truck) biological reactor foam and clarifier scum. Sometimes, a cationic polymer can be used to correct a foam problem (Shao et al., 1997). Once removed, this material must be separated so that it will not recycle through the facility.

4.2.2.2 Microthrix parvicella *Foaming*

Microthrix parvicella is a long, thin, irregular coiled filament that does not have branching and does not have sheathing. *Microthrix parvicella* commonly occurs at facilities with tank configurations that involve different zones, such as anaerobic and aerobic zones. Recent studies have found *M. parvicella* grows in a range of 7.1 to 8.5 pH, and the optimum temperature for cell growth is 25 °C (77 °F) (Rossetti et al., 2005).

The following are some causes and controls for *M. parvicella*:

- Will result in foaming and bulking;
- Can occur at high SRTs;
- Can occur at low dissolved oxygen conditions;
- Can occur at low F:M;
- Presence when operating with anoxic, anaerobic, and intermittently aerated zones—in some cases, continuous aeration can eliminate the filament;
- Control *M. parvicella* by reducing SRT; and
- Control *M. parvicella* by adding chlorine or PAX-14.

4.2.3 *Controlling Filamentous Bulking*

A variety of influent and facility conditions can encourage growth of filamentous organisms that cause filamentous bulking. Table 7.1 is a general guide that relates bulking by specific filamentous organisms to various facility conditions.

TABLE 7.1 Filament Types as Indicators of Conditions Causing Activated Sludge Bulking (Jenkins et al., 2003)

Suggested causative condition	Indicated filament types
Low dissolved oxygen concentration (for applied organic loading)	*Sphaerotilus. natans*, Type 1701, and H. *hydrossis*
Low organic loading rate (F:M) in completely mixed biological reactors	*M. parvicella, Nocardia* spp., *Haliscomenobacter hydrossis*, and Types 021N, 0041, 0675, 0092, 0581, 0961, and 0803
Septic wastes and sulfides	*Thiothrix* spp., *Beggiatoa* spp., and Type 021N
Nutrient deficiency: nitrogen or phosphorus (industrial and mixed industrial–domestic wastes)	*Thiothrix* spp. and Types 021N, 0041, and 0675
Low pH (less than 6.5 pH)	Fungi

A practical approach to control filamentous bulking includes the following:

- Confirming that the problem is indeed caused by filaments,
- Identifying the filaments involved, and
- Determining the specific set of appropriate remedies (both short- and long-term) for the filaments involved, as follows:
    - Short-term solutions involve treating symptoms by changing SRT, changing influent feed points, changing RAS rates, adding settling aids, and chlorinating or treating with other chemicals such as PAX (note that PAX will not work on all filaments) and
    - Long-term solutions involve treating the cause by controlling mixed liquor pH, controlling influent septicity, adding nutrients, changing aeration rates, and changing SRT.

Chlorination neither satisfactorily controls poor settling because of viscous activated sludge caused by nutrient deficiency nor controls dispersed-growth bulking. Chlorination exposes activated sludge to chlorine levels that damage filaments extended from the floc surface while leaving organisms within the floc largely untouched. Filamentous and floc-forming bacteria do not seem to significantly differ in their chlorine susceptibility (Richard, 1989).

One approach is to set a target SVI (or other measure of settleability) for satisfactory operation of secondary clarifiers and solids-processing units and chlorinate only when the target is significantly and consistently exceeded (Jenkins et al., 2003). However, this approach treats the symptom, not the underlying problem.

Location of the chlorine application point is critical. The point should be located where there is excellent mixing, the sludge is concentrated, and wastewater concentration is at a minimum (to reduce unwanted reactions with chlorine) (Richard, 1989). The three common application points are in the RAS stream, directly in the biological reactor at each aerator, and in an installed sidestream that recirculates mixed liquor within the biological reactor.

Chlorine dose and the frequency at which organisms are exposed to chlorine are the two most important parameters. The dose is adjusted so that concentrations are lethal at the floc surface but not within the floc. The chlorine dose should be based on solids inventory in the process (biological reactors plus clarifiers), which is called the *overall chlorine mass dose*. Effective dosages are in the range of 1 to 10 g/kg·d MLVSS (1 to 10 lb chlorine/day/1000 lb MLVSS). The dose must be accurately measured, preferably by using a chlorinator dedicated to this purpose. Initially, start at a lower dose and gradually increase the dose until it is effective (Richard, 1989).

Frequency of exposure is a measure of how often the entire solids inventory is subjected to an excessively high chlorine dosage. Required frequency of exposure depends on relative growth rates of filamentous and floc-forming organisms and the effectiveness of each dose. While the frequency of exposure is facility-specific, three times or more per day were reported as sufficient (Jenkins et al., 2003), and success has been reported at frequencies as low as once per day (Richard, 1989). The actual dose applied at each exposure should simply be the total calculated dose per day divided by the number of exposure events.

Control tests should be performed during chlorination to assess the effects of chlorine on both filamentous and floc-forming organisms. Tests should measure settleability (e.g., SVI), effluent quality (turbidity), and activated sludge quality (microscopic examination) (Jenkins et al., 2003). An adequate chlorine dose should start to improve settleability within 1 to 3 days (Richard, 1989). A turbid, milky effluent and a reduction in BOD removal are signs of overchlorination, although a small increase in effluent suspended solids and BOD concentrations are normal during chlorination for bulking control. Microscopically visible effects of chlorine on filaments include the following (in correct order):

- Intracellular sulfur granules (if present) disappear,
- Cells deform and cytoplasm shrinks, and

- Filaments break up and dissolve.

Chlorine does not destroy sheathed filaments, which cause poor sludge settling until they are wasted from the system. Chlorination should be stopped when only empty sheaths remain and should not be continued until SVI falls (Richard, 1989). Adding chlorine beyond this point may overchlorinate. Because the effects of chlorine on filaments can be detected microscopically before settleability improves, microscopic examinations can provide an early indication of filament control.

4.2.4 Filamentous Microorganisms Not Present

If few or no filamentous microorganisms are present, check the F:M and/or SRT to determine if the system is operating at a higher or lower than normal F:M or SRT. The presence of small, dispersed floc is characteristic of an increased F:M by lowering the SRT. If the F:M is higher than normal by 10% or more, the wasting rate should be decreased. The decrease in F:M should be reflected by the disappearance of dispersed floc over a period of 2 to 3 MCRTs.

The amount of turbulence and dissolved oxygen in the biological reactor is also important. In most facilities, dissolved oxygen concentrations greater than 4.0 mg/L indicate that excess air is being used and the aeration rate should be reduced to lower dissolved oxygen concentrations to the 1.5- to 3-mg/L range. If the facility uses fine-bubble diffusers or pure oxygen, higher dissolved oxygen concentrations can be safely maintained. Excessive turbulence (i.e., overaeration) in the biological reactor may hinder MLSS floc formation and result in pinpoint floc being carried over with clarifier effluent. Toxics such as industrial wastes may also cause dispersed-growth buildup.

5.0 SECONDARY CLARIFIERS

Operational problems with secondary clarifiers can be grouped into the following categories: sludge quality, equipment malfunction, hydraulic problems, solids overload, biological activity within the clarifier, and temperature and density currents.

5.1 Sludge Quality

Clarifier performance should be compared to settleometer performance and is further described in Chapter 6, "Process Control Tests." If clarifier effluent and settleometer supernatant are of comparable quality for total suspended solids (TSS) and/or turbidity, then the clarifier is performing as well as it can with the sludge quality that is being produced in the biological reactor. Pinpoint floc, straggler floc, and filamentous bulking are fundamentally sludge quality problems that must be solved by making adjustments in biological reactors to improve sludge settleability. The 30-minute settling test provides useful information that operators must consider when running this test. Operators should not just fill the graduated cylinder and record the reading after 30 minutes. This test is a useful troubleshooting tool for determining how the activated sludge process is operating because this test represents some of the conditions that are occurring in the secondary clarifier. When running this test, operators must observe the following:

- Is the supernatant (water above the sludge) clear or cloudy?
- How well is the sludge settling?
- Are floc particles well-formed or are they small and granular or nonexistent?
- Does a large portion of the sludge rise to the surface in a short period of time?
- Is there ashing or a greasy layer on the surface of the water?
- Is there a large amount of pin floc in the water?
- Are there any unusual smells such as gasoline or other substances?
- Is there any color in the water?

If an issue exists, then determine what course of action must be taken to troubleshoot and correct the problem. For example, a cloudy supernatant may indicate that the facility received a toxic material, the process is organically overloaded, or there is another issue.

5.1.1 Pinpoint Floc

The appearance of small, dense, pinpoint floc particles suspended in the clarifier is a common problem seen in WRRFs operating near the lower end of the loading range (between conventional and extended aeration). Typically, this problem is related to a sludge that settles rapidly but lacks good flocculation characteristics.

Some probable causes of pinpoint floc include the following:

- Operating at excessively long SRTs;
- The process is being operated at an F:M near or in the extended aeration range, resulting in old sludge with poor floc formation characteristics; and

- Excessive turbulence (overaeration) in the biological reactor shears floc formations.

The following measures can be taken, when applicable, to correct the problem:

- If sludge-settling characteristics observed during the mixed liquor settleability test indicate a sludge settling too rapidly with poor floc formation, clarifier effluent quality can be improved by gradually lowering the SRT by increasing the WAS rate. If nitrification is required, caution must be used to avoid decreasing nitrification by wasting too much sludge.
- If the settleability test indicates good settling with clear supernatant above the settled sludge, check for proper aeration and mixing in the biological reactor. If the average dissolved oxygen concentration in the reactor is more than 4 mg/L, consider reducing the aeration rate until the dissolved oxygen concentration is between 1.5 and 3 mg/L.
- If there are facilities for feeding settling aids such as alum, ferric chloride, or polymer, determine the required dose and start feeding a settling aid as a temporary measure while correcting the source of the problem.

5.1.2 Straggler Floc

The appearance of small, almost transparent, very light, fluffy, buoyant sludge particles rising to the clarifier surface near effluent weirs and discharging over the weirs is a problem often seen when MLSS concentration is too low. When this type of straggler floc occurs while the secondary effluent is otherwise exceptionally clear and, particularly if it prevails even during relatively low surface overflow rates, typically, the problem is related to a low MCRT (high F:M). At some facilities, floc is particularly noticeable during early morning hours. In many facilities, this problem is present but has no significant effect on effluent quality. Straggler floc is not a concern unless it causes an effluent quality problem.

The following are some probable causes of this problem:

- The biological reactor is being operated at an MLSS concentration that is too low. (Typically, this would occur during process startup until the proper MLSS concentration is reached.) A WAS rate that is too high will result in low SRT, low MLSS, and a high F:M.
- Sludge is being wasted on a batch basis during the early morning hours, resulting in a shortage of microorganisms to handle daytime organic loading.

The following measures can be taken, when applicable, to correct the problem:

- Decrease the WAS rate to increase SRT, which also means that MLSS concentration will also increase.
- If wasting sludge on a batch (or intermittent) basis, avoid wasting when the BOD load is increasing. Also, increase the RAS rate as peak flows start. All of the organisms are needed at this time to handle the daily increase in organic loading. Decrease the RAS rate at night.

5.1.3 Denitrification

When clumps or rafts of sludge appear on the surface of a secondary clarifier, this indicates a condition known as *denitrification.* This condition occurs when there is a shortage of oxygen for microorganisms in the secondary clarifier. Once free oxygen runs out, microorganisms search for other forms of oxygen; that oxygen is found in nitrogen forms. Microorganisms use oxygen from nitrate-nitrogen and produce nitrogen gas. Nitrogen gas bubbles gather in the sludge blanket, and these bubbles attach to sludge and cause clumps of sludge to rise to the surface of the clarifier.

The following measures can be taken to correct this problem:

- Increase the sludge withdrawal rate from the clarifier to get microorganisms back to the biological reactor.
- Check dissolved oxygen leaving the biological reactor and in the clarifier. If dissolved oxygen is low leaving the biological reactor, then increase the airflow.

5.2 Equipment Malfunction

Inspection of all equipment specific to the clarification process is important to ensure proper operation. Equipment malfunctions can cause solids washout, short-circuiting, poor sludge collection, poor sludge removal, and so on. This could lead to solids flowing over secondary clarifier weirs. Solids flowing over weirs can result in effluent violations. Specifically, check the following:

- Flowmeters
    - Clarifier influent meter or effluent meter, if provided. If there is an influent or effluent meter that is not reading properly, then that clarifier could receive an excessive amount of flow that could result in a solids washout from that clarifier.

  - Return sludge flow meter. If the return sludge meter is not reading properly, then the proper quantity of solids will not be removed for the secondary clarifier.
  - If return sludge flow is being flow-paced, verifying accuracy of the influent flowmeter is important.
- Return activated sludge pumps and lines
  - Verify that pump(s) are operational.
  - Are the pumps removing the correct amount of sludge?;
  - Are there any mechanical issues with the pumps, such as unusual noises, excessive leaking from the seals, and so on?
- Sludge- and scum-collection equipment
  - Look for broken drives or support members.
  - Look to see if the collector mechanism is operating smoothly and not hanging up on anything.
  - Look for uneven blanket depth at several locations in the clarifier. An uneven blanket may indicate a plugged suction collector or broken plough/scraper.
  - When the clarifier is empty, check the seal around the center column to ensure that it is intact.
  - Check microswitches and shear pin.
  - Look to see if the scum trough is clear and the flushing system is operating properly.
- Baffles and skirts
  - Look for broken welds, bolts, or supports,
  - Look for holes in the center well skirt wall,
  - Look for missing skirt panels on the center well, and
  - Look for missing or damaged scum baffles.
- Weirs
  - Look for unbalanced flow over the weirs and
  - If weirs on one side of the clarifier are more deeply covered than those on the other side, they should be adjusted to an equal elevation.

5.3 Hydraulic Issues

A secondary clarifier can be hydraulically overloaded because of two different situations, and it is important that, when operators are troubleshooting these problems, they recognize each situation. The first hydraulic overload situation would be if the facility is subjected to excessive and infiltration and inflow or combined sewer flows. During this situation, all secondary clarifiers could be washing out, which is a facility (process) problem, so operators must focus on flows entering the facility. The second hydraulic overload situation would occur when only one secondary clarifier is washing out solids. For this situation, operators must focus on what is occurring for that one clarifier. In this situation, the influent gate to that clarifier could be opened too much, causing a flow imbalance to the secondary clarifiers.

Check the hydraulic loading on each clarifier by either measuring the flow to each clarifier or by estimating the flow balance among multiple units, as indicated by the depth of flow over the weirs in each clarifier. Differences in depth of flow, however, will be hard to measure unless differences in flow are excessive. If headloss to all clarifiers is the same, clarifier overflow weirs should all be set at the same elevation. If not, weirs should be adjusted to equalize headloss to them. Overloading can result from excessive flow or unevenly distributed flow among multiple units.

The problem should be approached as follows:

- Determine if flows are being distributed equally to biological reactors and secondary clarifiers. Weirs at the effluent end of the reactors must be adjusted to equally distribute flow to each secondary clarifier if these two processes are paired together. Weirs, valves, or gates on the RAS distribution lines must be adjusted to properly distribute RAS flows. Weirs, weir gates, or gates at clarifier distribution structures must be adjusted to uniformly distribute forward flow into each secondary clarifier.
- Inspect clarifier effluent weirs to make sure that flow is moving over them equally.
- After checking and correcting weir elevations where needed, determine clarifier surface overflow rate at both average and peak flows. Compare actual calculated surface overflow rate with the design value. If actual overflow rate exceeds the design value, clarifiers are hydraulically overloaded and additional clarifiers may need to be placed into operation, if available. Also, perform an SVI test to check settling characteristics. If results are satisfactory and all clarifiers are operating, facility expansion or flow equalization should be considered.

To illustrate by way of an example, the clarifier surface overflow rate at average and peak flows can be calculated as follows:

Given

$$\text{Surface area, each clarifier} = 729\ \text{m}^2\ (7850\ \text{sq ft}),$$
$$\text{Total surface area, two clarifiers} = 1458\ \text{m}^2\ (15\,700\ \text{sq ft}),$$

Facility flow, average = 28 400 m^3/d (7.5 mgd or 7 500 000 gpd), and

Facility flow, peak 2 hour = 56 800 m^3/d (15.0 mgd or 15 000 000 gpd).

The solution in International Standard units is as follows:

$$\text{Surface overflow rate} = \frac{\text{Facility flow}}{\text{Clarifier surface area}}$$

$$\text{Average surface overflow rate} = \frac{28\ 400\ \text{m}^3/\text{d}}{1458\ \text{m}^2}$$

$$= 19.48\ \text{m}^3/\text{m}^2{\cdot}\text{d}$$

$$\text{Peak surface overflow rate} = \frac{56\ 800\ \text{m}^3/\text{d}}{1458\ \text{m}^2}$$

$$= 38.9\ \text{m}^3/\text{m}^2{\cdot}\text{d}$$

The solution in U.S. customary units is as follows:

$$\text{Average surface overflow rate} = \frac{7\ 500\ 000\ \text{gpd}}{15\ 700\ \text{sq ft}}$$

$$= 478\ \text{gpd/sq ft}$$

$$\text{Peak surface overflow rate} = \frac{15\ 000\ 000\ \text{gpd}}{15\ 700\ \text{sq ft}}$$

$$= 955\ \text{gpd/sq ft}$$

5.4 Solids Overload

Clarifier solids overloading is when solids enter the secondary clarifier faster than they can be settled. Solids overloading is related to facility flow, return flow, and MLSS concentration. It becomes apparent when sludge settleometer readings are near normal, but clarifier blanket levels are high, return rates are high, and high effluent solids are observed (or anticipated).

Adjusting the return rate should quickly verify whether a solids-overloading problem exists. If the return rate is increased and the blanket drops, the problem is not solids overloading and increasing the return rate has helped. If an increased return rate raises the blanket and keeps it up, the clarifier is probably overloaded with solids and higher return rates will only aggravate the problem. The return rate can then be reduced by 20%, but only if the blanket level, effluent and MLSS concentrations, and mixed liquor settleability are closely watched. The sludge blanket in clarifiers should be measured after 30 minutes and again after 60 minutes to check the effect. It should then be checked three times per shift (about once every 3 hours) to ensure that the blanket does not build up to an excessive depth. This may only work for a short time because storing solids in the clarifier is not effective with high blankets. To reduce the sludge blanket in secondary clarifiers, the wasting rate must be increased.

For example, the solids loading rate can be determined as follows:

Given

$$\begin{aligned}\text{Surface area, each clarifier} &= 729\ \text{m}^2\ (7850\ \text{sq ft}),\\ \text{Total surface area, two clarifiers} &= 1458\ \text{m}^2\ (15\ 700\ \text{sq ft}),\\ \text{Facility flow, } Q &= 28\ 400\ \text{m}^3/\text{d}\ (7.5\ \text{mgd}),\\ \text{Return flowrate, } R &= 14\ 400\ \text{m}^3/\text{d}\ (3.8\ \text{mgd}), \text{ and}\\ \text{MLSS} &= 2500\ \text{mg/L}.\end{aligned}$$

The solution in International Standard units is as follows:

$$\text{Solids loading rate} = \frac{(Q+R)(\text{MLSS})(\text{Conversion factor})}{(24\ \text{hours/day})(\text{Clarifier surface area})}$$

$$\text{Solids loading rate, kg/m}^2\cdot\text{h} = \frac{\left(28\ 400\dfrac{\text{m}^3}{\text{d}} + 14\ 400\dfrac{\text{m}^3}{\text{d}}\right)\left(2500\dfrac{\text{mg}}{\text{L}}\right)(0.001)}{\left(\dfrac{24\ \text{h}}{\text{d}}\right)(1458\ \text{m}^2)}$$

$$\text{Solids loading rate} = 3.1\ \text{kg MLSS/m}^2\cdot\text{h}$$

The solution in U.S. customary units is as follows:

$$\text{Solids loading rate} = \frac{(Q+R)(\text{MLSS})(\text{Conversion factor})}{(24\ \text{hours/day})(\text{Clarifier surface area})}$$

$$\text{Solids loading rate, lb/sq ft/hr} = \frac{(7.5\ \text{mgd} + 3.8\ \text{mgd})\left(2500\dfrac{\text{mg}}{\text{L}}\right)(8.34)}{\left(\dfrac{24\ \text{hr}}{\text{d}}\right)(15\ 700\ \text{sq ft})}$$

$$\text{Solids loading rate} = 0.63\ \text{lb/hr/sq ft}\ (15.12\ \text{lb/d/sq ft})$$

For peak flow conditions, solids loading rates of approximately 7.08 to 9.28 kg/m^2·h (1.45 to 1.90 lb/hr/sq ft or 35 to 46 lb/d/sq ft) are practical upper limits for nitrification and normal secondary clarifier operations, respectively. For average flow conditions, a solids loading rate of 6.10 kg/m^2·h (1.25 lb/hr/sq ft or 30 lb/d/sq ft) is a practical upper limit for normal secondary clarifier operation. However, it is important to know each facility's design value because a number of secondary clarifiers are only designed for maximum solids loading rates of 1.45 lb/hr/sq ft (35 lb/d/sq ft).

Also, an interstitial volume dilution test reportedly can determine the clarifier safety factor (CSF) and the minimum return rate required. If the actual return rate is more than the calculated minimum and the CSF is more than 1.5, the facility may not be overloaded (Wilson, 1983).

If calculated values and observations suggest an overload, the following approaches can be tried:

- Place another clarifier online, if possible.
- Place another biological reactor online, if possible, to reduce MLSS concentrations.
- Reduce MLSS concentration by increasing wasting.
- Change to sludge reaeration or contact stabilization, if possible.
- If necessary, temporarily add a settling aid such as polymer or alum until MLSS is reduced. This will not help if the problem is strictly the result of solids overload, but it may help if poor settling is a factor.
- Try to implement a step-feed operating strategy, if possible.

Using additional clarifiers increases the clarifier surface area, reducing the solids loading rate. This is a low-cost option and, typically, is easy to do. Bringing another biological reactor on line makes it possible to reduce the MLSS concentration without changing the F:M. The extra volume provided by additional biological reactors makes it possible to have the same volatile solids inventory with a lower MLSS concentration, which effectively reduces the solids loading rate. Placing another biological reactor on line may increase power costs; therefore, other corrective actions should be considered first.

If previous solutions do not improve the situation, MLSS concentration should be decreased by significantly increasing the wasting rate. Be aware that the F:M is increasing and SRT is decreasing during this procedure because solids inventory will decrease. If this is the last resort to reduce solids overload on the clarifier, reduce the MLSS concentration 20% in 1 day, reduce an additional 15% the second day, and hold it at that level. If no improvement occurs during any of these adjustments, high effluent suspended solids concentrations are most likely not the result of solids overloading. Try to improve sludge settleability by changing process conditions through wasting or environmental changes. Adding polymer, alum, or iron salts will help if poor settleability is a factor by shortening the response time for improvements.

Temperature currents occasionally occur in large, deep clarifiers in colder climates, most often in the spring and fall. A temperature profile of the clarifier will identify the presence of temperature currents. The temperature probe on a dissolved oxygen meter is an excellent tool for this procedure. To make the profile, measure and record the temperature at the head, one-quarter, one-half, three-quarters, and tail end of a rectangular or square clarifier, or at quarter points across a circular clarifier. At each point, measure the temperature at the surface and quarter points down to the reactor bottom. Be careful that the temperature probe and wires do not get entangled with sludge-collection equipment.

If deeper temperatures are consistently cooler by 1 to 2 °C (2 to 4 °F) or more, this is an indication that temperature currents are present. If temperature differences are caused by surface warming from the sun, reducing the number of clarifiers on line or rigging temporary covers over clarifiers may help. Installing baffles may improve settling by breaking up currents and stopping turbulence.

Temperature also affects the settleability of activated sludge. Colder temperatures are commonly associated with poorer sludge settleability and vice versa for warmer temperatures. This relationship has been attributed to temperature-induced changes in the density of water; the density of water increases with decreasing temperature over temperature ranges typical for WRRFs.

5.4.1 Clumping/Rising Sludge

Occasionally, sludge that has otherwise good settling characteristics may rise to the clarifier surface (Figure 7.10). This can occur even if settling aids such as polymer or alum are being fed as a coagulant. Floating clumps of solids and fine, rising bubbles will be observed. The cause of this phenomenon is denitrification, which requires nitrification to occur first (unless industrial wastes are contributing nitrates).

For significant nitrification to take place, the mixed liquor temperature must be high and/or SRT must be long. Denitrification occurs in settled sludge when there is little to no oxygen and in the presence of nitrate. Nitrogen gas is released and becomes trapped in the sludge mass. If enough gas is produced, portions of the sludge mass become buoyant and rise to the surface, typically in dark clumps, where they eventually flow over the weir.

FIGURE 7.10 Rising/Clumping Sludge (Denitrification) (Reprinted with permission from Kenneth Schnaars)

For denitrification to take place, not only must the activated sludge process be nitrifying (unless industrial wastes are contributing nitrates), but the following three conditions must also be present within the sludge blanket:

1. Low dissolved oxygen concentration (less than 0.5 mg/L). Removing the sludge too slowly from the clarifier may drive the dissolved oxygen down to this concentration.
2. High nitrate concentration (typically more than 5 mg/L).
3. Residual BOD_5 (typically more than 10 mg/L).

A long SRT in the clarifier can result in denitrification. Thick sludge blankets (caused by low RAS rate) and low mixed liquor dissolved oxygen concentrations are contributing factors. Rise time in the mixed liquor settleability test is a key parameter to watch during warm weather operations. When rise times drop below 2 hours, be alert for rising sludge problems.

High respiration rates on samples of mixed liquor taken at the clarifier inlet may also indicate that conditions are sufficient for denitrification in clarifiers. The following are some probable causes of sludge clumping:

- The process is being operated at a low F:M or high SRT and, consequently, the process has slipped partly or completely into the nitrification zone.
- Sludge is being held too long in the clarifier and, consequently, all available dissolved oxygen has been used by microorganisms.
- There is a higher than normal wastewater temperature, resulting in a higher rate of microorganism activity that causes the process to nitrify at a higher F:M. A higher rate of microorganism activity will also more quickly deplete dissolved oxygen in settled sludge and, consequently, pose a greater potential for denitrification.

The following measures can be tried, when applicable, to correct the sludge clumping problem:

- Increase the RAS rate to reduce the detention time of sludge in the clarifiers. A periodic measurement of the clarifier sludge blanket depth will help determine the proper return rate.
- If a suction-type sludge collector is used, check that all suction tubes are flowing freely with a fairly consistent suspended solids concentration. Some of the suction tubes may be improperly adjusted or plugged, resulting in coning in some areas and sludge blanket buildup in other areas.
- Where possible, increase the sludge collector speed.
- Reduce the number of clarifiers on line to shorten SRT.
- If nitrification is not required, gradually increase the WAS rate to shorten SRT and increase the F:M to stop nitrification. Initially, the solids inventory should be decreased by 20% to 30% over 1 week, and then process operation must be observed to determine if nitrite and nitrate levels are dropping. If, after approximately 2 weeks, they have not dropped to below 5 mg/L, reduce the inventory by another 20% to 30%. As a guide, calculate the F:M or SRT and compare it to published values for nitrification. In very warm waters, it may be necessary to drop the SRT to approximately 2 days (F:M greater than 0.6) and MLVSS to less than 1000 mg/L.
- In severe cases, and if nitrification is not required, chlorinate RAS at low rates to stop nitrification.
- When nitrification is required, gradually decrease the WAS rate to increase SRT and lower the F:M to ensure that nitrification is complete and soluble BOD concentrations are low.

5.4.2 Cloudy Secondary Effluent

During periods of high effluent suspended solids concentrations, the mixed liquor settleability test should be run immediately and followed up with additional tests several times a day until the problem is identified and corrected. When mixed liquor in the settleability test settles poorly, leaving a cloudy supernatant, the next step is to examine mixed liquor and RAS under a microscope. One of the important purposes of this examination is to determine the presence of protozoa and the status of their health.

When protozoa are present but appear to be inactive, it frequently indicates that a slug of toxic material, such as heavy metals, has recently entered the treatment system. Also, determining the type of protozoa and the number of each type are important in troubleshooting the problem.

Check the respiration rate. If it is low, this is a further indication of a toxic sludge. Reduce the WAS rate but, otherwise, maintain normal operation until the material passes through the treatment system, if it has not already. Then, try to identify whether toxics were present in sludge by determining the presence of, for example, metals. Whenever evidence of toxics is found, thoroughly follow up to identify the pollutant and remove its source.

If protozoa appear normal and active (i.e., healthy), but the effluent is cloudy, the activated sludge floc may be dispersed as a result of excessive turbulence (i.e., overaeration) in the biological reactor or excessive floc shearing at the clarifier inlet. Also, there could be too many flagellates or *Amoeba*, which indicate a young sludge that may not settle well and leave a cloudy effluent.

If few or no protozoa are present, there are two possibilities. First, the F:M could be too high and the system may be overloaded. In this case, the following measures can be taken, when applicable, to correct the problem:

- Calculate the F:M and compare it to F:Ms for periods of satisfactory operation. If the current F:M is greater than these values, reduce the WAS rate to increase the solids inventory.
- Increase the RAS flow to lower the sludge blanket level in the clarifier to the minimum. Increased RAS flow will increase solids inventory by transferring MLSS stored in the clarifier to the biological reactor.
- Consider letting the blanket build up. Sometimes, this can produce a blanket-filtering effect, although this can be a complicated way to operate.

Second, the F:M may be lower than or in the normal range. This condition is frequently associated with one of the following:

- Low dissolved oxygen concentration in the biological reactor—if the average dissolved oxygen concentration measured at several locations is less than 0.5 mg/L, aeration should be increased until the dissolved oxygen concentration is between 1.5 and 3 mg/L.
- A toxic waste in the treatment system—toxic waste adversely affects the health of activated sludge. If metals are present in mixed liquor, consider increasing the WAS rate for approximately 1 week to purge the system. A short-term solution would then be to add large quantities of healthy seed sludge from another facility to build up the volatile solids inventory. The long-term solution requires conducting an industrial waste survey to identify the source of the toxic material and strictly enforcing the sewer use ordinance.

Table 7.2 shows U.S. EPA findings for inhibition of various pollutants in the activated sludge process. There are many cases in which concentrations shown in the table overlap; that is, the concentration for "no effect" may be more than the concentration for "inhibitory." This occurs because of the number of variables leading to a wide range of concentrations for observed effects. Key variables not accounted for in this table include treatment process mode (i.e., high-rate, complete-mix), pH, temperature, synergism, acclimation, and full-scale or pure culture at bench scale. Research reported in the literature indicates that activated sludge systems may become acclimated to heavy metal levels considerably higher than those shown.

Except for the data in Table 7.2, information on some organic chemicals is limited. Information on shock load sensitivity is also limited but can be addressed in general qualitative terms. If other factors are equal, the inhibitory effect should be a function of concentration in relation to the active biomass (expressed as weight of pollutant per weight of biomass). If this assumption is used, the effect of a given influent pollutant concentration (expressed as milligrams per liter) can be evaluated for each treatment mode. For plug flow with zero intermixing, the shock load effect would be most intense. Systems using higher MLSS concentrations would be affected less because the influent pollutant would be spread over more biomass. Similarly, the complete-mix system is considered more resistant to shock loads.

5.4.3 Ashing

The appearance of small, ashlike particles floating on the surface of the secondary clarifier (Figure 7.11) is commonly referred to as *ashing*. The ash may be particles of dead cells as well as normal sludge particles and grease. Some probable causes of ashing are denitrification beginning or occurring

TABLE 7.2 Concentrations of Inorganic Pollutants Affecting Activated Sludge Processes

Pollutant	Concentration, mg/L		
	No Effect (Less Than)	Inhibitory	Upset
Ammonia	100.0	—	480.0
Arsenic	0.1	0.1	0.1
Borate	0.1	1.0	0.1
Cadmium	—	1.0–100	50.0
Chromium (+6)	—	1.0–10	—
Chromium (+3)	—	50.0	—
Copper	0.2	0.5*	1.0
Cyanide	0.1	0.3–100	200.0
Iron	—	1.0–1000	1000.0
Lead	0.1	0.1	—
Manganese	10.0	20.0	60.0
Mercury	—	0.1	200.0
Nickel	—	1.0	20.0
Silver	—	9.0	30.0
Sulfide	—	20.0	—
Zinc	0.3	0.5	5.0

*The nitrification process would be inhibited at 0.05 mg/L.

in the clarifier, extremely low F:M and long SRTs beyond normal extended aeration (less than 0.05), and unusually high grease content in the mixed liquor.

The ashing problem should be approached and solved as follows. First, stir the solids that float in the mixed liquor settleability test. Next, proceed as follows:

- If floating solids release bubbles and then settle, this indicates that denitrification has begun. See Section 5.4.1 Clumping/Rising Sludge earlier in this chapter for a solution.

FIGURE 7.11 Ashing (Reprinted with permission from Kenneth Schnaars)

- If stirred floating solids do not settle, there may be an overoxidized mixed liquor that has many higher microorganisms, few lower microorganisms, and many dead cells. This type of mixed liquor may settle quickly, but it will not flocculate well and leaves many particles behind. Some of the particles of dead cells that do not settle will accumulate on the water surface, forming an ash. If ashing is significant enough to affect effluent quality by increasing effluent suspended solids concentrations, WAS rates can be increased by not more than 10% per day to increase the F:M and lower SRT to optimum values.
- If floating solids do not settle, there may also be excessive amounts of grease in the sludge. Perform a grease analysis. If the grease content exceeds 15% by weight of MLSS, the problem may be one of the following:
    - The primary tank scum baffles are malfunctioning because of hydraulic overloading or mechanical failure. Specific attention should be given to scum baffles and the scum-collection system in the facility.

- Too much grease is being put into the sewer by an industrial or commercial discharger. If too much grease is in the raw wastewater, an industrial waste survey must be conducted to identify the discharger and have the problem corrected.

5.4.4 Floating Sludge Caused by Deep Tank Aeration

Facilities that have space restrictions have had to construct deep biological reactors (more than 6-m [20-ft] deep or deeper) to meet the necessary aeration volume to treat wastewater. A number of these deep reactors have resulted in problems, causing MLSS to become supersaturated with gases (oxygen and nitrogen) and floating MLSS in the secondary clarifiers (see Figure 7.12). When mixed liquor leaves biological reactors and enters secondary clarifiers, gases are released, causing sludge to float to the surface (see Figure 7.13), similar to a DAF thickener. Studies conducted in the United States and Japan found that floating sludge caused by mechanical aeration systems occurred both in pilot tests and full-scale operations (Fujii & Fukuda, 1974). Floating sludge from deep biological reactors also occurs at facilities using diffused air systems such as New York City's North River Facility (Lu et al., 1990).

If a WRRF is experiencing floating sludge on its secondary clarifiers because of deep biological reactor systems and not filamentous organisms, there are several alternatives for correcting the problem. The key is to strip out gases from mixed liquor before they reach the clarifiers. This can be accomplished by the following:

FIGURE 7.12 Sludge From Deep Aeration Tank: (a) Initial Removal and (b) After 15 minutes (Reprinted with permission from Kenneth Schnaars)

FIGURE 7.13 Floating Solids on the Secondary Clarifier Because of Supersaturated Gases From Deep Tank Aeration (Reprinted with permission from Kenneth Schnaars)

- Stripping gases in an aerated mixed liquor channel to clarifiers by increasing solids agitation using coarse-bubble or fine-bubble diffusers in the mixed liquor channel (fine-bubble diffusers would be more efficient in stripping off gases). It is recommended that there is a minimum of 10 minutes residence time through the fine-bubble diffuser stripping section.
- If the tank is equipped with draft tube mechanical mixing and air spargers, then installing fine-bubble diffusers in biological reactors at the tank midpoint liquid elevation can help strip gases.
- Installing a weir in the mixed liquor channel near the biological reactor or at the effluent end of each biological reactor would help release entrained gases (in some cases, this may not be enough agitation to strip away saturated gases).
- Installing a shallow (3 to 5 m [10 to 15 ft]) biological reactor with a short hydraulic retention time (HRT) (15 to 30 minutes) after deep tanks can release gases. However, this is only possible if there is sufficient area available at the WRRF.

6.0 BIOLOGICAL NUTRIENT REMOVAL SYSTEMS

6.1 Introduction

Regulatory agencies are requiring WRRFs to produce cleaner effluents with fewer pollutants and nutrients such as phosphorus and nitrogen. More WRRFs are being converted or constructed as BNR facilities that are designed to control effluent phosphorus, nitrogen, or both phosphorus and nitrogen. There are many variations to biological phosphorus and nitrogen processes currently used at WRRFs, and these variations are described in other sections of this manual as well as *Standards for Operation of Water Resource Recovery Facilities* (2023, WEF 11), *Standards for Operation of Nutrient Removal Facilities* (2024, WEF 37), *Nutrient Removal* (2011, WEF Manual of Practice No. 34), and *Standards for Design of Water Resource Recovery Facilities* (2023, WEF 8).

Specialized processes for both phosphorus and nitrogen control use biological reactors that have baffles for creating zones with different environmental conditions. Several terms are used to describe these zones. *Aerobic*, or *oxic*, indicates a zone that contains dissolved oxygen. *Anoxic* refers to a zone that has a dissolved oxygen of less than 0.5 mg/L but has an input of nitrate-nitrogen. An *anaerobic* zone has an input of organic substrate, contains no dissolved oxygen, and has no input of nitrate-nitrogen.

Other terms used are *RAS recycle* and *internal recycle*. Return activated sludge recycle is the return flow of biomass from the underflow of the secondary clarifier to biological reactors. Internal recycle is flow directed from one reactor zone to another (e.g., the internal recycle flow from the effluent end of the aerobic zone [after nitrification] to the anoxic zone to recirculate nitrate-nitrogen from the aerobic zone to the anoxic zone for denitrification). This internal recycle flow can be in the range of 100% to 400% of the forward flow.

Processes have been developed for biological phosphorus removal only, nitrogen removal by biological nitrification and denitrification only, and removal of both nutrients in dual or combined systems. Depending on effluent residual requirements and reliability attainment goals, some chemicals may have to be added to supplement these processes.

This section highlights general troubleshooting aspects of these processes. Standard troubleshooting guidelines for activated sludge systems can be used for some aspects of BNR processes; however, these systems have unique concerns that must also be addressed, such as BOD-to-phosphorus ratios, nitrate concentrations in the anaerobic zone, internal recycle rates, and other process controls related to a particular BNR process.

6.2 Biological Phosphorus Removal Processes

The basic concept of a biological phosphorus removal system is to cycle activated sludge microorganisms through an anaerobic zone and then an aerobic zone. Typically, the HRT through the anaerobic zone is 2 to 4 hours and the HRT through the aerobic zone is 4 to 6 hours. In addition, this system requires the BOD-to-phosphorus ratios to work. Typically, these ratios are in the range of 20 to 25 mg/L BOD to 1 mg/L phosphorus to achieve a 0.5-mg/L effluent total phosphorus concentration.

This process can also be affected by the type of in-facility sidestreams that are returned to the treatment process. Because this is a biological process, operators must watch various environmental conditions such as dissolved oxygen concentration in the aerobic zone, alkalinity levels, toxic materials entering the system, temperature, and other conditions that could affect microorganisms.

If the nitrate-nitrogen (NO_3-N) concentration gets higher than 10 mg/L as N in the anaerobic zone, it would inhibit the biological phosphorus process. The nitrate concentration in the anaerobic zone is controlled by the RAS flowrate. The addition of an anoxic zone helps control nitrate that would be returned to the anaerobic zone. It is best to keep the nitrate-nitrogen concentration below 1 mg/L as N for efficient biological phosphorus removal.

Table 7.3 presents a troubleshooting guide for biological phosphorus removal processes. This guide will enable operators to help identify problems and their solutions.

6.3 Biological Nitrogen Removal

Biological nitrogen removal involves two processes: nitrification and denitrification. When designed properly, BNR can be achieved in standard-designed biological reactors, oxidation ditches, sequencing batch reactors, and other process modifications.

To determine the success of BNR processes, clarifier effluent is tested for ammonia-nitrogen, total nitrogen, and nitrate-nitrogen. Typically, ammonia-nitrogen should be less than 1.0 mg/L, total Kjeldahl nitrogen (TKN) at approximately 5 mg/L, and nitrate-nitrogen between 7 and 10 mg/L.

Typically, BNR processes are controlled by the following:

- Adjusting dissolved oxygen in the aerobic zone for nitrification. Conventional processes use dissolved oxygen concentrations ranging between 0.5 and 2.0 mg/L.
- Maintaining the absence of dissolved oxygen in the anoxic zone will trigger microorganisms to use NO_2–N and NO_3–N and complete the

TABLE 7.3 Biological Phosphorus Removal Troubleshooting Guide

Observations	Solutions
Effluent soluble phosphorus >1 mg/L, total phosphorus >2 mg/L, and BOD_5 <20 mg/L	Ensure that dissolved oxygen concentration in the anaerobic zone is <0.2 mg/L and that nitrate-nitrogen concentration in the anaerobic zone is <5 mg/L as nitrogen
Effluent soluble phosphorus <1 mg/L, total phosphorus >2 mg/L, BOD_5 >25 mg/L, and suspended solids >30 mg/L	Check suspended solids frequently; use polymer or metal salt to control suspended solids; find reason for high suspended solids, such as high flowrates, high RAS rate, or bulking
Effluent soluble phosphorus <0.2 mg/L, total phosphorus <1 mg/L, and BOD_5 >30 mg/L	Facility may be organically overloaded; check F:M
Effluent soluble phosphorus >1 mg/L, total phosphorus >2 mg/L, and BOD_5 <30 mg/L	May not be enough low-molecular-weight acids in anaerobic zone; consider adding anaerobic digester supernatant to this zone
Thin-looking MLSS, effluent total phosphorus >2 mg/L, and BOD_5 <30 mg/L	Sludge may be forming a blanket in the secondary clarifier; check sludge blanket depth; adjust recycle rate if necessary
Effluent total phosphorus >2 mg/L, BOD_5 <25 mg/L, and suspended solids <25 mg/L	Start or increase dose of metal salt to assist in precipitating total phosphorus

denitrification process. Therefore, it is critical to minimize the dissolved oxygen concentration in the recycle stream before it enters the anoxic zone. Measuring the oxidation–reduction potential (ORP) in the anoxic zone is one troubleshooting method that can be used by operators to determine how well the anoxic zone is performing. The ORP reading in the anoxic zone should be in the range of −50 to 50 mV.

- Adjusting the return sludge and internal recycle rates is also important to monitor the denitrification process. The internal recycle rate from the aerobic zone to the anoxic zone is important to return nitrates back into the anoxic zone for microorganisms. Typically, the internal recycle rate will be in the range of 100% to 400% of incoming flow.

It is important that proper conditions be maintained in the anaerobic, anoxic, and aerobic zones of the treatment process. Conversion between carbonaceous and nitrification treatment modes depends on the system used. The two basic types of nitrification systems are a single-stage system, in which both carbonaceous and nitrogenous uptake occur in the same tank, and a two-stage system, which is essentially two complete activated sludge processes in series. In the two-stage system, the first stage, which would have intermediate clarifiers after the first-stage biological reactors, provides carbonaceous removal, while the second stage provides ammonia conversion.

In either system, the basic feature enabling nitrification is a long SRT (or low F:M) of sufficient duration to allow more slowly growing nitrifiers to develop. The necessary SRT or F:M values are temperature dependent. For single-stage systems, the SRT ranges from 8 to 10 days during the summer, when the wastewater temperature is 16 to 21 °C (60 to 70 °F), but may climb into the 12- to 20-day range as the temperature drops. When wastewater temperature drops below 10 °C (50 °F), reestablishing a nitrifying culture, once lost, may not be possible, although typically, it can be maintained once it is developed. Typically, operating nitrifying systems at low temperatures is more difficult and requires considerable extra capacity in the biological reactor to provide extended detention times needed to achieve acceptable nitrification levels. Therefore, increasing SRT during the winter months is a method for improving nitrification performance.

Dissolved oxygen in nitrifying systems must start at a higher concentration than in carbonaceous activated sludge systems because of the aerobic nature of nitrifying organisms. A dissolved oxygen concentration of approximately 2.0 mg/L should prevent inhibition of nitrifiers. It is possible to nitrify under low dissolved oxygen conditions, but research indicates that gradual adjustments to dissolved oxygen level must be made to allow the population of nitrifiers to adjust to the conditions and avoid process upsets.

If a nitrification process is not capable of achieving the desired effluent ammonia-nitrogen concentration, there are various parameters to check and different controls that can be implemented to correct the process. Nitrifying organisms are sensitive to reactor conditions; as such, slight changes in their environment will cause them to become inactive. Nitrifying organisms are affected by the following:

- Toxic material,
- Temperature changes,
- pH, and
- Solids retention time.

An indication that there is a problem with the nitrification process is if the chlorine demand on facility effluent increases dramatically. This is an indication that nitrite is present and, therefore, that nitrification is incomplete. The nitrification process is only complete when ammonia-nitrogen is converted first to nitrite and then to nitrate. Again, the dissolved oxygen concentration in the biological reactor, SRT, and other such parameters must be checked.

Nitrification cannot go to completion if pH is significantly below 7.0. The optimum operating pH for nitrification is in the range of 7.0 to 8.3. Nitrification produces acid, thereby reducing alkalinity and pH. It requires 7.14 mg of alkalinity for each 1 mg of ammonia-nitrogen converted to nitrate-nitrogen. In contrast, denitrification of 1 mg of NO_2–N or NO_3–N produces 3.57 mg of alkalinity. This affects process pH only if alkalinity is insufficient. The nitrification reaction can be inhibited by low pH; therefore, it is necessary to ensure that there is adequate alkalinity in wastewater. Alkalinity may have to be added to the influent end of the biological reactor if there is an insufficient concentration of alkalinity for the nitrification process.

The following example illustrates the procedure for checking if sufficient alkalinity is present and, if it is not, determining how much to add. The example is given to assist the operator in determining how to calculate the amount of additional alkalinity required for the nitrification process. As stated earlier in this chapter, operators can add magnesium hydroxide, lime, and other alkaline chemicals to increase the alkalinity concentration to achieve complete nitrification. Careful control of alkalinity is important. By the time the pH decreases to less than 7.0, inhibition of nitrifying organisms has already occurred.

Given:

Influent TKN concentration = 20 mg/L as N,
Influent alkalinity = 130 mg/L $CaCO_3$, and
Facility flow = 37 800 m^3/d (10 mgd)

1. Calculate the amount of alkalinity required as follows:

   Alkalinity to maintain optimum pH = Alkalinity for ammonia oxidation + Residual alkalinity

   Alkalinity for ammonia oxidation = (20 mg/L NH_3–N) (7.14 mg $CaCO_3$ required/mg NH_3–N oxidized) = 143 mg/L $CaCO_3$

   Alkalinity required = 143 mg/L $CaCO_3$ required + 50 mg/L of residual alkalinity = 193 mg/L $CaCO_3$

2. Calculate the amount of alkalinity needed as follows:

   Alkalinity needed = Alkalinity required − Alkalinity available = 193 mg/L alkalinity required − 130 mg/L available in influent = 63 mg/L $CaCO_3$

3. Calculate the amount of alkalinity to be added as follows:

   In International Standard units:

   Alkalinity to add = (Alkalinity needed) (Facility flow) (Conversion factor)

   Alkalinity to add = (63 mg/L $CaCO_3$) (37 800 m^3/d) (0.001) = 2381 kg/d $CaCO_3$

   In U.S. customary units:

   Alkalinity to add = (Alkalinity needed) (Facility flow) (Conversion factor)

   Alkalinity to add = (63 mg/L $CaCO_3$) (10 mgd) (8.34) = 5254 lb/d $CaCO_3$

It is important that operators maintain a residual alkalinity concentration at the effluent end of the biological reactor between 50 to 60 mg/L to maintain the optimum pH range.

If a problem arises with the nitrification process, attempt to identify the cause of the problem:

- Monitor the dissolved oxygen and ORP in the aerobic and anoxic zones.
- Calculate the SRT and adjust wasting rates to move the SRT into the range necessary to develop nitrifying organisms (4 to 15 days, depending on temperature).
- Monitor influent and effluent pH (nitrifiers operate best in a range of 7.0 to 8.3 pH).
- Monitor influent alkalinity and ensure that there is an adequate concentration to complete the nitrification process.
- Monitor influent and effluent ammonia, nitrite, and nitrate-nitrogen concentrations.

Table 7.4 presents a troubleshooting guide for BNR processes to assist operators with identifying problems and their solutions.

TABLE 7.4 Biological Nitrogen Removal Potential Issues

Observation	Solutions
Low dissolved oxygen levels in the aeration zone, <2.0 mg/L and ammonia-nitrogen >1.0 mg/L	Increase dissolved oxygen levels to a minimum of 2.0 mg/L; dissolved oxygen may be reduced gradually if required by the process to limit oxygen in downstream anoxic or anaerobic zones
Low alkalinity, <7.14 mg of alkalinity as $CaCO_3$ per 1.0 mg of ammonia-nitrogen, and <50–60 mg/L of alkalinity as $CaCO_3$ in secondary effluent	Add chemicals such as magnesium hydroxide, lime, or other alkaline chemicals
Low pH inhibiting nitrification, <7.0 pH	Add chemicals such as magnesium hydroxide, lime, or other alkaline chemicals; if alkalinity is high, sodium hydroxide may also be used
High pH inhibiting nitrification, >8.3 pH	Increase aeration or add acidic chemicals such as carbonic acid or hydrochloric acid
High ORP >50 mV and/or dissolved oxygen >0 mg/L in the anoxic zone, and/or nitrate-nitrite levels >10 mg/L	Gradually reduce aeration upstream of anoxic zone and/or reduce internal recycle rates from aeration zones
Chemical inhibition, for example, industrial wastes or sidestreams can introduce chemicals such as disinfectants, metals, cyanides, and sulfur-containing compounds	Identification and pretreatment of industrial wastes and sidestream.
Inert solids from chemical addition, for example, the addition of ferric or alum can increase inert solids in the reactor can lead to overwasting and washout of the nitrifiers in a MLSS controlled process	Maintain SRT within the nitrifier acceptable range (4–15 days depending on temperature) and proper chemical addition rates

7.0 DESIGN AUDIT

Various studies and publications have found that many factors limiting WRRF performance were attributed to inadequate facility design. These include deficiencies in sludge wasting capability, process flexibility, process controllability (inability to measure or adjust the process), secondary clarifier design, sludge treatment capacity or operating ease, and aeration capability.

Before troubleshooting a problem, it is important to determine the facility's capabilities and limitations. Design limitations may make it impractical to fully implement some of the more typical solutions to facility problems. Although operator ingenuity has overcome many minor facility limitations, certain design conditions will be difficult or impossible to correct without facility modifications. In some cases, design limitations may be so severe that proper operation under normal conditions will be difficult. It is important, therefore, to make a design audit of the facility. Some problems can be avoided if the WRRF is designed properly and with operators and maintenance personnel in mind. Therefore, it is important that the engineer and facility staff work closely together through the design phase of the project.

Questions that must be asked include the following:

Wastewater characteristics

- Is the facility protected against shock loads (e.g., with a flow equalization basin)? If not,
    - What type of shock load is experienced: organic, toxic, or hydraulic?
    - How often does a shock load occur?
    - What is its duration?
    - What is its source?
- Is there industrial waste? If so,
    - What percentage of the influent flow is industrial waste?
    - What percentage of influent BOD_5 and TSS load is from industrial waste?
    - Does the city or agency have industrial waste pretreatment regulations and are they enforced?
    - Are industrial waste flows sampled?
    - How often are flows sampled?
    - Are industrial waste flows measured, recorded, and totalized?
    - Is influent flow free of toxic concentrations of industrial waste?
    - Was industrial waste considered in the facility design?
    - Are there provisions for adequate pH control?
    - Are there provisions to feed supplemental nutrients if needed?
- Do septic waste trucks discharge to the facility? If so,
    - Is the septage analyzed for pH, COD, and BOD (and TSS if there are no primary clarifiers)?

    - Is there a septage-handling facility to control discharge to the facility?
    - Is septage a significant load on the activated sludge process (more than 15% of BOD_5)?
    - Can septage flows be measured?
    - Is there a program to regulate receiving septage?
- Is in-facility recycled flow routinely measured for flow, total phosphorus, nitrogen forms (ammonia, nitrate, etc.), BOD, and suspended solids concentrations? If so,
    - Is it high strength?
    - Is it variable rate?
    - Where is the in-facility recycle flow returned to within the facility? For example, supernatant from an anaerobic digester, centrate from a centrifuge, and sidestreams from other types of solids-handling processes could contain high concentrations of solids, BOD, phosphorus, and/or nitrogen.
    - Is it a significant load on the activated sludge process (more than 15% of BOD_5)?
    - Is the facility recycle flow affecting the facility process, such as nitrification?

Facility layout

- Are there two or more of each biological reactor, secondary clarifier, and pump systems for operating flexibility?
- If the process tank is covered, can it be readily inspected for process and/or maintenance considerations?
- Do biological reactors have the flexibility to operate either as parallel process trains or as a single process train?
- Can flow be accurately split and measured to parallel reactors?
- Can return sludge flow be split to each biological reactor for process control?
- Is the influent sampler, if located outside, weather protected?
- Is the facility easily accessible during bad weather?
- Is there adequate piping flexibility so a biological reactor does not have to be shut down if a clarifier is out of service?
- Is there enough room around pumps and other equipment to perform maintenance?

- Are there bypasses around individual treatment units, such as biological reactors and secondary clarifiers? If so,
  - Where do these bypasses discharge?
  - How are they controlled?
  - Are these bypass flows measured?
  - Are bypasses listed in the discharge permit?

Instrumentation

- Is the flowmeter properly sized for low and peak flows?
- Is the meter free from any obstructions to avoid inaccurate readings?
- If a magnetic flowmeter, is it installed properly (flowing full, proper upstream and downstream straight pipe, etc.)?
- Does the flowmeter have a local chart recorder or light-emitting diode trend recorder that is transmitted to a programmable logic controller (PLC) or SCADA so that flow variations can be adequately displayed and recorded?
- Is the flow-meter unaffected by upstream and downstream hydraulic conditions?
- Are flow-meters and recorders routinely and properly calibrated?
- Are recycle flows included in the facility flow measurement? If so, are they subtracted from influent flow?
- If a Parshall flume is used, is it free of grit deposits?
- Are there provisions for any other control-type instrumentation, such as solids meters, ORP meters, nitrogen probes, and so on?

Biological reactors

- Do biological reactors have adjustable weirs?
- Do outlet pipes remain free of rags?
- Is there adequate piping to operate in different modes of operation, such as step-feed or contact-stabilization modes as well as plug flow mode?
- Can influent flow be accurately split and measured among multiple biological reactors?
- Can RAS flow be accurately split and measured among multiple biological reactors?
- Is there a toxic element sensing device ahead of, or in, the biological reactor? If so,

  - Where?
  - What type?
  - What is the flow travel time between the sensor and biological reactor?
- Do in-facility recycle sidestreams discharge at the headworks rather than directly to the biological reactor? Also,
  - Where do flows originate?
  - Are they measured for flow, phosphorus, nitrogen, pH, and BOD and suspended solids concentrations?
  - Are there facilities for adjusting the pH?
- Are biological reactors sized for secondary influent plus recycle loads?
- Is the tank designed so it does not trap foam?
- Is there an ability to add chlorine or another type of chemical into return sludge or biological reactor to assist with the reduction of filaments?

Aeration equipment

- Are blowers properly sized?
- Are there leaks in air piping?
- Do air diffusers need only normal cleaning and can they be cleaned with reasonable effort?
- What is the design air rate per diffuser? It is good practice to design diffusers so that there can be some turn-up capacity during higher loading rates.
- Do surface mechanical aerators operate without overheating or shutting off under increased flows?
- Can the speed of mechanical aerators be varied?
- Do cables on floating aerators withstand intermittent operation?
- Are submerged turbine aerator bearings and shafts reliable?
- Is there adequate freeboard to contain splashing from surface mechanical aerators?
- Are surface mechanical aerators free of icing problems? If not, what is the frequency and extent of the problems?
- Are surface mechanical aerators free of rag problems? If not,
  - What is the frequency and extent of the problem?
  - Are there upstream screening devices?

  - Can aerators operate in reverse to help clear off rags?
- Are there provisions for removing rags from surface mechanical aerators?
- How is mixed liquor dissolved oxygen concentration controlled?
- Do aeration devices meet their design oxygen-transfer efficiencies?
- Can dissolved oxygen concentrations of at least 1.0 mg/L be maintained under all conditions?

Secondary clarifiers

- Is there accurate flow-splitting and measurement among multiple clarifiers?
- Are clarifier side water depths (SWDs) sufficient?
  - Tank diameters of up to 12 m (40 ft) should have a SWD of 3.7 m (12 ft). Note that a number of state regulatory agencies stipulate that the minimum secondary clarifier depth shall not be less than 3.7 m.
  - Tank diameters from 12 to 21 m (40 to 70 ft) should have SWDs of 3.7 m (12 ft or deeper).
  - Tank diameters from 21 to 41 m (70 to 160 ft) should have SWDs of 4 m (13 ft) or deeper.
- Are solids retained even during peak flows?
- Do clarifiers remain free of ice during cold weather?
- Is scum removed efficiently from the process and disposed of properly?
- Is the scum trough accessible from the bridge for maintenance processes. Also, is there a flushing mechanism on the scum trough to keep it clean?
- Are overflow weirs unaffected by downstream hydraulic restrictions?
- Is short-circuiting prevented by adequate inlet baffle construction?
- Are weirs on a single launder balanced to pull evenly from each side?
- Does the center column contain a seal around the lower end of the column and, if so, is that seal in place?

Return sludge system

- Are RAS pumps properly sized?
- Can RAS flows be varied over a wide range (25% to 100% of influent flow)?
- Is RAS pumping capacity adequate to control sludge blanket depth?

- Is RAS flow visible?
- Is RAS flow measured?
- Does the RAS flow-meter respond to flow changes?
- Can RAS flows be easily balanced with multiple biological reactors and secondary clarifiers?
- Do telescoping valves remain free of rags even at lower flows?
- Is piping symmetrical on multiple clarifiers to avoid imbalances in RAS flows, both from clarifiers and to biological reactors?
- Is RAS returned near the inlet of biological reactors?
- Is the valve controlling air to an airlift pump the regulating type rather than the shutoff type?
- Do valves used for RAS flow control remain clear from obstructions? Try to avoid using butterfly valves because rags can wrap around the valve and not close.
- Can RAS be chlorinated?

Waste sludge system

- Is WAS pumping system adequately sized to prevent wasting restrictions?
- Can WAS flowrate be varied?
- Can WAS flowrate and volume be measured?
- Can sludge be wasted continuously?
- Can the WAS pumping rate be reliably used to calculate waste volume if necessary?
- Does the WAS pump have taps for measuring pressure if checking the pumping rate with a pump curve?
- Can WAS be pumped to different locations?

Miscellaneous

- Are process tanks adequately sized, with recycle flows included?
- Are actual peak flows, including infiltration and inflow, less than or equal to design peak flows?
- Is there a reasonable design hydraulic gradient so that tanks are not surcharged and weirs are not submerged?
- Are there facilities for feeding settling aids such as alum, ferric chloride, and polymer? If so, are they operating properly?

- Are there facilities for feeding chemicals for adjusting the pH?
- Are there facilities for feeding supplemental nutrients or BOD?
- Can facility influent be chlorinated?
- Are there automated control systems, such as PLCs, SCADA, and distributed control systems?

Laboratory facilities

- Is there an on-site process control laboratory?
- Is there sufficient equipment to run necessary operational control and National Pollutant Discharge Elimination System tests?
- Is the balance unaffected by vibrations?
- Do sampling point locations allow representative samples to be taken?
- Does the heating, ventilation, and air conditioning system provide comfortable working conditions and the proper environment for equipment?
- Is there good lighting?
- Is there sufficient flow and bench space?

8.0 PRACTICE CASE HISTORY TROUBLESHOOTING PROBLEM

The example troubleshooting problem is based on an actual case history of a WRRF that was having issues remaining in compliance. The following problem provides the background of the facility, the problems the facility was experiencing, and the information provided by operators. Provide solutions and then check them against what was actually done to bring this facility back into compliance.

8.1 Background

As a new facility superintendent at a 18.9-ML/d (5-mgd) average design flow (379-ML/d [100-mgd] maximum flow) WRRF, one must quickly understand the facility's operation and process controls. The facility consists of facility influent screening, velocity-controlled grit removal, two large single-pass aeration tanks, two secondary clarifiers, chlorine disinfection, and sulfur dioxide dechlorination. There are no primary settling tanks. Solids-handling facilities consist of two aerobic digesters and liquid hauling of digested sludge. The facility receives domestic, commercial, and industrial wastewater.

Facility data are as follows:

- Influent facility flowrate—15 ML/d (4 mgd)
- Number of aeration tanks in service—1
- Volume per aeration tank—13 853 m^3 (3.66 mil. gal)
- Type of aeration—Surface aerators
- Number of secondary clarifiers in service—two
- Secondary clarifiers—23-m (75-ft) diameter, 3.6-m (12-ft) liquid depth
- Volume per secondary clarifier—1500 m^3 (396 347 gal)
- Sludge blanket depth—1 m (3 ft) (during dry weather)
- Number of RAS pumps—3 (only 1 operable)
- Mixed liquor suspended solids concentration—3500 mg/L
- Mixed liquor volatile suspended solids—49%
- Return activated sludge concentration—5000 mg/L
- Waste activated sludge rate—0 m^3 (0 gpm)
- Aeration tank organic loading— 48 $kg/m^3{\cdot}d$ (3 lb BOD/d/1000 cu ft)
- Mean cell residence time—30 days

8.2 Problem

The facility washes out solids during rain events. As a new superintendent, you question the facility staff to see what the facility policies were before your arrival. What are at least five questions you would ask the facility staff?

1. ______________________________

   ____

2. ______________________________

   ____

3. ______________________________

   ____

4. ______________________________

   ____

5. ______________________________

   ____

Based on facility data, what course of action would you take, if any, if you were placed in charge of this facility?

8.3 Corrective Action Items

As the new superintendent, what are some of the corrective steps, if any, that you would take based on the facility data shown above?

1. ______________________________
2. ______________________________
3. ______________________________
4. ______________________________
5. ______________________________

After you write down your questions and corrective actions, see the actual operator responses provided by the facility staff.

8.4 Operator Responses

The following is some of the information you obtained from facility staff after questioning them:

- They like to keep a high mixed liquor concentration between 3500 and 4000 mg/L in case they receive a high organic load from an industry.
- They turned off the RAS pumps when it started to rain so they do not send too many solids back to the aeration tanks.
- They have not wasted sludge for months because MLSS have been remaining between their target numbers and it goes down slightly during rain events.
- Dissolved oxygen is approximately 1 mg/L in the aeration tank.
- Most of the aerators are out of service because of maintenance issues, which is okay because they like to keep the back half of the aeration off so not as many solids enter the secondary clarifiers and remain in the aeration tanks.

8.5 Actual Corrective Action Taken

1. The MLVSS in the aeration tank is too low. The volatile solids content is more like an aerobic digester. Start wasting immediately and bring down the MLSS to 1200 to 1500 mg/L. Lowering the MLSS and producing a healthier biomass will increase volatile solids to the 70% to 80% range.

The idea of maintaining a high mixed liquor concentration with a very low MLVSS concentration to handle an industrial waste is incorrect thinking because low volatile solids content indicates that approximately one-half of the material in the aeration tank is inert and would not reduce organics.

2. Fix surface aerators and turn on aerators throughout the aeration basin.
3. Fix return activated sludge pumps.
4. Keep RAS pumps operating at all times and even during rain events.
5. Maintain a lower sludge blanket in the secondary clarifiers because the secondary clarifiers are shallow.
6. Lower the MCRT and set facility personnel up on an SRT process control scheme for their activated sludge process.

These changes improved the facility effluent quality and eliminated its solids washouts during rain events. Lowering the amount of solids in the aeration tank meant fewer solids entering the secondary clarifiers, so the blanket rose at a slower rate. In addition, by keeping RAS pumps running, facility operators pulled solids out of the clarifier and did not allow them to build up in the secondary clarifier.

9.0 PRACTICE QUESTIONS

1. A WRRF with an activated sludge process is experiencing white foam in its aeration tank. Which of the following would not be the cause?
   a. High F:M
   b. The facility wasted too many solids
   c. Low F:M
   d. The facility received a high hydraulic load because of a rainstorm

2. While an operator was making rounds, the operator noticed that one of their three secondary clarifiers was losing solids. Why?
   a. The facility received a toxic load
   b. The clarifiers are all receiving the same hydraulic load
   c. All clarifiers are in service
   d. The clarifier mechanism is not moving

3. *Thiothrix* is typically associated with what process conditions?
   a. Low pH and high F:M
   b. Increased BOD and nitrite

c. Nutrient deficiency or septic conditions
d. High dissolved oxygen and reduced wasting

4. A WRRF has four secondary clarifiers and two clarifiers are currently in service. Each clarifier is 36.5 m (120 ft) in diameter with a 4-m (14-ft) SWD. The facility flow is 265 ML/d (70 mgd), the return sludge flowrate is 60% of the influent, and the aeration tank mixed liquor concentration is 3000 mg/L. Based on the solids loading rate, as an operator what process modifications would you make?
   a. Take a secondary clarifier out of service
   b. Place a secondary clarifier in service
   c. Do nothing
   d. Adjust the facility influent flow

5. Which of the following observations would tell an operator that they are seeing *Nocardia* filaments under a microscope?
   a. The filament looks like tree branches
   b. The filament is long and stringy
   c. The filament is segmented and looks like barrel shapes
   d. Solids are dispersed

6. An operator notices that some of the secondary clarifiers have rafts of sludge floating on the liquid surface of the clarifiers. What is the problem?
   a. There is an excessive amount of filamentous organisms in the activated sludge process
   b. Secondary clarifiers are experiencing a solids washout because of excessive flows entering the clarifiers
   c. Sludge is not being removed quickly enough from clarifiers, causing denitrification to occur in tanks
   d. Sludge is floating because the SVI of the activated system is at 300

7. An operator looks at a mixed liquor sample under a microscope and sees fine floc particles with no large grouping of floc particles. This is describing what?
   a. *Nocardia*
   b. Denitrification
   c. Dispersed floc
   d. Good-settling sludge

8. A facility has five secondary clarifiers and four of the secondary clarifiers are in service. Each secondary clarifier is 30 m (98 ft) in diameter

with a SWD of 4 m (14 ft). The clarifiers receive a facility flow of 57 ML/d (15 mgd) and an RAS flow of 23 ML/d (6 mgd). The mixed liquor concentration entering the secondary clarifiers is 3200 mg/L. What is the solids loading rate of these clarifiers? Show formula and all math calculations here next to the answers.

a. 63 kg/m^2·d (13 lb suspended solids/d/sq ft)
b. 93 kg/m^2·d (19 lb suspended solids/d/sq ft)
c. 142 kg/m^2·d (29 lb suspended solids/d/sq ft)
d. 405 kg/m^2·d (53 lb suspended solids/d/sq ft)

9. Based on the calculated results in Question No. 8, as an operator what process change would you make to improve the solids loading rate?

   a. Place the fifth secondary clarifier in service
   b. Decrease the RAS rate
   c. Increase wasting
   d. Do nothing because the clarifier is at an acceptable loading

Answer key:

1. c, 2. d, 3. c, 4. b, 5. a, 6. c, 7. c, 8. b, 9. d

10.0 REFERENCES

Cha, D. K., Jenkins, D., Lewis, W. P., & Kido, W. H. (1992). Process control factors influencing *Nocardia* populations in activated sludge. *Water Environment Research*, *(64)*1: 37–43.

Fujii, H., & Fukuda, H. (1974). Improvement of the aeration tank: The deep aeration tank. *Water Supply Wastewater, 18*, 959, in Japanese.

Jenkins, D., Richard, M. G., & Daigger, G. T. (2003). *Manual on the causes and control of activated sludge bulking and foaming* (3rd ed.). Lewis Publishers.

Lu, L., Turcotte, A., Donnellon, J., & Wilson, T. E. (1990, October 7–11). Deep tank aeration experience at New York City's North River Plant. Paper presented at the 63rd Annual Water Pollution Control Federation Technical Exposition and Conference, Washington, DC, United States.

Parker, D., Geary, S., Jones, G., McIntyre, L., Oppenheim, S., Pedregon, V., Pope, R., Richards, T., Voigt, C., Volpe, G., Willis, J., & Witzgall, R. (2001, October 13–17). Making classifying selectors work for foam control in the activated sludge process. *Proceedings of the 74th Annual*

Water Environment Federation Technical Exposition and Conference [CD-ROM], Atlanta, GA, United States. Water Environment Federation.

Parker, D., Bratby, J., Esping, D., Hull, T., Kelly, R., Melcer, H., Merlo, R., Pope, R., Shafer, T., Wahlberg, E., & Witzgall, R. (2014). A critical review of nuisance foam formation and biological methods for foam management or elimination in nutrient removal facilities. *Water Environment Research*, *86*(6), 483–506.

Richard, M. (1989). *Activated sludge microbiology.* Water Environment Federation.

Rossetti, S., Tomei, M. C., Nielsen, P. H., & Tandoi, V. (2005) *"Microthrix parvicella"*, a filamentous bacterium causing bulking and foaming in activated sludge systems: A review of current knowledge. *FEMS Microbiology Reviews*, *29*(1), 49–64.

Shao, Y. J., Starr, M., Kaporis, K., Kim, H. S., & Jenkins, D. (1997). Polymer addition as a solution to Nocardia foaming problems. *Water Environment Research, 69,* 25.

U.S. Environmental Protection Agency. (1977). *Process control manual for aerobic biological wastewater treatment facilities.* EPA-430/9-77-006.

Water Environment Federation. (2011). *Nutrient removal* (Manual of Practice No. 34). Water Environment Federation.

Water Environment Federation (2023) *Standards for operation of water resource recovery facilities,* (8th ed., WEF 11).

Water Environment Federation. (2023). *Standards for design of water resource recovery facilities* (7th ed., WEF 8).

Water Environment Federation. (2024). *Standards for operation of nutrient removal facilities* (2nd ed., WEF 37).

Wilson, T. E. (1983). Application of ISV test to the operation of activated sludge plants. *Water Research, 17,* 707.

11.0 SUGGESTED READINGS

Arasmith, E. E. (1981). *Operational control tests for wastewater treatment facilities: Depth of blanket.* Linn-Benton Community College. http://eric.ed.gov/?id=ED221391

Bell, K. Y., & Abel, S. (2011). Optimization of WWTP aeration process upgrades for energy efficiency. *Water Practice Technology*, 6(2).

Bentivogli, D., & Smith, D. A. (1998). Creative financing helps Buffalo cut energy use. *Utility Executive, 1,* 2.

Brassard, M., & Ritter, D. (1994). *The memory jogger.* Goal/QPC.

California State University. (2022). *Operation of Wastewater Treatment Plants* (8th ed., Volume II). California State University.

Davies, W. J., Le, M. S., & Heath, C. R. (1998). Intensified activated sludge process with submerged membrane microfiltration. *Water Science & Technology, 38,* 4/5, 421.

Drury, D., Dezham, P., Torres, B., Ooten, K., & Gabb, D. (1995). Novel nitrogen removal. *Water Environment & Technology*, *7*(9), 74.

Ekama, G. A., Barnard, J. L., Günthert, F. W., Krebs, P., McCorquodale, J. A., Parker, D. S., & Wahlberg, E. J. (1997). Secondary settling tanks: Theory, modelling, design and operation. *International Water Association.*

Engelhardt, N., Firk, W., & Warnken, W. (1998). Integration of membrane filtration into the activated sludge process in municipal wastewater treatment. *Water Science & Technology, 38,* 4/5, 429.

Finger, R. (1983). Evaluation of ZULLIG dissolved oxygen probes. Internal Memo to William Burwell and Warren Uhte. Municipality of Metropolitan Seattle, Washington.

Gabb, D. M. D., Ekama, G. A., Jenkins, D., & Marais, G. v. R. (1989). The incidence of *Sphaerotilus natans* in laboratory-scale activated sludge systems. *Water Science & Technology, 21,* 29.

Gabb, D. M. D., Still, D. A., Ekama, G. A., Jenkins, D. &, Marais, G. v. R. (1991). The selector effect on filamentous bulking in long sludge age activated sludge systems. *Water Science & Technology, 23,* 867.

Grady, C. P. L., Jr., Daigger, G. T., Love, N. G., & Filipe, C. D. M. (2011). *Biological wastewater treatment* (3rd ed.). CRC Press.

Hammer, M. J., & Hammer, M. J., Jr. (2011). *Water and wastewater technology* (7th ed.). Prentice Hall.

Hartley, K. (2013). *Tuning biological nutrient removal plants.* IWA Publishing.

Henze, M. (Ed.). (2000). The activated sludge models (1, 2, 2d and 3). International Water Association.

Hermann, J. D., & Jeris, J. (1992). Estimating parameters for activated sludge plants. *Pollution Engineering,* December.

Holt, J. G. (Ed.). (1994). *Bergey's manual of determininative bacteriology* (9th ed.). Lippincott Williams and Wilkins Co.

Juretschko, S., Timmermann, G., Schmid, M., Schliefer, K., Pommerening-Roser, A., Koops, H., & Wagner, M. (1998). Combined molecular and convention analysis of nitrifying bacterium diversity in activated sludge: *Nitrosococcus mobilis* and *Nitrospira*-like bacteria as dominant population. *Applied Environmental Microbiology, 64,* 3042.

Kelly, R., Melcer, H., Smith, D., & Kresge, R. (2009). Evaluating the West Boise secondary clarifiers: CFD modeling and field verification of performance. Paper presented at the Pacific Northwest Clean Water Association Conference, Boise, ID, United States.

Metcalf & Eddy, Inc. (1972). *Wastewater engineering: Collection, treatment, disposal.* McGraw-Hill.

Metcalf & Eddy, Inc./AECOM. (2014). *Wastewater engineering: Treatment and resource recovery* (5th ed.). McGraw-Hill, Inc.

Mino, T., Van Loosdrecht, M. C. M., & Heijnen, J. J. (1998). Microbiology and biochemistry of the enhanced biological phosphorus process. *Water Research, 32,* 3193.

Mueller, J. A., Krupa, J. J., Shkreli, F., Nasr, S., & FitzPatrick, B. (1996, October 5–9). Impact of a selector on oxygen transfer—A full scale demonstration. *Proceedings of the 69th Annual Water Environment Federation Technical Exposition and Conference*, Dallas, TX, United States. Water Environment Federation.

Nagaoka, H., Yamanishi, S., & Miya, A. (1998). Modeling of biofouling by extracellular polymers in a membrane separation activated sludge system. *Water Science & Technology, 38,* 4/5, 497.

New York State Energy Research and Development Authority. (2008). *Statewide assessment of energy use by the municipal water and wastewater sector.* Final Report 08-17.

Norris, D. P. (1982). Energy savings for the plant operator. Paper presented at Pacific Northwest Water Environment Association Southern Oregon Regional Wastewater Conference.

Porter, J. E. (1917). The activated sludge process of sewage treatment—A bibliography on the subject. General Filtration Co., Inc.

Rittmann, B. E., & McCarty, P. L. (2001). *Environmental biotechnology: Principles and applications.* McGraw-Hill.

Secondary treatment, 40 CFR § 133.102 (1993).

Smith, W. G. (1981). Energy utilities cost control: findings from previous studies at operating treatment plants. Paper presented at the 53rd Annual Conference of the California Water Pollution Control Association, Long Beach, CA, United States.

U.S. Environmental Protection Agency. (1974). *Operational control procedures for the activated sludge process, part II control tests.* EPA-330/9-74-0016.

U.S. Environmental Protection Agency. (1976). *Selected applications of instrumentation and automation in wastewater treatment facilities.* EPA-600/2-76-276.

U.S. Environmental Protection Agency. (1987). *Summary report—The causes and control of activated sludge bulking and foaming.* EPA-625/8-87-012. Office of Research and Development.

U.S. Environmental Protection Agency. (1987). *Design manual—Phosphorus removal.* EPA-625/1-87-001, Center for Environmental Research Information.

U.S. Environmental Protection Agency. (1989). *Design manual—Fine pore aeration systems*. EPA-625/1-89-023. Center for Environmental Research Information.

U.S. Environmental Protection Agency. (2010). *Evaluation of energy conservation measures for wastewater treatment facilities.* EPA 832/R-10-005.

U.S. Environmental Protection Agency. (2010). *Nutrient control design manual.* EPA-600/R-10-100.

Van Loosdrecht, M. C. M., & Jetten, M. S. M. (1998). Microbiological conversions in nitrogen removal. *Water Science & Technology, 38*(1), 1.

Vinton, R. H., & Mace, G. R. (1998). Accurate aeration. *Operations Forum, 15*(6), 21.

Wang, L. K., Pereira, N. C., Hung, Y.-T., & Shammas, N. K. (Eds.). (2009). Biological treatment processes. In L. K. Wang, &M.-H. S. Wang (Eds.), *Handbook of environmental engineering* (Volume 8). Springer.

Water Environment Federation. (2005). *Clarifier design* (2nd ed., Manual of Practice FD-8).

Water Environment Federation. (2014). *Wastewater treatment process modeling* (2nd ed., WEF Manual of Practice No. 31).

Water Environment Federation. (2021). *Energy in water resource recovery facilities* (2nd ed., Manual of Practice No. 32.

Water Environment Federation. (2024). *Membrane bioreactors* (2nd ed., WEF Manual of Practice No. 36).

Water Environment Federation. (2024). *Standards for automation of water resource recovery facilities* (WEF 21).

Wheeler, D. J. (1993). *Understanding variation, The key to managing chaos.* SPC Press Inc.

8

Advanced Concepts in Nutrient Removal

1.0 INTRODUCTION AND NITRIFICATION BASICS

Ammonia nitrification, total nitrogen removal, and total phosphorus removal are crucial processes in domestic water resource recovery facilities (WRRFs) because they help mitigate environmental pollution and protect aquatic ecosystems. Ammonia nitrification converts toxic ammonia into less harmful nitrate, preventing oxygen depletion and toxicity in receiving waters. Total nitrogen removal, which includes nitrification and denitrification, reduces overall nitrogen content, preventing eutrophication that can lead to harmful algal blooms and dead zones. *Eutrophication* is the enrichment of surface waters with excessive nutrients. Similarly, total phosphorus removal is essential to control eutrophication, as phosphorus is a key nutrient that promotes excessive algal growth. Together, these processes ensure the treated effluent is safe for discharge to natural water bodies, maintaining ecological balance and public health.

Many WRRFs must provide enhanced removal of either nitrogen or phosphorus because excessive nutrients overstimulate growth of plants and algae in surface water and, when this plant and algae growth dies off, it creates a food source for bacteria that use dissolved oxygen in the water consuming this food, resulting in an oxygen-deficient condition in the water known as *hypoxia*. Typically, nitrogen is the limiting nutrient in marine systems and, typically, phosphorus is the limiting nutrient in freshwater systems. Many WRRFs must also remove nitrogen to protect drinking water supplies that must meet the drinking water standard of 10 mg/L nitrate-nitrogen. Frequently, regulators will require a permit limit of 10 mg/L total nitrogen because it is assumed the nonnitrate forms of nitrogen may be converted to nitrate in the environment.

Nitrogen exists in the water environment as ammonia (NH_3), ammonium (NH_4^+), nitrite (NO_2^-), nitrate (NO_3^-), organic nitrogen, and nitrogen gas (N_2). Before any nitrogen removal can take place, ammonia/ammonium must first be oxidized to nitrate, a process known as *nitrification*. Nitrification is required at most WRRFs with some occasional exceptions, like those of WRRFs discharging to large surface waterbodies with high dilution factors, because the un-ionized form of ammonia (NH_3) is toxic at the low concentration of 0.02 mg/L as nitrogen. The fraction of un-ionized ammonia depends on the pH and temperature of the water. Figure 8.1 provides the maximum concentration of ammonia/ammonium (ammoniacal nitrogen in the figure) in the water to stay below this 0.02-mg/L NH_3–N limit. At a normal pH of 7 and temperature of 20 °C, 99.6% of what we refer to as "ammonia" is actually ammonium, whereas at a pH of 8.5 and temperature of 30 °C, the ammonium fraction drops to 80%.

Nitrogen from the collection will primarily consist of NH_3/NH_4^+ and organic nitrogen, which can be measured together as a parameter called total Kjeldahl nitrogen (TKN). Typically, domestic wastewater contains 40 to 70 mg/L of TKN, consisting of 25 to 45 mg/L of ammonia-nitrogen and 15 to 25 mg/L of organic nitrogen. During secondary treatment, two reactions take place that affect nitrogenous oxygen demand in the activated sludge process:

- All but 1 to 3 mg/L of organic nitrogen is converted to ammonia through a process known as *ammonification*, and this is why it is important to always measure TKN in influent wastewater.

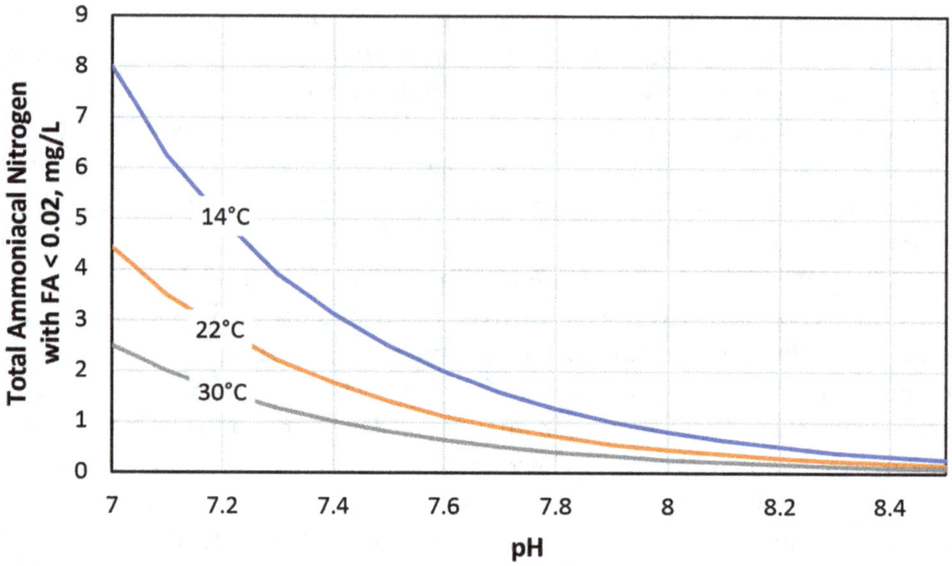

FIGURE 8.1 Maximum Concentration of Total Ammoniacal-Nitrogen (NH_3-N + NH_4^+-N) to Remain Below 0.02 mg/L NH_3–N at Various Water Temperatures

- Approximately 20% of influent nitrogen is removed by assimilation into cell mass and then removed though wasting of cells as waste activated sludge (WAS). Nitrogen is an essential nutrient and component of bacterial cells; the simplistic formula for a cell is $C_5H_7O_2NP_{0.1}$ (mixed liquor volatile suspended solids). Ammonium is the preferred form of nitrogen for biomass growth, although a few microorganisms can use nitrate or nitrite.

To remove nitrogen to a greater extent than assimilation into cells, two microbially-mediated processes called nitrification and denitrification are most commonly used. Nitrification is an aerobic process that converts ammonia-nitrogen to nitrite-nitrogen (nitritation) and then nitrate-nitrogen (nitratation). Denitrification is an anoxic process that converts the produced nitrite or nitrate-nitrogen to nitrogen gas that is then released to the atmosphere (see Section 2.0). Each of these processes requires different organisms and different environmental conditions within biological reactors that optimize microorganisms' growth and function. Most influent organic nitrogen is converted to ammonia-nitrogen in the treatment process; however, approximately 1 to 3 mg/L of organic nitrogen is *recalcitrant* (unreactive) *dissolved organic nitrogen* (rDON), passing through the WRRF until it is discharged in the effluent. Because it is recalcitrant, the effect of rDON on the environment is minimal. The presence of rDON in final effluent continues to be researched and discussed, especially in sensitive receiving waters with extremely low total nitrogen limits below 2 mg/L. The recalcitrant nature of rDON suggests that limits below 2 mg/L total nitrogen may have limited environmental benefit.

1.1 Stoichiometry and Kinetics

Nitrification is a two-step process in which toxic ammonium-nitrogen is oxidized to nonoxic forms of nitrogen: first to nitrite-nitrogen (NO_2-N) and then further to nitrate-nitrogen (NO_3-N). Nitrification involves two specific groups of autotrophic bacteria, ammonia-oxidizing bacteria (AOB) and nitrite-oxidizing bacteria (NOB). Autotrophic bacteria differ from heterotrophic bacteria in that autotrophic bacteria use inorganic carbon (CO_2) for cell synthesis, whereas heterotrophic bacteria use organic carbon (carbonaceous oxygen demand [COD]/biochemical oxygen demand [BOD]) for cell synthesis. Ammonia-oxidizing bacteria carry out the oxidation of ammonia to nitrite and NOB carry out the oxidation of nitrite to nitrate. Oxidation of ammonia or nitrite provides the energy needed for autotrophic bacteria cell synthesis. These bacteria are referred to as *obligate aerobes* because they grow only when dissolved oxygen is available.

The proportion of reactants and products involved in this process of biochemical reactions is referred to as *stoichiometry*. Understanding stoichiometry is important because it defines the basic inputs and outputs for each of the steps in the process and can determine which of these inputs will limit the reaction. The stoichiometric equation that defines molar ratios for the oxidation of ammonia to nitrite by AOB is given as follows:

$$NH_4^+ + 1.5\ O_2 \rightarrow NO_2^- + H_2O + 2H^+ \qquad (8.1)$$

Similarly, the stoichiometric equation that describes the oxidation of nitrite to nitrate by NOB is as follows:

$$NO_2^- + 0.5\ O_2 \rightarrow NO_3^- \qquad (8.2)$$

The overall reaction for the two-step nitrification process combines Equations 8.1 and 8.2, as follows:

$$NH_4^+ + 2\ O_2 \rightarrow NO_3^- + 2\ H^+ + H_2O \qquad (8.3)$$

The overall stoichiometric equation that incorporates both ammonia oxidation and cell synthesis reactions is given by the following:

$$\begin{gathered} NH_4^+ \leftrightarrow NH_3 + H^+ + 1.863\ O_2 + 0.098\ CO_2 \rightarrow \\ 0.0196\ C_5H_7O_2N + 0.98\ NO_3^- + 1.88\ CO_2 + H_2O + 1.88\ H^+ \qquad (8.4) \end{gathered}$$

Based on the stoichiometry of the overall energy reaction, 2 mol of oxygen is required to oxidize 1 mol of ammonia to nitrate, which is equivalent to 4.57 g O_2/g NH_4^+-N oxidized.

During nitrification, 2 mol of hydrogen ion are also formed. The formation of 1 mol of hydrogen is equivalent to the consumption of 1 mol of alkalinity or 0.5 mol of $CaCO_3$ (50 g $CaCO_3$). The consumption of alkalinity during nitrification may decrease the pH of the wastewater if there is insufficient alkalinity available and can adversely affect the nitrification process, as discussed later in this chapter. Therefore, sufficient alkalinity is needed to consistently nitrify ammonia with approximately 7.14 mg/L of alkalinity consumption for each mg/L of ammonia oxidized. Table 8.1 presents the typical oxygen and alkalinity relationships of the nitrification process. Because of the consumption of alkalinity in the nitrification process, operators must ensure that there is sufficient alkalinity in incoming wastewater.

The growth rate of microorganisms is referred to as *kinetics*. The first step in nitrification (i.e., conversion of ammonia to nitrite) is the rate-limiting step because, typically, AOB have lower specific growth rates than NOB. *Rate-limiting* means that the second reaction can occur only as fast as the first reaction. Growth of AOB and NOB is the result of the oxidation of ammonia and nitrite, respectively. Both ammonia and dissolved oxygen are substrates for AOB growth. The concentrations of either or both of these substances could be low enough to limit specific AOB growth rate in wastewater treatment systems.

At temperatures lower than 25 °C, the growth rate of NOB is significantly higher than that of AOB (see Figure 8.2). As a result, typically, the growth rate of AOB controls the overall rate of nitrification (i.e., as soon as nitrite is produced by AOB, it is used by NOB). However, facilities that transition in and out of nitrification or have dissolved oxygen limitations may experience conditions where the NOB growth rate is unable to keep up with the AOB growth rate, resulting in elevated effluent nitrite concentrations. This condition is known as *nitrite lock* and results in a significant increase in effluent chlorine demand caused by high levels of nitrite in the effluent (5.1 mg of chlorine is consumed per milligram of nitrite, thus increasing the amount of chlorine needed to meet disinfection requirements by 5:1).

TABLE 8.1 Key Nitrification Relationships

Parameter	Coefficient units
Oxygen use	4.57 g O_2/g NH_4-N
Alkalinity consumption	7.14 g alkalinity (as $CaCO_3$)/g NH_4-N

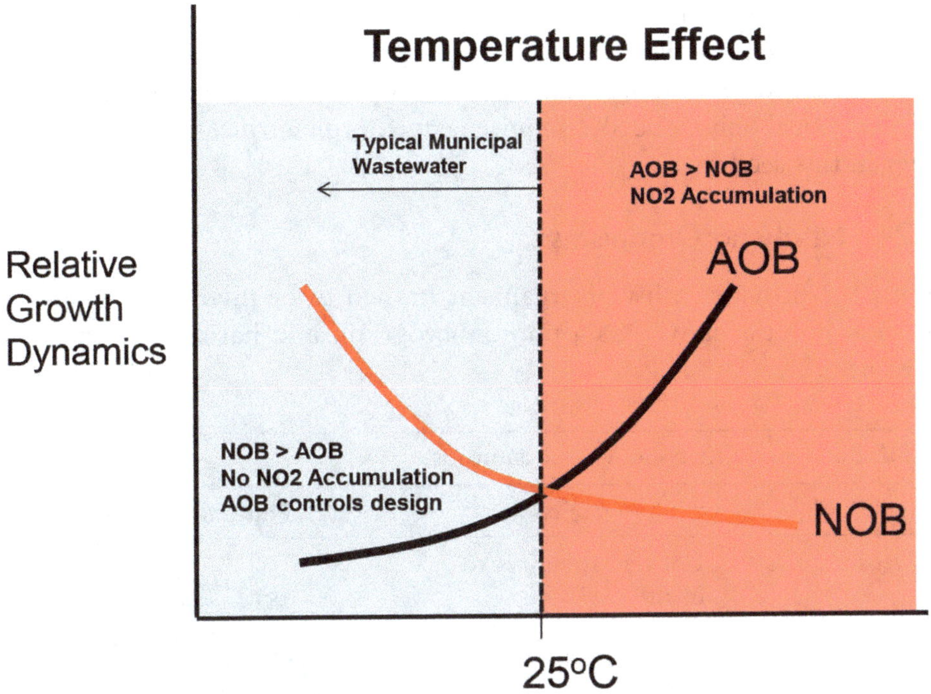

FIGURE 8.2 Temperature Effect of AOB and NOB Growth Rates

Typically, the overall kinetic nitrification reaction is expressed as follows:

$$\mu_n = \left(\frac{\mu_{nm} N}{K_n + N}\right)\left(\frac{DO}{K_o + DO}\right) - K_{nd} \tag{8.5}$$

Where

μ_n = specific growth rate of nitrifying bacteria (g new cells/g cells d),
μ_{nm} = maximum specific growth rate of nitrifying bacteria (g new cells/g cells d),
N = nitrogen concentration (g/m^3),
K_n = half-velocity constant (g/m^3),
K_{nd} = endogenous decay coefficient for nitrifying organisms (g volatile suspended solids [VSS]/g VSS·d),
DO = dissolved oxygen concentration (g/m^3), and
K_o = half-saturation coefficient for dissolved oxygen (g/m^3).

Bacterial growth rate of AOB, defined by the time required to double their population, is 10 to 20 times slower than that of aerobic heterotrophic bacteria used in BOD removal. Because the aerobic solids retention time (SRT) of an activated sludge nitrification system is inversely proportional to the net growth rate of the slowest growing nitrifiers (AOB), the aerobic SRT needed to effectively nitrify must be significantly longer than the typical SRT needed to maintain a stable heterotrophic bacterial population (i.e., BOD removal only). The SRT requirements for full nitrification are discussed later in this section.

1.2 Nitrifying Organisms

Table 8.2 provides a list of organisms known to be involved in the nitrification process as well as a photomicrograph and basic function of the organism.

TABLE 8.2 Key Nitrifying Microorganisms

Name	Function	Photomicrograph
Nitrosomonas	Convert ammonia to nitrite (AOB)	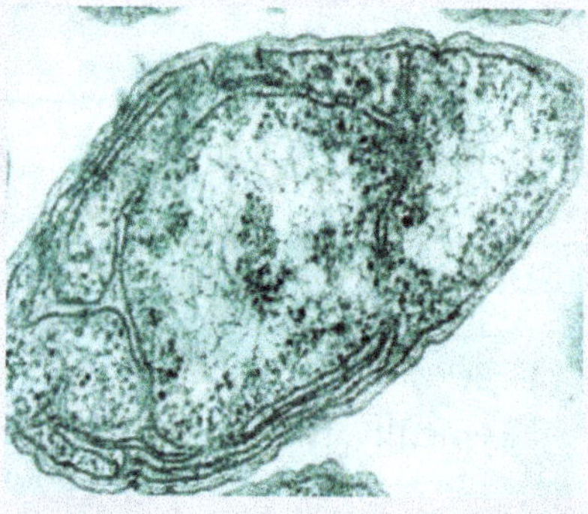
Nitrobacter	Convert nitrite to nitrate (NOB)	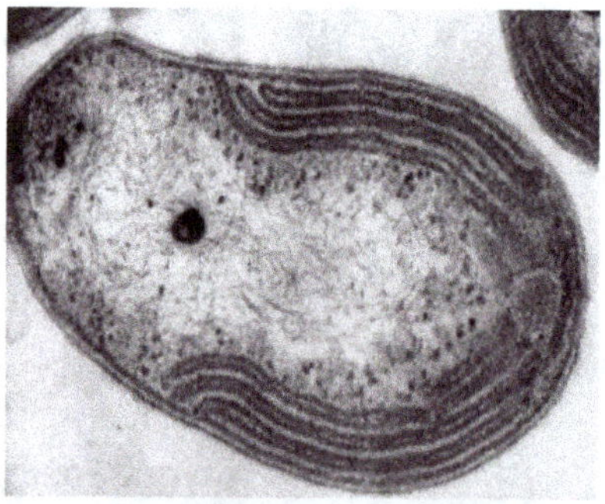
Nitrospira	Convert ammonia to nitrate (one step)	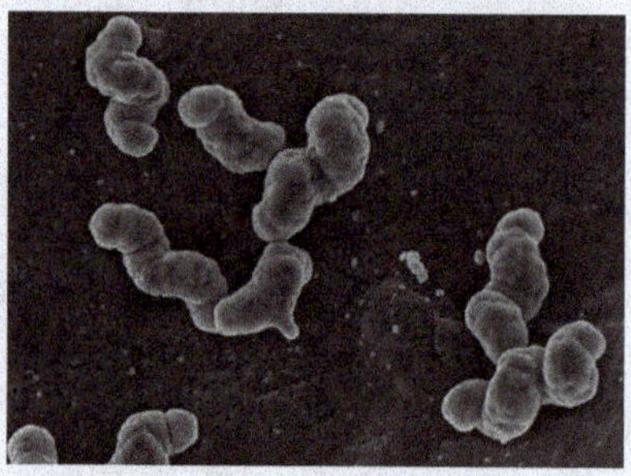

There is increasing evidence that *Nitrospira* plays an important role in nitrification, particularly with the tendency toward operating aeration basins at lower dissolved oxygen concentrations to facilitate nutrient removal. *Nitrospira* that can accomplish complete nitrification in a single step (NH_4^+ to NO_3^-) are called *comammox* and can nitrify at dissolved oxygen concentrations lower than 2 mg/L.

1.3 Process Control Variables

A number of environmental factors affect the nitrifier growth rate significantly, including but not limited to temperature, dissolved oxygen, pH and alkalinity, toxic substances, and variations in flows and loads. These factors and their effects on the process are discussed here.

1.3.1 Solids Retention Time and Temperature

Aerobic SRT and temperature of the mixed liquor are the two most important process control variables in the nitrification process. Aerobic SRT is defined as follows:

$$SRT_{aerobic} = \frac{Volume_{aerobic} \times MLSS}{(Flow_{WAS} \times TSS_{WAS}) + (Flow_{EFF} \times TSS_{EFF})} \tag{8.6}$$

When it comes to nitrification, only aerobic SRT should be used because nitrifying organisms only survive and metabolize ammonia in an aerobic environment. The SRT required to fully nitrify is highly dependent on the temperature of the mixed liquor, as shown by Equation 8.7, which is a function of the nitrifier's maximum specific growth rate ($\mu_{Aut,T}$) and decay rate ($b_{Aut,T}$). Figure 8.3 shows the relationship between SRT and temperature, including temperature-dependent equations for $\mu_{Aut,T}$ and $b_{Aut,T}$.

$$SRT_{min,aerobic} = \frac{1}{\mu_{Aut,T} - b_{Aut,T}} \tag{8.7}$$

Whereas nitrification in ideal conditions can be initiated at relatively low SRT, as indicated by the safety factor = 1 curve in Figure 8.3, WRRFs are designed using various degrees of "safety factors" to account for variability in several factors, including aeration basin configuration, influent nitrogen loadings, temperature, wasting schedules, and toxic compounds. The safety factor chosen is a matter of engineering judgment based on the known variability of these factors, but the single most important variable is the degree of compartmentalization or plug flow pattern of the aeration basin. Whereas a completely mixed basin might require a safety factor of 2

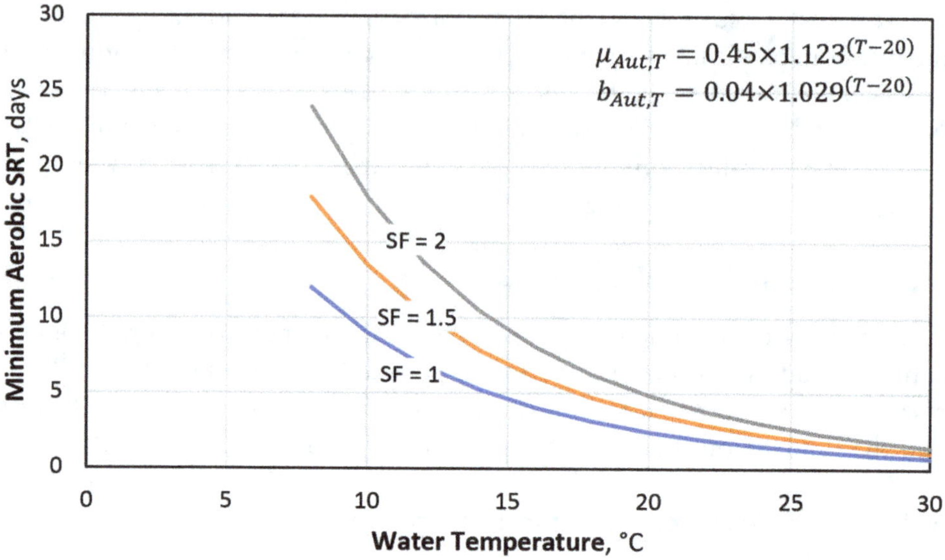

FIGURE 8.3 Solids Retention Time Required for Full Nitrification (Chen et al., 2023) (SF = safety factor; $\mu_{Aut,\,T}$ = maximum specific growth rate; $b_{Aut,\,T}$ = decay rate)

or more, a decreased safety factor can be used for a plug flow basin with a high food-to-microorganism ratio (F:M) gradient because of the proven benefits of plug flow to nitrification kinetics. Another important temperature relationship of operational importance is relative growth dynamics between AOB and NOB. A schematic representation of this effect was shown in Figure 8.2. Whereas slow-growing AOB dictate a required SRT at temperatures lower than 25°C, at higher temperatures, AOB growth can outpace NOB growth and the unwanted nitrite lock condition is more likely to occur at higher temperatures. Therefore, any condition inhibiting NOB growth, such as limited dissolved oxygen or inadequate SRT, may result in excessive nitrite in the effluent. This is particularly troublesome for WRRFs using chlorine for disinfection because nitrite has an extremely high chlorine demand.

1.3.2 Hydraulic Retention Time

Typically, hydraulic retention time (HRT) is less significant when it comes to nitrification. In a nitrifying wastewater treatment aeration basin, SRT is often more critical than HRT because nitrifying bacteria have relatively slow growth rates, so maintaining an adequate SRT ensures that these bacteria remain in the system long enough to effectively convert ammonia to nitrite and then to nitrate. On the other hand, HRT primarily affects the time wastewater spends in the basin, which is important but less directly tied to biological processes of nitrification. While most aeration basins are designed with an HRT of 6 hours or more, it has been proven that, with adequate SRT, complete nitrification can be achieved at an HRT as few as 3 hours. When HRT does become a factor is in a step-feed WRRF, where influent is allowed to be distributed at various points in the aeration basin. When significant amounts of raw wastewater or primary effluent are fed at the back end of the reactor, HRTs of 1 hour or less can result in a significant amount of ammonia in the effluent.

1.3.3 Dissolved Oxygen

Traditionally, it has been believed that maintaining a dissolved oxygen concentration of 2 mg/L or more has been necessary to ensure a stable nitrification process. This has been based on the effect of dissolved oxygen on AOB and NOB growth rates, as indicated in Figure 8.4.

As mentioned previously, many WRRFs can nitrify completely while operating much of the aerated portion of the bioreactor at a dissolved oxygen concentration of less than 2 mg/L. In fact, research and full-scale applications show that many facilities are able to completely nitrify at much lower dissolved oxygen concentrations (0.5 to 1 mg/L) after improving dissolved oxygen control and acclimating microbial activity to lower dissolved oxygen conditions. The growing understanding and successful application of process designs involving *anammox* and *comammox* bacteria will continue

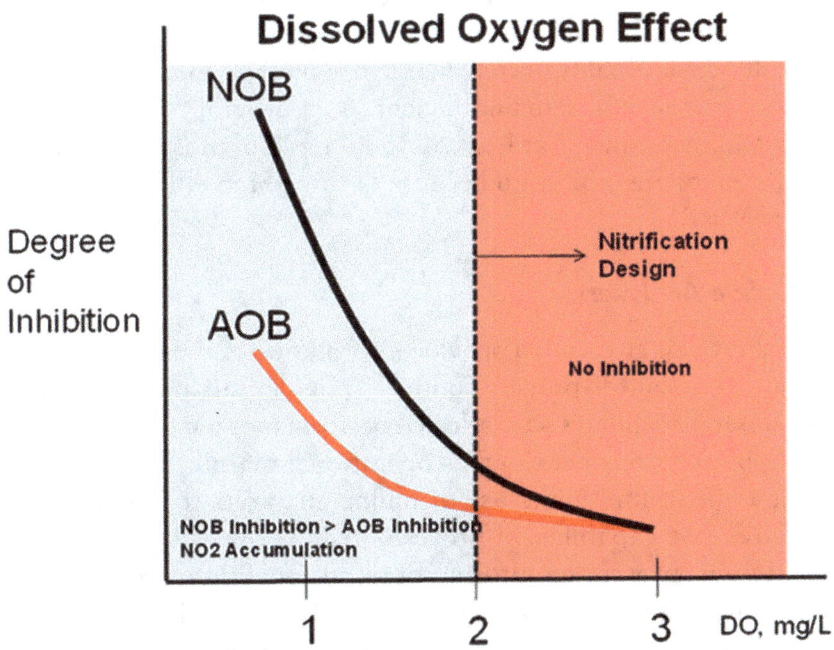

FIGURE 8.4 Effect of Dissolved Oxygen on AOB and NOB Growth Rates

to influence the design and operation of future WRRFs. Increasing emphasis is being placed on operating at lower dissolved oxygen levels for several reasons: cut operating costs, reduce carbon footprint, and improve nutrient removal. This is achieved through the application of specialized microbial strains, by enhancing simultaneous nitrification/denitrification in the aeration basin, and minimizing dissolved oxygen carryover in recycle streams to anaerobic and anoxic zones used for enhanced biological phosphorus and nitrogen removal.

1.3.4 Alkalinity/pH

As indicated in Table 8.1, 7.14 parts of alkalinity are consumed by every part of ammonia oxidized. For a typical influent of 40 to 60 mg/L of total nitrogen, 290 to 420 mg/L of alkalinity is required in raw wastewater. Therefore, the stability of the nitrification process at the WRRF is highly dependent on the amount of alkalinity in the water supply of the customers being served by the WRRF. Alkalinity in water supplies can vary between 50 and 400 mg/L. This stresses the importance of incorporating denitrification to the design of a nitrifying WRRF because of the return of alkalinity involved with the denitrification process, as explained later in this chapter. For facilities with low influent alkalinity that are not designed to denitrify, the feeding of supplemental alkalinity is frequently required.

Nitrification will consume alkalinity, resulting in decreased pH within the mixed liquor. Once the pH goes below 7, nitrification rates begin to decrease significantly with increasing hydrogen ion activity (Chen et al., 2023). If sufficient alkalinity is available to buffer against pH values lower than 7, the process will remain efficient. As a general rule, monitoring of alkalinity and maintaining at least 50 to 80 mg/L of alkalinity in effluent is preferred to mitigate inhibition because of low pH in effluent of alkalinity-challenged WRRFs.

1.3.5 Online Analyzers

Analysis of process streams to provide information and data on operational conditions is critical for process control. Though grab and composite sampling for laboratory analysis on selected performance parameters is certainly a viable approach for operators, as operational margins shrink because of tightening effluent limits, the use of online analyzers to provide real-time data and trending capability of process streams should be considered. The near-continuous process monitoring of online analyzers provides real-time understanding of time-of-day, day-of-week conditions that may contribute to performance variability and enable more timely operational decisions for optimized performance.

Although online analyzers present opportunity for improved data collection and operational understanding and control, proper implementation is critical to enable representative and reliable data collection. Each WRRF is unique, so the use of online instruments must be tailored to fit the application. Key considerations for online analyzer selection include flow and load variability, location of and conditions at the desired monitoring point, technology selection for the application/level of needed accuracy, potential interfering agents in the process stream, and equipment maintenance. Because of differing site-specific conditions, pilot testing of online analyzers should be considered before implementation to confirm performance response and reliability for the particular application.

To optimize nitrification, there are several types of analyzers that are currently used to support process control decisions and, in some cases, these are integrated with automatic control systems for equipment operation. These analyzers are presented in the following sections.

1.3.5.1 Dissolved Oxygen

Typically, dissolved oxygen sensors are divided into electrochemical sensors and optical sensors. Electrochemical sensors measure dissolved oxygen based on the electrochemical reduction of oxygen at a cathode. The current measured in the sensor is proportional to the oxygen concentration in the basin. Optical sensors use a luminescent dye that emits light when excited by a light source. Typically, when compared to electrochemical sensors, optical sensors are more stable and require less maintenance; however, they have slower response times.

Maintaining dissolved oxygen sensors in WRRFs can be challenging because of harsh conditions in the wastewater environment. Membranes on electrochemical sensors can become fouled or damaged and, often, require frequent calibration and membrane replacement to ensure accuracy. While more robust, optical sensors can also suffer from fouling; periodic cleaning can help mitigate instrument drift. Both types of sensors can be affected by temperature fluctuations, which requires compensation to obtain reliable measurements. Dissolved oxygen sensors are crucial for optimizing the aeration process and maximizing operational efficiency. Routine and proper maintenance and calibration of these sensors is required to maintain data integrity that guides operational decisions.

1.3.5.2 Ammonia

There are various online ammonia (ammonium) analyzers that make use of different sensing technologies, including ion-selective electrode, spectrophotometry, and gas-sensing electrode. These analyzers can be grouped into two general types: console/wet chemistry units and insertion probe analyzers. Typically, probe-type analyzers are more easily implemented but may not provide accuracy or reliability, particularly at lower ammonium concentrations. Wet chemistry analyzers can measure more consistently to lower concentrations but require sample delivery and/or filtration ahead of the analyzer and, typically, are higher maintenance systems.

Selection of the right analyzer/technology is application-based. If online instruments are used to supplement process control data for influent ammonia loading and nitrification performance conditions, then probe-type systems could be favored. If accuracy and reliability are essential for automated equipment control, such as for aeration system/aeration blower control, then wet chemistry analyzers may offer a higher level of reliability and consistency needed for the application.

1.3.5.3 pH/Alkalinity

The measurement of pH can be accomplished using standard pH/conductivity-type probes. However, if ammonium probes are considered, some manufacturers offer pH in combination with the ammonium probe analyzer.

Online alkalinity measurement can be beneficial for process control when influent alkalinity is insufficient to meet nitrification alkalinity demand, necessitating the addition of supplemental alkalinity. These analyzers are based on wet chemistry and require sample delivery to the units. Two basic analytical methods are used: phenolphthalein titration method and direct measurement based on CO_2 gas generation.

Titration analyzers are colorimetry-based and are sensitive to sample quality and consistency. Solids and air bubbles in the sample stream will cause inconsistent performance; therefore, sample filtration and care to eliminate bubbles in the sample delivery tubing is essential.

The alkalinity direct measurement method is based on CO_2 derived from the analysis of total inorganic carbon (TIC) using acid digestion and infrared analysis. The alkalinity concentration of that sample is calculated based on stoichiometric relationships of alkalinity, TIC, and CO_2.

1.3.5.4 Total Suspended Solids

Total suspend solids (TSS) analyzer probes are useful in a number of locations in the process for ongoing monitoring of solids distribution in the system and managing return activated sludge (RAS), WAS, and SRT conditions for optimum nitrification. The TSS probe can be immersed in the aeration basin or inserted to process piping (RAS, WAS). The TSS concentration is derived from measurement of infrared light absorbance.

1.4 Operational Considerations

1.4.1 Temperature

Although temperature certainly affects the kinetics of nitrification, operators have minimal control over operating temperature conditions. Flow equalization may provide a measure of near-term relief from cold stormwater flow or ice melt conditions but, as seasonal conditions trend towards lower or higher temperatures, operators must track operational conditions and adjust biomass to balance responsive performance. Understanding the conditions and trends of flows, loads, pH, dissolved oxygen, SRT, and temperature that deliver optimum nitrification provides the operator knowledge and understanding of the relationships in this balance, how changes in the balance influence performance, and when operational adjustments are needed to maintain balance.

In locations where temperatures begin to drop in the fall, the typical operational response is to reduce WAS to increase SRT (sustain the slower-growing nitrifier population). However, this can lead to excessive SRT and, possibly, aggravate foaming and/or settleability issues.

A better strategy, while providing adequate alkalinity/pH and dissolved oxygen, is to track conditions that affect nitrification within the facility and how those are affected with changing temperature. In complete-mix activated sludge systems, trend influent loads, SRT, and aeration basin effluent temperature and ammonia to develop the understanding of the lowest SRT for the temperature and loading conditions that provide adequate stable nitrification. In plug flow basins, track the characteristics of the facility's fluctuating flows and loads compared to the point along the plug flow path where adequate nitrification has been completed. As temperatures drop, this nitrification completion point will move further downstream toward the basin effluent. With this information, as temperatures drop (or as they increase in spring), SRT can be adjusted based on data to maintain system balance while sustaining adequate nitrifiers to achieve nitrification performance.

Online instruments are ideal for tracking operational balance, trending performance, and providing operators tools to proactively address process adjustments.

1.4.2 Supplemental Alkalinity Addition

In many areas, the alkalinity content of influent wastewater is inadequate to supply the alkalinity demands for nitrification. Also, the alkalinity in dewatering sidestreams from anaerobic digesters is limiting when compared to ammonia loads returned. In these cases, supplemental alkalinity is often required to provide adequate pH buffering (minimize pH decline) and prevent nitrification inhibition.

There are a number of alkalinity/pH control supplements that may be used; common supplements are magnesium hydroxide [$Mg(OH)_2$], calcium hydroxide [$Ca(OH)_2$, lime], sodium hydroxide (NaOH, caustic soda), and sodium carbonate (Na_2CO_3, soda ash). The use of these supplements may be dependent on a number of criteria, including performance, cost, operator preference, safety considerations, and storage and feed system capability. It is important to note that each facility situation is dynamic, and the control application is best determined for each facility through full-scale testing to determine the most cost-effective chemical for the application. Jar testing may not provide an equivalent evaluation of the various control chemicals.

Magnesium hydroxide slurry (milk of magnesia) is increasingly becoming an alkalinity supplement in many applications because of the stability of the alkalinity delivered and the measured pH response, reduced safety concerns, and ease of application. Lime (calcium hydroxide) is another option for alkalinity addition but, typically, requires higher dosage per pound of alkalinity delivered than magnesium hydroxide. Lime can also create localized, elevated pH, depending on feed location, and increases sludge production because of precipitation of calcium carbonate and magnesium carbonate as a result of its softening effect while reducing carbonate alkalinity and the buffering capacity of the water. If lime is procured as a powder rather than a slurry, powdered material handling and slurry preparation require extra equipment and operation and maintenance.

Sodium hydroxide (caustic soda) is a liquid and more directly controls pH. It requires careful handling and feed considerations. It reacts quickly and immediately elevates the pH of the receiving liquid stream. Depending on the feed location, localized pH may reach 10 or higher, which is detrimental to the affected mixing zone. This elevation in pH condition is dependent on the chemical dosage necessary to overcome the magnitude of alkalinity deficiency.

Other less commonly used alkalinity/pH control supplements are calcium oxide (CaO, quick lime) and sodium bicarbonate ($NaHCO_2$, baking soda). Of the commonly used chemicals, only soda ash and baking soda add carbonate alkalinity, whereas the other chemicals provide hydroxide alkalinity. Only carbonate alkalinity adds to the buffering capacity of the water, so this needs to be considered when dealing with low alkalinity waters.

1.4.3 Special Inhibitory Conditions and Compounds

Nitrifying bacteria are more susceptible to inhibition than heterotrophic bacteria. Substrate inhibition can occur when elevated dissolved ammonia concentrations greater than 10 mg/L occur at conditions higher than 8.5 pH. This inhibition condition may arise if nitrification performance declines because of aeration system issues causing inadequate dissolved oxygen for nitrification and pH increases. Under these conditions, pH must be decreased to 7 to 7.5 to revive nitrification and operators must be ready to supplement alkalinity upon nitrification restart.

If inadequate alkalinity is available for nitrification, pH will decline and, as pH drops, nitrification slows as nitrifiers are inhibited. Additionally, nitrous acid formation from nitrite develops as pH approaches 6.0, which further inhibits the conversion of nitrite to nitrate.

Excessive loads of carbonaceous BOD or drops in ammonia loading (increases in the BOD-to-TKN ratio) may inhibit nitrification because heterotrophic activity with its relative nutrient requirements and air demand will predominate, limiting nitrifier action.

Beyond substrate-type inhibition, there are many compounds that can cause inhibition of nitrifiers, including heavy metals and organic and inorganic chemicals like quaternary ammonium compounds. For more information on toxicity to nitrifiers, refer to *Industrial Water Quality* (Eckenfelder et al., 2008).

1.4.4 Sidestream Loads

Sidestream flows from anaerobic digester/biosolids-dewatering operations can deliver significant ammonia loads and inadequate alkalinity back to the biological system. Care should be taken to avoid excessive loads from sidestreams that overwhelm the operational and control capabilities of the aeration system, alkalinity feed, and SRT of the system, and cause pass-through of ammonia to facility effluent. Strategies to control sidestream loads include managing dewatering activities, time of day, duration and dewatering flowrates, and storage/equalization of sidestream flow to manage return during lower diurnal flow/loading periods to level the loading to the aeration basin.

If sidestream ammonia loads cannot be accommodated by the treatment system by physical load control/management, sidestream treatment and/or improvements to the existing treatment system to accommodate this loading would be needed. If centrifuges are used for dewatering, consideration should be given to full-time operation because of the operational issues associated with starting and stopping centrifuges.

1.4.5 Endogenous Release

In WRRFs with low ammonia limits, operators must be aware of the potential for ammonia release from settled activated sludge in the secondary clarifiers because of endogenous bacterial activity and cell lysis. To minimize this condition, RAS pumping rates should be adjusted to minimize residence time in the secondary clarifiers. This will result in minimum sludge blanket depths (<25% of the clarifier's side water depth). Sludge blankets should be monitored regularly to provide understanding of sludge blanket conditions and the effects from adjustments to RAS and WAS rates. Additionally, the SRT must be managed to promote generation of a well-settling, manageable mixed liquor suspended solids (MLSS).

These minor ammonia releases can be beneficial when the WRRF is practicing chlorination for effluent disinfection. When the nitrification process in the bioreactor results in inadequate concentration of ammonia to form a stable monochloramine residual, this small ammonia release in the final clarifier can be significant and avoid the need to feed supplemental ammonia, as some fully nitrifying WRRFs have found the need to do. Typically, practicing breakpoint (free chlorine) disinfection on unfiltered WRRF effluent requires high dosages of chlorine and chloramination is preferred.

2.0 DENITRIFICATION BASICS

2.1 Stoichiometry and Kinetics

In denitrification, bacteria reduce nitrate to nitrogen gas. Nitrogen gas is relatively insoluble in water, so it is released to the atmosphere. The atmosphere naturally consists of more than 70% nitrogen, so emissions released from denitrification at WRRFs do not adversely affect the environment.

The stoichiometry of the denitrification process depends on the carbon source used:

Wastewater

$$C_{10}H_{19}O_3N + 10\ NO_3^- \rightarrow 5\ N_2 + 10\ CO_2 + 3\ H_2O + NH_3 + 10\ OH^- \quad (8.8)$$

Methanol

$$5\ CH_3OH + 6\ NO_3^- \rightarrow 3\ N_2 + 5\ CO_2 + 7\ H_2O + 6\ OH^- \quad (8.9)$$

Acetate

$$5\ CH_3COOH + 8\ NO_3^- \rightarrow 4\ N_2 + 10\ CO_2 + 6\ H_2O + 8\ OH^- \quad (8.10)$$

Equation 8.11 is typically used to describe the kinetics associated with the denitrification process:

$$r_{su} = \left(\frac{kXS}{K_s + S}\right)\left(\frac{NO_3}{K_{s,NO_3} + NO_3}\right)\left(\frac{K'_o}{K'_o + DO}\right)\eta, \tag{8.11}$$

Where:

K'_o = dissolved oxygen inhibition coefficient for nitrate reduction (mg/L) (0.1 to 0.2 mg/L),

K_{s,NO_3} = half-velocity coefficient for nitrate-limited reaction (mg/L) (0.1 mg/L),

DO = dissolved oxygen, and

η = fraction of denitrifying bacteria in the biomass (g VSS/g VSS).

Understanding biochemical reactions associated with denitrification is important when operating WRRFs. For example, reducing nitrate-nitrogen eliminates some of the BOD demand in the aerobic zone because some of the BOD is removed in the anoxic zone. Theoretically, 2.86 mg of oxygen is recovered for every 1.0 mg of nitrate-nitrogen reduced to nitrogen gas. This is more than 60% of the oxygen needed for nitrification, so this recovered oxygen is valuable in offsetting aeration costs in nitrifying WRRFs. This is why it is good practice to always incorporate denitrification in nitrifying WRRFs when possible.

Denitrification also results in new bacterial cells. The cell yield depends on the carbon source. For example, if methanol is the carbon source, then the cell yield is approximately 0.5 mg VSS per 1.0 mg of nitrate-nitrogen removed. If BOD is the source, then the cell yield is approximately 1.5 mg VSS/1.0 mg of nitrate-nitrogen removed. Another advantage of denitrification is that cell yield per kilogram (pound) of BOD removed is approximately 80% of aerobic cell yield, so fewer waste biosolids are produced in denitrifying facilities.

In addition, approximately 3.57 mg of alkalinity (as calcium carbonate) is produced for each 1.0 mg of nitrate-nitrogen removed. So, approximately 50% of the alkalinity lost during nitrification can be recovered during denitrification. Denitrifiers are less sensitive than nitrifiers, so if the WRRF environment does not inhibit the nitrifiers, then denitrifiers should have no problem functioning at optimal growth rates.

Therefore, in addition to the environmental benefit of producing an effluent lower in total nitrogen, as a result of these three operational benefits (oxygen credit, alkalinity credit, and lower biomass yield), operators should always strive to maximize the amount of denitrification in the WRRF.

2.2 Denitrifying Organisms

Denitrification is carried out by *facultative heterotrophic microorganisms*, that is, organisms that use organic carbon for food and energy (referred to as an *electron donor*) and oxygen or nitrate for cell synthesis (referred to as an *electron acceptor*). Heterotrophic organisms spend less energy on synthesis than autotrophic organisms do, so they grow more quickly and yield more cell mass. Numerous heterotrophic bacteria can denitrify wastewater. Bacteria that reduce nitrate are referred to as *denitrifiers* and are "facultative" bacteria, meaning they can function in both aerobic and anoxic environments. *Anoxic* is defined as less than 0.3 mg/L dissolved oxygen, but with either nitrate or nitrite available. Denitrifiers prefer to use molecular oxygen, but if the environment contains less than 0.3 mg/L of dissolved oxygen, they will cleave oxygen from nitrate-nitrogen molecules to synthesize carbon compounds (e.g., BOD), thus producing nitrogen gas.

2.3 Process Configurations

2.3.1 Preanoxic Denitrification

In preanoxic denitrification, the anoxic zone(s) precedes the aeration basin and allows the use of raw wastewater or primary effluent BOD to drive the denitrification reaction. Nitrified mixed liquor is recycled from the end of the aeration basin to the anoxic zone(s) at a rate of 200% to 700% of the influent flow. Because 100% of influent flow cannot be recycled, this configuration is used to achieve total nitrogen limits of 6 to 10 mg/L.

2.3.2 Postanoxic Denitrification

In postanoxic denitrification, the anoxic zone is located downstream of the aeration basin. This is done for WRRFs with total nitrogen limits of less than 6 to 8 mg/L to further denitrify the forward flow that cannot be captured by the nitrified mixed liquor recycle. Postanoxic denitrification can be designed to use carbon released from endogenous decay of the mixed liquor or a supplemental carbon source. Using a supplemental carbon source allows for significant downsizing of the postanoxic zone.

2.3.3 Postsecondary Denitrification

Postsecondary denitrification is used to achieve "limits of technology" in nitrogen removal, or less than 3 mg/L. Supplemental carbon is added to tertiary treatment processes, typically filters.

2.3.4 Simultaneous Nitrification/Denitrification and Sludge Granulation

There is rapidly increasing interest in simultaneous nitrification/denitrification (SND) as greater emphasis is being placed on operating WRRFs at lower dissolved oxygen to save energy costs and enhance nutrient removal. With the discovery of low dissolved oxygen nitrifying organisms such as *Nitrospira* and *anammox*, we are now capable of achieving significant total nitrogen removal in the aeration basin. One of the most rapidly developing fields of research today is the enhancement of granular flocs in activated sludge. By various manipulations of the process, such as using high substrate (food-to-mass ratio) gradients along the bioreactor, surface wasting and/or physical separation (e.g., hydrocyclone), only the heaviest flocs are retained in the process and dense granules form within the floc. These granules are able to harbor layers of nitrifying and denitrifying bacteria in the same floc, similar to fixed-film processes. This can lead to the process known as "nitrite shunt," where ammonia is oxidized only to nitrite before denitrifiers complete denitrification cycles, resulting in savings of both oxygen required for nitrification and carbon required for denitrification. Also, if conditions are favorable, particularly in warmer climates, this can lead to the development of a red-colored bacteria known as *anammox* that are able to anaerobically metabolize ammonia and convert it to nitrogen gas. One commercially available system utilizing the nitrifying organism *Anammox* to convert ammonium and nitrite into nitrogen gas is the Anammox® process. The Anammox® process requires less energy and less oxygen than other processes, and no additional carbon source is needed. Because of the energy and capital cost benefits, sludge granulation and SND may soon become the preferred method of operating WRRFs.

2.3.5 Alternating Aeration

Alternating aeration is used in sequencing batch reactors (SBRs), some flow-through processes, and some low dissolved oxygen systems with inadequate blower turndown to maintain low dissolved oxygen operations. Aeration is turned on and off using timed cycles. If operated properly, alternating aeration can achieve lower total nitrogen limits than staged processes by exploiting both the purely oxic/anoxic and SND principles.

2.4 Process Control Variables

2.4.1 Carbon-to-Nitrogen Ratio

The amount of nitrate that can be removed from a given wastewater is a function of organic carbon availability and environmental and operating conditions and is primarily determined as the ratio between the electron donor and acceptor (i.e., COD-to-nitrate-nitrogen ratio). Studies have reported a wide range of COD:nitrogen required for satisfactory denitrification. For a typical municipal wastewater and a Modified Ludzack–Ettinger process with an anoxic zone volume equal to one-third of the aerated volume (oxic SRT of 10 days, anoxic SRT of 3.3 days), inhibition of denitrification is seen at an influent COD:TKN of less than 10:1 and complete denitrification is achieved at an influent COD:TKN of 13:1. If there is insufficient carbon in the wastewater, then an external source is required. When external carbon sources are used, these COD:TKN are greatly reduced because of the highly biodegradable nature of these carbon sources, with a dose of 5 to 6 g supplemental carbon COD/g of NO_3 removed being sufficient.

2.4.2 Carbon Sources

A large variety of supplemental carbon sources have been used to increase denitrification rates. Typically, when a supplemental carbon source is required, the first action would be to survey local industries, particularly food-processing industries that might be producing a high COD waste product. This not only provides a low-cost supplemental carbon source for the municipality, but it can reduce operating costs for the industry. When local carbon is not readily available, methanol is commonly used as a supplement. Although methanol is dangerous to ship, store, and handle, it is still the most cost-effective carbon source available today. However, because of the hazards associated with methanol, many municipalities are turning to alternative carbon sources that provide a better work environment for their operators. Table 8.3 provides some basic physical characteristics for the most commonly used carbon sources.

2.4.3 Solids Retention Time

The degree of denitrification, although primarily dependent on COD:nitrogen, is also a function of anoxic SRT. Typically, denitrifying organisms require a 2- to 4-day SRT to produce 0.5 mg of VSS per 1.0 mg of BOD removed; however, given carbon limitations in many wastewaters, longer anoxic SRTs are frequently used.

2.4.4 Oxidation–Reduction Potential

Oxidation–reduction potential (ORP) is a measure of the ionic charge in water in millivolts. A positive ORP is indicative of oxidizing conditions and a negative ORP is indicative of reducing conditions. Online ORP meters can assist the operator in determining the ionic charge in the mixed liquor at various points in the aeration basin using real-time data. Oxidation–reduction

TABLE 8.3 Physical Characteristics of Commonly Used Supplemental Carbon Sources

Property[a]	Methanol	Ethanol	MicroC® 1000[b]	MicroC® 2000[b]	Acetic acid
Typical strength, % lb chemical/lb solution	100%	95%	100%	100% (71% glycerine)	56%
Bulk density at 25 °C, lb solution/gal solution	6.6	6.6	10.2	10.2	8.71
COD dose required per NO_3-N, lb COD/lb N[c]	4.82	6.36	6.45	6.45	6.09
COD available in chemical, lb COD/lb chemical	1.5	1.98	0.54	0.90	0.98
Dose required, gal solution/lb NO_3-N	0.48	0.46	1.15	0.70	0.71
Freezing point, °C	−98	−114	−7	−35	16.6

[a]1 lb/gal = 0.12 g/cm^3 and 1 gal/lb = 8.33 cm^3/g.
[b]Proprietary chemical with product name subject to change.
[c]WERF Report NUTRIR06b.

potential is more accurate than dissolved oxygen readings in determining denitrification conditions. Optimum denitrification is achieved at an ORP from −50 to +50 mV. An ORP of less than −50 mV is indicative of anaerobic conditions. At −50 mV, odorous sulfide formation occurs, and this is an indication that the nitrate recycle rate needs to be increased. Many WRRFs are oversized and have excessive anoxic volume; for these facilities, once maximum nitrate removal is achieved, the ORP can drop to less than −50 mV in the latter stages of the anoxic zones. At −100 mV, volatile fatty acid formation begins, resulting in conditions required for enhanced biological phosphorus removal (EBPR). Frequently, these underloaded WRRFs will achieve EBPR despite not having a dedicated anaerobic zone; this is known as "incidental" EBPR. An additional discussion of ORP is provided in Section 3.3.4.

2.4.5 Online Nitrate Analyzers

Online nitrate analyzers use ion-selective electrode probes, spectrophotometry units, and UV measurement probe technologies. The technology used needs to be selected based on the specific application. Selection criteria/considerations include sampling location, wastewater character/facility conditions, level of accuracy/consistency desired, and data use (process control monitoring vs equipment control).

Upon evaluation of the technology that is determined best to meet the conditions of need and use, pilot testing of the selected technology is recommended to confirm acceptable performance for the application.

2.4.6 Internal Mixed Liquor Recycle Ratio

Internal mixed liquor recycle (IMLR) refers to staged denitrification facilities where nitrified mixed liquor is recycled using low-head pumps to the anoxic zones at the beginning stages of the bioreactor or in oxidation ditches where a portion of the recirculated flow is diverted to the anoxic zone(s) using an adjustable gate. This allows for the use of raw wastewater or primary effluent to drive the denitrification reaction. The optimum IMLR rate for most WRRFs ranges from 200% to 400% of influent flow. One of the most useful aspects of mechanistic treatment process models is determination of optimum IMLR rates. Typically, higher IMLR rates result in increased denitrification, but the benefit rapidly decreases at IMLR rates >400%. There are two problems with high IMLR rates: (a) excessive dissolved oxygen being recycled to the anoxic zone (known as "oxygen poisoning") and (b) the energy associated with IMLR pumping may offset the nitrate oxygen credit benefits. Frequently, modeling can predict significant gains in denitrification at IMLR rates exceeding 400%, but these high IMLR rates are rarely used because of high pumping costs. An exception to this case has been found in oxidation ditch facilities where high IMLR rates can be achieved by simply adjusting the internal recycle gate position. Figure 8.5 provides a typical example of expected effluent nitrate concentrations for three different influent TKN concentrations.

2.5 Operational Considerations

2.5.1 Temperature

Denitrification performance, as with other biological processes, is affected by changes in temperature; denitrification rates will increase or decrease as temperatures rise and fall. However, denitrification performance is not as limited with respect to temperature as nitrification. Therefore, with

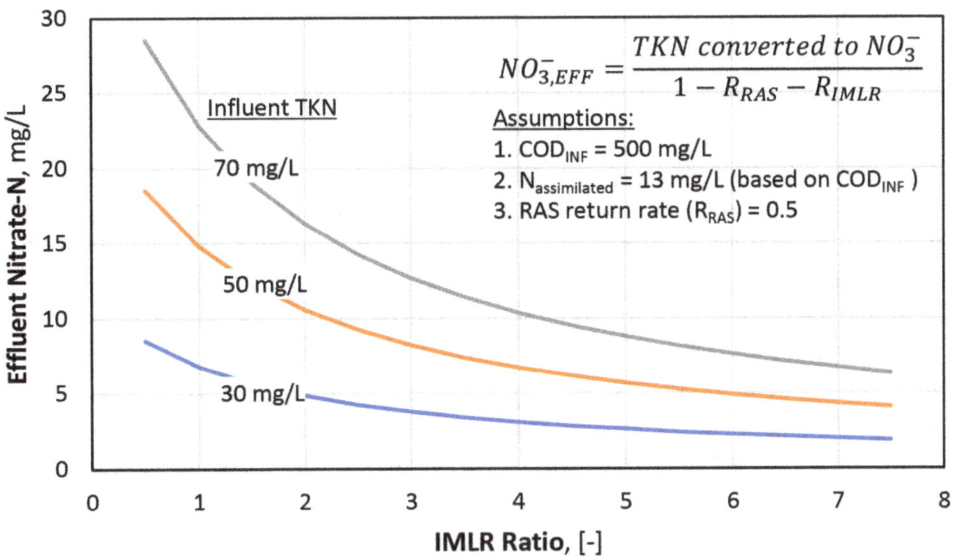

FIGURE 8.5 Effect of IMLR on Effluent Nitrate

temperature as the variable, denitrification performance will continue as long as the nitrification process produces nitrate.

2.5.2 Alkalinity/pH

As alkalinity is consumed by the nitrification process (7.1 mg alkalinity consumed/mg ammonia nitrified), the denitrification process returns alkalinity (3.5 mg alkalinity returned/mg NO_3–N denitrified). This approximately 50% alkalinity credit can be critical for maintaining adequate alkalinity in the system for pH stability. Also, if supplemental alkalinity is being purchased to ensure adequate alkalinity for nitrification, return of this "free" alkalinity can result in significant chemical cost savings.

2.5.3 Dissolved Oxygen Levels

Optimum use of anoxic zone volume is critical for optimum denitrification control and efficiency. Dissolved oxygen admittance to the anoxic zone reduces the capacity of the anoxic zone by reducing the effective anoxic HRT and reduces the carbon available for denitrification because facultative bacteria will use the available dissolved oxygen preferentially in consuming the carbon. Likewise, if supplemental carbon is being added to support denitrification, it will also be consumed aerobically, resulting in increased chemical costs.

Sources of dissolved oxygen in the anoxic zone may be raw feed, transfer at the liquid surface, and IMLR. Dissolved oxygen in raw influent feed should be minimal as it is consumed in the collection system. The transition of influent through preliminary/primary treatment process units to the anoxic zone may introduce some dissolved oxygen to the process, so dissolved oxygen should be checked periodically to ensure that resulting dissolved oxygen levels are minimal.

Excessive mixing/turbulence in the anoxic zone creates more contact time at the liquid–air interface and transfer of dissolved oxygen to the liquid. Although this should be minor, it can contribute to dissolved oxygen loading and resulting inefficiencies.

The most significant point for introduction of dissolved oxygen to the anoxic zone is from IMLR, which transfers nitrified flow from the end of the aeration basin back to the anoxic zone. Minimizing dissolved oxygen by managing aeration conditions at the pump suction location is important. Some WRRFs have zones at the end of the aeration tank to enable control of aeration rates, along with mechanical mixing to help reduce dissolved oxygen levels in IMLR flow.

As facility flows increase toward design capacity, preserving effective anoxic zone capacity becomes increasingly important to enable optimum denitrification performance.

2.5.4 Sidestream Loads

Sidestream flows from dewatering of biosolids produced from anaerobic digesters can affect the carbon-to-nitrogen balance in the treatment system and cause issues with nitrification and denitrification processes. Typically, these sidestreams deliver high levels of ammonia and disproportionately lower levels of alkalinity for nitrification and carbon for denitrification.

Flow management of the return of sidestreams may help reduce effects to treatment processes, but supplemental carbon for denitrification may be necessary to overcome sidestream-induced carbon deficiency.

2.5.5 Denitrification in Secondary Clarifiers

Denitrification in the secondary clarifier sludge blanket should be minimized to avoid negative effects to final effluent quality. Denitrification in secondary clarifiers is promoted by elevated levels of nitrate in MLSS flow from the aeration basin, elevated SRT, long clarifier HRT, excessive sludge blanket levels, and warmer water temperatures. Under these conditions, carbon needed for denitrification becomes available through endogenous activity and cell lysis.

The early stage of denitrification in secondary clarifiers is characterized by fine, ashlike solids accumulating on the clarifier surface. In this stage, fine TSS particles are lifted to the surface by near-microscopic nitrogen gas bubbles. This is a common occurrence and is controlled so as not to become excessive to the point of affecting downstream treatment processes. If denitrification levels in secondary clarifiers increase, larger solids are lifted by coalescing nitrogen bubbles within the sludge blanket and, in advanced denitrification, clumps and cakes of settled solids are lifted to the surface. This flotation of once settled sludge is termed *rising sludge*.

Typically, reducing the nitrate load to clarifiers, minimizing sludge blankets via RAS control, and managing the number of clarifiers in operation can help control inadvertent and excessive denitrification in secondary clarifiers. As a general rule, maintaining an effluent nitrate of less than 10 mg/L helps to minimize sludge blanket denitrification.

2.5.6 Nitrogen Gas Stripping/Release

Nitrogen gas contained in denitrified MLSS, particularly in postanoxic zones provided to meet low nitrogen limits, must be liberated from MLSS before secondary clarification to promote flocculation of solids and good settling conditions. Postaeration zones and/or aerated clarifier distribution channels can help liberate entrained nitrogen gas and promote improved settling conditions.

2.5.7 Carbon Addition

Supplemental carbon addition to enhance denitrification is often necessary because of inadequate/disproportionate levels of carbon in influent flows. One approach to gain additional carbon is to increase carbon in influent flow. If an industrial source of carbon is available, consideration should be given to relaxing pretreatment requirements to exploit this "free" carbon source. A more direct approach for carbon supplementation is to procure products and byproducts such as glycerin, sugars, and methanol for controlled feed as required by process conditions.

Determining the type of product and the magnitude and frequency of the need to supplement carbon is important because overfeeding is costly and may raise effluent BOD enough to violate permit limits and underfeeding may result in inadequate denitrification. Therefore, supplemental carbon feed requirements may fluctuate. Feed may be necessary only during certain times of the day, week, and seasonally based on flows, loads, and permit conditions. Based on requirements of a particular facility, decisions on the type of chemical, feed system, and manual versus automatic feed control must be determined.

Carbon sources found in raw influent and sugars provide a level of complexity to the organic carbon menu that can promote biodiversity and enhance process stability. Carbon sources such as methanol, ethanol, acetate, and glycerin can narrow denitrification population diversity, creating more of a dependency on a particular substrate.

Flow and load consistency at a particular facility determines the level of sophistication needed in a carbon feed system. A manually controlled feed system may be sufficient for facilities with consistent/predictable flows and loads, but, if significant fluctuation is common, automated feed systems that track and respond to varying carbon demands may be a cost-effective approach. Online nitrate analyzers and feed control systems can be used effectively in these cases to optimize carbon feed and denitrification performance.

An additional means to supplement carbon includes managing carbon at a facility but perhaps is in a different location at the facility than needed. Those internal carbon management strategies could include bypassing primary clarifiers, changing the number of primary clarifiers in service, implementing primary sludge fermentation, implementing in situ mixed liquor, and/or implementing RAS fermentation.

3.0 ENHANCED BIOLOGICAL PHOSPHORUS REMOVAL BASICS

This section considers the factors surrounding the use of a special consortium of microorganisms that have the capacity to uptake phosphorus in mass amounts in excess of what they would need for normal cellular growth. These special organisms, named *phosphorus-accumulating organisms* (PAOs), are capable of removing up to 80% of influent reactive phosphorus completely through biological metabolism. Because all organisms require some phosphorus for growth, this process is known today as EBPR. The basic reaction environments for effective EBPR are first anaerobic, where inorganic phosphorus (PO_4^{3-}) is released from the PAO cell into the bulk solution, followed by aerobic, where inorganic phosphorus is taken up into the PAO cell. Removal of phosphorus then occurs through the PAO cell being wasted via WAS. In domestic municipal wastewater, typically, pH is typically basic (7.0 to 8.0 pH), which means that dominant inorganic phosphorus is in the form of hydrogen phosphate (HPO_4^{2-}). For simplicity, this is written as PO_4 or o-PO_4 for “phosphate” or “orthophosphate”, respectively. A schematic of the EBPR process is shown in Figure 8.6.

3.1 Stoichiometry and Kinetics

Stoichiometry is a physical description of those things in wastewater that react with one another to form some type of byproduct. In biological systems, these reactions are mediated by bacterial activity. So, another way to describe stoichiometry is the removal of various constituents present in the

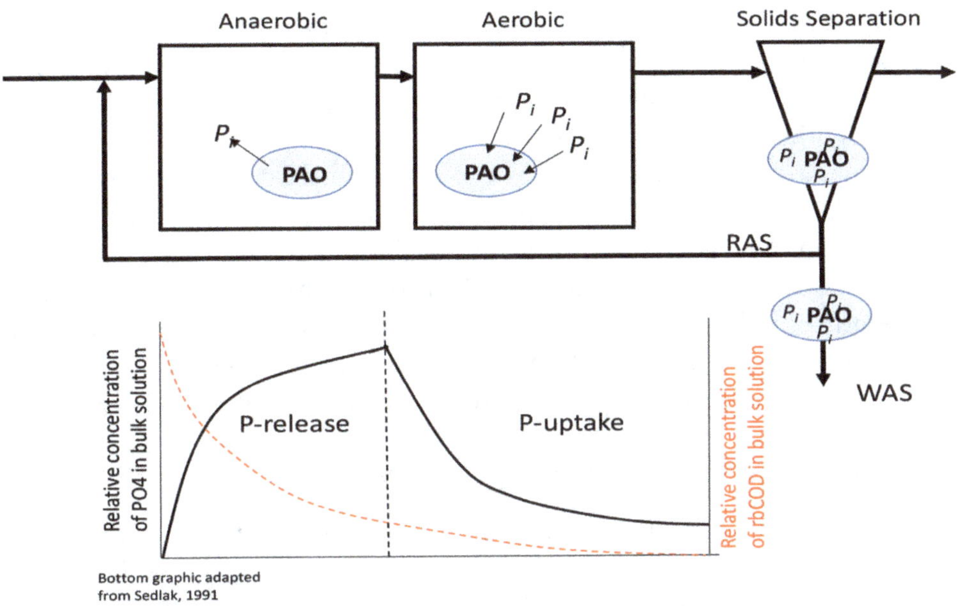

FIGURE 8.6 Basic EBPR Process Schematic (Bottom graphic adapted from Sedlak, 1991) (rbCOD = readily biodegradable COD)

waste stream by microorganisms. Simplistically, to support their growth, PAOs in an EBPR system remove organic carbon, phosphorus, and metal cations. The relative ratios of these constituents to one another in wastewater determines the overall efficiency of the EBPR process. Because municipal wastewater typically contains more than sufficient amounts of cationic metals (e.g., K^+, Mg^{2+}, Ca^{2+}), typically, EBPR stoichiometry focuses on the relative ratio between influent forms of COD and phosphorus. Forms of COD of interest, in order of increasing breakdown and increasing direct influence on EBPR, are total COD (the sum of the soluble and particulate fractions), soluble COD (sCOD), readily biodegradable COD (rbCOD), and volatile fatty acids (VFAs).

Each of these forms is present in wastewater. Under anaerobic conditions (such as in an influent pipeline), some COD can ferment or break down into more viable forms. As a result, in a typical municipal wastewater, there will be some VFAs, rbCOD, and sCOD. The most important constituents for effective EBPR are VFAs. The predominant VFA form from fermentation is acetate, followed by propionate. In general, approximately 1.0 g of acetate is produced from 1 g of rbCOD. This VFA is used by PAOs for growth and, because this growth takes place in an anaerobic environment, the overall yield is low, in the range of 0.45 g VSS/g acetate consumed by PAOs. By these facts, the general stoichiometric range of an EBPR system ranges between 8 and 20 g phosphorus removed/g rbCOD by PAO biomass. Obviously, because phosphorus is required by all organisms, additional phosphorus will be taken up for synthesis of ordinary heterotrophic and autotrophic microorganisms, resulting in the total removal by the total biological treatment system being the sum of that taken up by PAO biomass and that taken up by all other microbial synthesis.

While stoichiometry describes those things that will react with one another, kinetics describe the rates, or speeds, with which those reactions take place in the reaction environment. Furthermore, kinetics are dependent on the ambient reaction environment; factors such as dissolved oxygen, pH, temperature, and the presence/absence of inhibitory substances; and operational parameters such as HRT and SRT. In general, PAOs have similar rates of reaction driving their growth to that of ordinary heterotrophic organisms.

3.2 Phosphorus-Accumulating Organisms

Table 8.4 lists the dominant species of microorganisms present and active in the EBPR process. The two classes of microorganisms affecting the performance of an EBPR process are PAOs and glycogen-accumulating organisms (GAOs). Whereas PAOs have the ability to store polyphosphate for energy during VFA uptake, GAOs do not. Because both PAOs and GAOs thrive in the anaerobic environment, they compete directly for available VFAs. In an anaerobic environment in which the GAO is the favored competitor, phosphorus-removal performance is greatly diminished. So, it is important that the anaerobic environment be created and sustained whereby PAOs actively outcompete GAOs for available VFA substrate.

3.3 Process Control Variables

3.3.1 Carbon-to-Phosphorus Ratio

As previously stated, the effectiveness of the EBPR process is highly dependent on the type and quantity of readily available carbon present in the anaerobic zone, specifically VFA in the form of acetate and propionic acid,

TABLE 8.4 Important Enhanced Biological Phosphorus Removal Microorganisms

Phosphorus-accumulating organisms	Glycogen-accumulating organisms (GAOs)
Accumulibacter phosphatis–Type 1 (able to use NO_2^- or NO_3^- as electron acceptors and, thus, classifies also as "denitrifying PAO")	*Competibacter* (all GAOs can use NO_3^- as an electron acceptor and, thus, classify also as "denitrifying GAOs")
Accumulibacter phosphatis–Type 2	
Acinetobacter	

as the primary substrate for PAO growth and driver for release of stored polyphosphate into solution. Although VFA : phosphorus is the fundamental parametric necessary for a reliable prediction of EBPR performance, most operators do not have immediate access to test methods for determining VFA concentration. Table 8.5 provides a set of ratios for VFA and other conventional descriptions of organic carbon that can be referenced to predict similar reliability.

It should be noted that these ratios all must increase if the overall process is operated with long SRTs (in excess of 15 days) or if there are residuals of dissolved oxygen and/or nitrate present that enter the anaerobic zone. This is because these parameters will consume VFAs, leaving less available for optimal PAO growth, resulting in lower phosphorus-removal efficiency. Typically, when the ratios shown in Table 8.5 are present, an effluent phosphorus concentration of 1.0 mg/L is achievable.

3.3.2 *Volatile Fatty Acids*

It cannot be emphasized enough how important is the role of VFAs in the EBPR process. Volatile fatty acids are short-chained carbon (fewer than 6 linked carbon atoms per molecule) compounds derived from the fermented breakdown of large, organic, readily biodegradable COD compounds such as polysaccharides, carbohydrates, sugars, and fats and oils. For the purposes of EBPR operations, VFAs can be produced in pipelines upstream of the facility, if those lines are anaerobic. If the influent carbon-to-phosphorus ratio is favorable, VFA formation is assumed to occur in the anaerobic zone of the EBPR system.

The type of VFA present is critical to the effective performance of the EBPR process. Typically, the biodegradable organic carbon present in municipal wastewater, when fermented, produces VFA fractions of approximately 75% acetate and 25% propionate. This is fortunate because PAOs can metabolize both acetate and propionate, whereas GAOs can only metabolize

TABLE 8.5 Minimum Ratios for Achieving Total Phosphorus Effluent Concentrations of Less Than 1.0 mg/L Phosphorus (WEF, 2018)

Substrate type	Recommended minimum ratio	Notes
$cBOD_5$: TP	20	Rough initial estimate
COD : TP	45	More accurate than $cBOD_5$
rbCOD : TP	15	More accurate than COD
VFA : TP	7–10	Most accurate and representative

acetate. Therefore, if propionate is present in reasonable amounts, typically, PAOs will outcompete GAOs for survival.

Where influent carbon-to-phosphorus ratios are not favorable for EBPR, operators can consider fermentation of primary sludge to acquire deficient VFAs. Primary sludge fermentation can be accomplished within the blankets of primary clarifiers or in a sidestream "reactor," such as a gravity thickener. The gravity thickener option is preferred because there is better control of the sludge blanket depth and it is not influenced by influent diurnal fluctuations. For information on primary sludge fermentation techniques and process designs, refer to *Standards for Design of Water Resource Recovery Facilities* (WEF, 2023).

Because GAO proliferation in the anaerobic zone must be minimized to ensure effective phosphorus-removal efficiency, factors other than readily degradable carbon are equally important. Attention to pH is critical. Phosphorus-accumulating organisms grow more favorably at pH values higher than 7.0, and experience shows that 7.5 pH is optimal because GAOs assimilate and metabolize available VFAs at much faster rates than PAOs at pH values lower than 7.0 (Filipe et al., 2001). Temperature is also an important factor. Typically, PAOs favor cooler water temperatures than GAOs and, at lower temperatures, PAOs are favored regardless of pH. At temperatures below 20 °C, growth rates for PAOs exceed GAOs, providing a competitive advantage for PAOs (Lopez-Vazquez et al., 2008). Unfortunately, the majority of municipal wastewaters fall within a range of 15 to 30 °C, a range within which GAOs compete effectively. Thus, it is critical that pH and a balanced VFA species condition is maintained. If this is done consistently, PAOs' dominance can be sustained.

3.3.3 Solids and Hydraulic Retention Time

The SRT for the overall biological nutrient removal (BNR) system tends to be controlled to maintain a high enough SRT for sufficient nitrifier growth, as discussed earlier in this chapter. Because phosphorus uptake in the aerobic environment is essential to the growth of PAOs, the SRT of the aerobic portion is also critical in EBPR. As mentioned previously, research has found that the specific growth rate for PAOs under controlled conditions is 0.95 g VSS/g VSS·d. This is significantly faster than that of nitrifiers, implying that nitrifiers will need a longer SRT than PAOs. In fact, as Mamais and Jenkins found from their research in 1992, for a conventional EBPR configuration (anaerobic selector followed by an aerobic zone), the minimum aerobic SRT for effective phosphorus removal was 2.5 days (for temperatures of at least 20°C). In a BNR system that is required to also remove nitrogen, this aerobic SRT is well below what would be needed for the nitrogen removal component. So, this minimum aerobic SRT is notable for biological phosphorus removal only systems where nitrogen removal is not required (e.g., the anaerobic/oxic [A/O] process). It should be noted that when long SRTs are imposed on reaction environments that are also lightly loaded, significant endogenous decay (lysing) of all organisms, including PAOs, occurs. This obviously depletes the PAO population. For those that remain in the mixed liquor, they deplete their store of carbon to remain viable. The reduction in PAO population, coupled with the reduced store of carbon, significantly reduce overall EBPR performance. As a general rule, then, experience shows that operating at an SRT just slightly above that needed to sustain nitrification is optimal for effective and sustained EBPR. This emphasizes the need for SRT control of combined nitrogen and phosphorus removal WRRFs to operate at minimum SRT for complete nitrification, depending on the mixed liquor temperature, to achieve best EBPR performance.

Experience has shown that, more critical to effective and efficient performance of the EBPR process, is HRT of the anaerobic zone. The uptake of VFA and subsequent release of inorganic phosphorus to the bulk solution is a relatively rapid reaction. Therefore, the time required for the PAO to complete these reactions is relatively short. In general, for municipal wastewater EBPR systems, the anaerobic zone is sized to allow for a nominal HRT of between 0.75 and 1.0 hours. This allows ample time for PAOs to take up available VFAs and expel their storage of phosphorus.

Some research has shown that this HRT can be as long as 3 hours. From an infrastructural point of view, this is excessive because HRT is a parameter by which tank volumes are selected; however, if there is a desire to repurpose existing infrastructure, whereby a tank volume is already fixed such that the HRT will be greater than 1 hour, it is helpful to know that some flexibility exists with this upper HRT limit. Nonetheless, the operator must be cognizant of the anaerobic HRT. Once PAOs consume available VFAs and have released their store of phosphate, if they continue to remain in the anaerobic environment, they will begin to use stored carbon to remain viable. The longer they remain in the anaerobic zone, the more of this carbon they will consume, and additional phosphorus can be released. However, that specific released fraction of the total release is done without the use of VFAs. Such a phenomenon has been labeled *secondary release*. Unfortunately, because PAOs have used stored carbon for viability during the extended anaerobic period, it means that less carbon will be available to use for the phosphorus uptake reaction when PAOs enter the aerobic zone. As a consequence, PAOs will end up taking in less phosphorus, leaving more phosphorus to be present in effluent and showing a gradual increase in effluent concentration over time. This results in overall decreased EBPR performance.

3.3.4 Oxidation–Reduction Potential

Effective performance of a BNR process requires sustaining critical reaction environments. These environments are aerobic, anoxic, and anaerobic. A means of measuring whether a reaction environment is any of these can be done by measuring the ORP of the solution. The ORP can be measured indirectly with an ORP probe, which measures the conductivity of the solution. Conductivity is related to the ionic strength of a solution as reflected by the concentration of ionized dissolved solids in solution. The units of the ORP probe are millivolts. Figure 8.7 illustrates the breadth of ORP across the biological reaction spectrum.

As is evidenced by Figure 8.7, reaction environments do not exist in isolation; rather, they function over ranges of ORP. For EBPR, typically, the anaerobic environment begins at approximately −50 mV because fermentation is possible at that ORP. In practice, however, most EBPR facilities consider anaerobic conditions beginning at less then −150 mV to be optimal for effective phosphorus release in the anaerobic zone, and at least +150 mV in the aerobic zone to ensure effective uptake.

3.3.5 Online Phosphorus Analyzers

Total phosphorus and soluble (ortho) phosphorus can be measured using online monitoring systems. For process control applications, online monitoring of orthophosphate is the most applicable system because orthophosphate levels present the functionality and performance of the EBPR system. Orthophosphate analyzer units use colorimetric/spectrophotometric

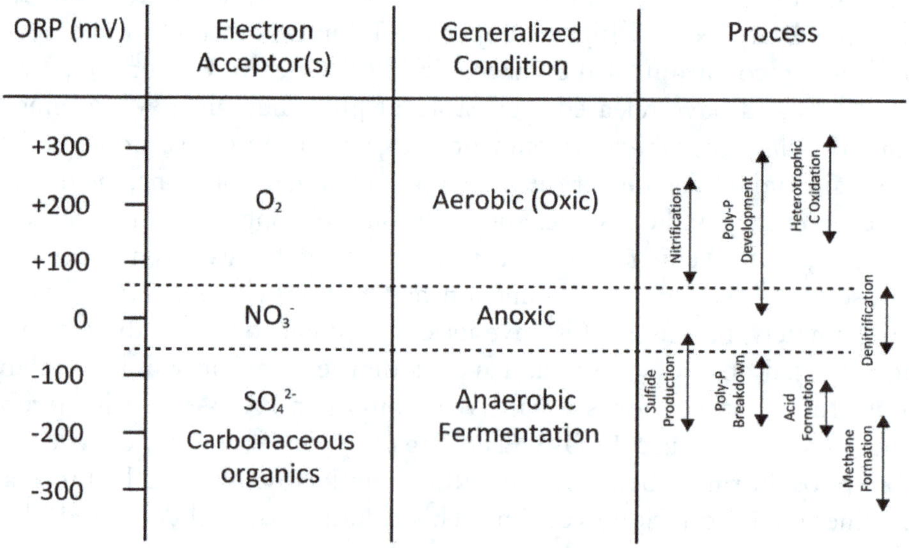

FIGURE 8.7 Oxidation–Reduction Potential Spectrum of Metabolic Processes

methods for analysis to deliver near-continuous phosphorus concentration data for the process stream sampled. Samples for analysis must be filtered and pumped to the unit for analysis. Many units have integrated sampling/filtering systems, but stand-alone sampling systems are available. Locations in EBPR systems for online orthophosphate analyzers include anaerobic zone influent, anaerobic zone, oxic zone, and secondary clarifier effluent.

3.3.6 Return Activated Sludge

In all EBPR processes, RAS is routed through the anaerobic zone (whether it be the anaerobic zone in the mainstream BNR process or through a side-stream phosphorus-release anaerobic tank). Because RAS flowrates are not insignificant (typically ranging between 50% and 100% of facility influent flowrate), RAS affects the localized HRT of the anaerobic zone. Because phosphorus release in the anaerobic zone is crucial to the effectiveness of EBPR and that release is dependent on the amount of time available in the anaerobic zone for release reactions to occur, RAS flowrate needs to be carefully controlled so as to not exceed the recommended HRT of the anaerobic zone with RAS flowrates that are too low or "short-circuit" the HRT of the anaerobic zone with RAS flowrates that are too high. An excessive HRT runs the risk of expending too much stored carbon, resulting in less phosphorus uptake in the aerobic zone. Conversely, a "short-circuited" HRT runs the risk of less-than-optimal phosphorus release occurring, again resulting in less-efficient uptake downstream. Operators should be aware of the design HRT of the anaerobic zone and adjust RAS accordingly.

3.4 Operational Considerations

3.4.1 Temperature

Typically, temperature effects on PAO organism activity follow trends of biological systems because temperature trends change during seasonal fluctuations. However, in many cases, overall EBPR system performance appears to benefit during winter operations. This is thought to be related to reduced competition from GAOs as temperatures decline. Experimental evidence suggests that GAOs tend to become stronger competitors with PAOs at temperatures higher than 20 °C (Filipe et al., 2001). This implies that competition by GAOs with PAOs in EBPR facilities may be more problematic in warm weather conditions and in warm climates.

3.4.2 Dissolved Oxygen

Optimal use of anaerobic zone volume is critical for optimal phosphorus release. Dissolved oxygen admittance to the anaerobic zone reduces the capacity of the anaerobic zone by reducing the effective anaerobic HRT and reduces carbon available for phosphorus release because facultative bacteria will use available dissolved oxygen preferentially in consuming carbon. Likewise, if supplemental carbon is being added to support phosphorus release, it will also be consumed aerobically, resulting in increased chemical costs.

Sources of dissolved oxygen in the anaerobic zone may be raw feed, transfer at the liquid surface, and RAS. Dissolved oxygen in raw influent feed should be minimal as it is consumed in the collection system. The transition of influent through preliminary/primary treatment process units to the anoxic zone may introduce some dissolved oxygen to the process, so dissolved oxygen should be checked periodically to ensure resulting dissolved oxygen levels in flow to the anaerobic zone are minimal.

Excessive mixing/turbulence in the anaerobic zone creates more contact time at the liquid–air interface and transfer of dissolved oxygen to the liquid. Typically, although this is minor, it can contribute to dissolved oxygen loading and resulting inefficiencies.

Dissolved oxygen may be contained in RAS flow to the anaerobic zone. Like surface dissolved oxygen transfer, typically, this is minimal but can contribute to dissolved oxygen loading and loss of efficiency.

As facility flows increase toward design capacity, preserving effective anaerobic zone capacity becomes increasingly important to enable optimum anaerobic zone performance.

3.4.3 pH

The EBPR performance can be influenced by system pH. Operating at slightly alkaline pH levels of 7.0 to 7.5 will tend to favor PAO growth over GAO growth, which leads to strengthening of EBPR performance. The optimum pH for a particular EBPR system should be evaluated and monitored over time to optimize performance while minimizing supplemental chemical addition for alkalinity/pH control.

3.4.4 Nitrate

Nitrate contained in RAS will further reduce the effective anaerobic HRT and reduce carbon available for phosphorus release because facultative bacteria will use the available nitrate preferentially in consuming the available carbon. Likewise, if supplemental carbon is being added to support phosphorus release, it will also be consumed anoxically if nitrate is present, resulting in increased chemical costs.

Nitrate levels in RAS can vary based on denitrification system performance and conditions in the secondary clarifier and, typically, is the same as in facility effluent. Therefore, care must be taken to control/reduce RAS return rates to enable ample MLSS settling/concentration, minimize the mass of nitrate returned, and preserve anaerobic zone HRT.

3.4.5 Sidestream Loads

Sidestreams from anaerobic digester decanting and biosolids dewatering will contain elevated levels of orthophosphate from PAO release and cell lysis during digestion, but disproportionally low levels of available carbon. Sidestream flows may overload EBPR system capacity and can result in pass-through of soluble phosphorus to facility effluent. In particular, intermittent sidestream flows may periodically overwhelm the capacity of the EBPR system. In this case, managing sidestream flows by selectively operating decanting and dewatering equipment or by flow equalization can be effective for balancing phosphorus loading to the system to stay within the capacity of the EBPR system. Additionally, sidestream treatment to reduce orthophosphate in the return flow is possible. Chemical addition to dewatering operations and nutrient harvesting from the sidestream are potential options.

3.4.6 Secondary Release

Secondary phosphorus release occurs under conditions without carbon storage (Barnard et al., 2017). This phosphorus release is not associated with PAO selection and this phosphorus will not be taken up in the aerobic zone. Therefore, if secondary release occurs, it can cause increases in effluent phosphorus. Conditions associated with secondary release include excessive detention times in anaerobic, anoxic or aerobic phases; RAS returned to primary clarifiers; excessive HRT/septic conditions in secondary clarifier sludge blankets; blending of primary and EBPR sludges; and failure to aerate stored EBPR sludges.

3.4.7 Carbon Addition

Supplemental carbon addition to promote anaerobic-zone phosphorus release is often necessary because of inadequate/disproportionate levels of carbon in influent flows and sidestreams. Determining the type of product and the magnitude and frequency of the need to supplement carbon is important because overfeeding can be detrimental and underfeeding may lead to inadequate performance and elevated effluent phosphorus. Therefore, supplemental carbon feed requirements may fluctuate. Feed may be necessary only during certain times of the day, week, and seasonally based on flows and loads and permit conditions. Based on the requirements of a particular facility, decisions on the type of chemical, feed system, and manual versus automatic feed control must be determined.

Carbon sources found in raw influent provide a level of complexity to the organic carbon VFA menu, promote a more robust PAO population, and enhance EBPR performance. Sidestream fermentation of primary sludge is a means to produce a mixture of VFAs for supplementing and balancing carbon demand with phosphorus-loading conditions. Pure supplemental carbon sources such as acetic acid can create a substrate dependency and, if fed under less-than-ideal pH and temperature conditions for PAOs, can give further advantage to GAO growth over PAO growth and reduce EBPR performance.

3.5 Process Configurations

3.5.1 Enhanced Biological Phosphorus Removal

3.5.1.1 Anaerobic and Oxic Configuration

Figure 8.8 illustrates the most basic EBPR reactor configuration in use today at WRRFs required to remove phosphorus to low levels. This process is traditionally known as the A/O process. (It should be noted that, in early literature sources, this process was known as the Phoredox process, but this terminology has seen less use in recent years). The anaerobic zone is first in the reaction sequence and receives the primary effluent commingled with RAS. The readily biodegradable carbon in the primary effluent is fermented to the appropriate VFA in the anaerobic zone, providing for biomass synthesis for PAOs and release of phosphorus. The A/O process is used when neither nitrification nor total nitrogen removal is required by the discharge permit. As such, typically, the aerobic SRT is designed to be below that which is necessary for nitrification, in the range of 2 to 5 days (for wastewater temperatures ranging from 10 to 20 °C). However, in climates where wastewater temperatures are chronically warm, nitrification can occur despite efforts by the operator to keep it from occurring. In these cases, nitrates produced during nitrification can be returned to the anaerobic zone (often termed an *anaerobic selector* in an A/O process). This "poisons" the anaerobic zone because the nitrate present will be denitrified, consuming

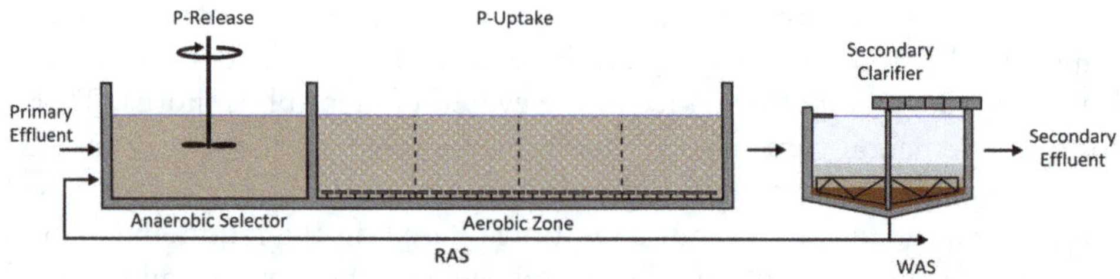

FIGURE 8.8 Anaerobic/Oxic Process

VFA carbon sources needed for EBPR. Because of this, implementation of the A/O process must be evaluated carefully.

3.5.1.2 Phostrip

The Phostrip process was developed to enhance the release of phosphorus from PAOs. Figure 8.9 shows an illustration of the Phostrip process. The process builds upon the configuration of the A/O process but relies on mostly phosphorus removal by chemical means rather than removal of phosphorus only via WAS.

The Phostrip process is more complex operationally than the A/O process because additional sidestream sequences are required. The key additions are first the anaerobic phosphorus stripper tank, which is essentially a gravity thickener. It receives a slip stream of WAS rich with PAOs. Typically, this tank has an HRT between 8 to 12 hours. This extended period of time promotes endogenous decay of a portion of the PAO-rich WAS, providing a VFA source for high levels of phosphorus release. Phosphate-rich supernatant from this tank is then conveyed to a second tank, where metal salts (or lime) are added to bind up phosphate into a thickened chemical sludge for disposal. Supernatant from this tank is returned to the head of the facility.

The fundamental objective behind process designs for EBPR is to maximize phosphorus release in the anaerobic environment so that an enhanced uptake of phosphorus by PAOs (far more than they need to live) can occur in the subsequent aerobic environment. This means that all use of VFAs present must be given top priority to PAOs. Any condition that limits PAOs' ability to favorably compete for VFAs will cause the EBPR performance to suffer. Therefore, process design is set up to make the anaerobic zone free

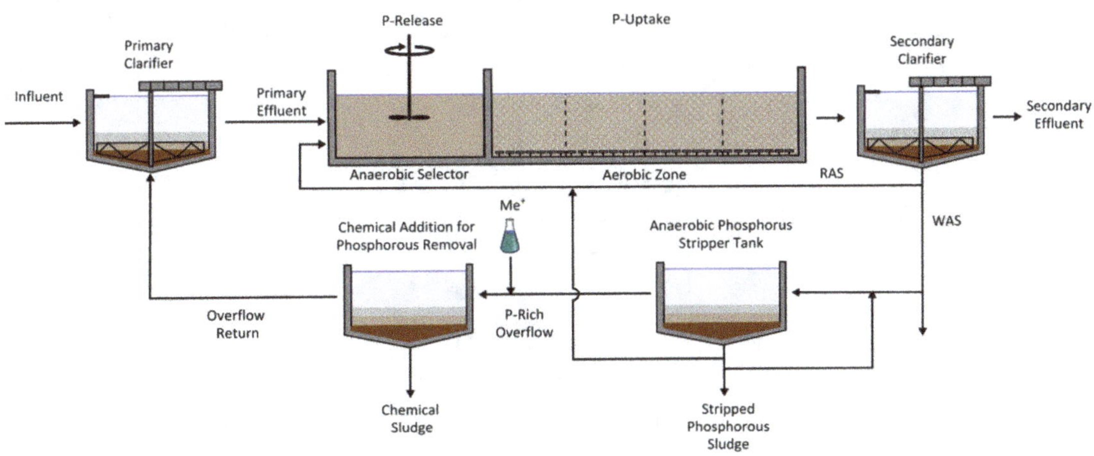

FIGURE 8.9 Phostrip Process

from residual dissolved oxygen and nitrate for introduction to the anaerobic zone via RAS. If either of these are present in the anaerobic zone, even at low concentrations, ordinary heterotroph organisms (OHOs) and denitrifiers in mixed liquor will immediately become active because they now have their required terminal electron acceptors present (oxygen for the OHOs and NO_3^- for denitrifiers). When this occurs, OHOs and denitrifiers will "steal" VFAs from PAOs, limiting PAO growth. In high-strength wastewater where the COD-to-phosphorus ratio is higher than 60 to 70, typically, there is enough rbCOD present to satisfy both residual dissolved oxygen and/or NO_3 demand and plenty still available for sustaining PAO growth. In weaker wastewater, this is not the and EBPR performance drops. For this situation, the process design is modified to prevent the RAS from entering the anaerobic zone directly. This keeps the anaerobic zone free of any residual dissolved oxygen and/or NO_3. It should be noted that when modification of the process is not physically feasible, supplemental carbon (VFA) sources, such as acetate, can be added to boost the COD-to-phosphorus ratio into more favorable levels.

3.5.1.3 Sidestream Enhanced Biological Phosphorus Removal

One of the primary struggles with developing robust biological phosphorus removal is ensuring stable VFA supply to the anaerobic zone. This is especially true if the collection system's production of VFAs is seasonal. To address this challenge, facilities have incorporated flow schemes that allow a portion of the mixed liquor or RAS to ferment and produce VFAs. This concept is termed sidestream EBPR (S2EBPR). Figure 8.10 illustrates three variations of S2EBPR.

Sidestream RAS fermentation involves redirecting a portion of RAS to a dedicated anaerobic reactor (or zone) with an extended SRT in this zone. For example, Danish facilities have implemented this configuration since the 1990s with good success by diverting 4% to 7% of RAS and achieving an

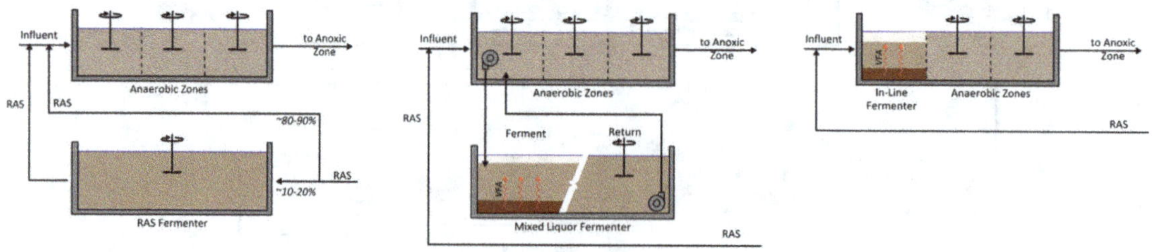

(a) Sidestream RAS Fermentation

(b) Sidestream MLSS Fermentation

(c) Sidestream Unmixed In-line MLSS Fermentation

FIGURE 8.10 Side-Stream EBPR Configurations

SRT of 30 to 40 hours within the sidestream RAS fermentation zone (Vollertsen et al., 2006). Sidestream MLSS fermentation can also be an effective method to produce VFAs, where MLSS is pumped into an anaerobic reactor (or zone) that is independent of the mainstream flow. Because MLSS is diluted when compared to RAS, this flow scheme may require a larger volume to achieve the same SRT and VFA production as RAS fermentation. To overcome this limitation, the MLSS fermenter is often unmixed for a portion of the day, allowing solids to settle and decouple SRT and HRT. A simplified version of MLSS fermentation is unmixed inline MLSS fermentation, where mixing within an anaerobic zone (or zones) is stopped for portion of the day, allowing solids to settle and ferment. Case studies of several facilities with successful MLSS fermentation are described by Barnard et al. (2017).

3.5.2 Combined Phosphorus and Nitrogen Removal

As stated previously, typically, processes designed for combined EBPR and nitrogen are best operated at the lowest SRT that achieves complete nitrification, depending on mixed liquor temperature, to ensure maximum stability of the EBPR process. Historically, SRT control has been based on curves such as those shown in Figure 8.3 that are based on "washout" SRT plus a safety factor. Because required safety factors vary from facility to facility, there is increasing interest in the use of online ammonia probes, such as those discussed in Section 1.3.5.2, to control the SRT to the minimum required for nitrification. These sensors are used to reduce SRT to the point where low concentrations of ammonia are detected in effluent and, in this way, the operator knows that the WRRF is running at the minimum SRT for nitrification and maximizing the stability of the EBPR process.

The anaerobic/anoxic/oxic (A^2O) process (Figure 8.11) is the most fundamental process for combined biological phosphorus and nitrogen removal. Typically, the process is capable of meeting moderately low limits (total nitrogen of 8 to 10 mg/L, total phosphorus of 1 to 1.5 mg/L). The disadvantage of this configuration is that nitrate in RAS can consume

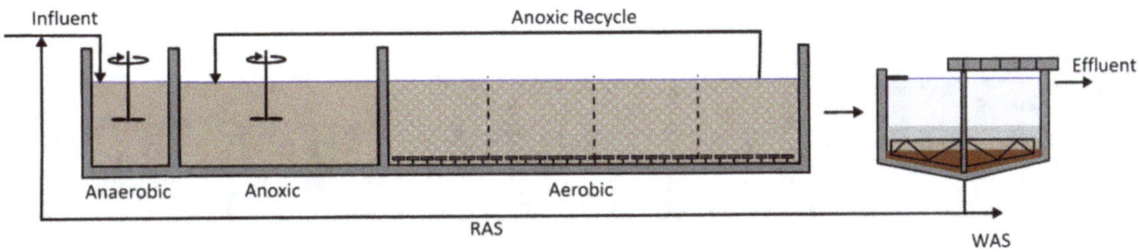

FIGURE 8.11 Anaerobic/Anoxic/Oxic Process

readily available carbon that is required in the anaerobic zone to initiate the phosphorus release/uptake cycle that is key to the EBPR process. The anoxic recycle is typically 2 to 5 times influent flow.

The five-stage Bardenpho process (Figure 8.12) is commonly used to meet lower nitrogen limits (3 to 8 mg/L). Because nitrate in the effluent is so low, nitrate effects on the anaerobic zone are minimized and phosphorus removal is enhanced as well.

In the University of Cape Town (UCT) process and Virginia Initiative Process (VIP) (Figure 8.13), denitrified mixed liquor is recycled from the anoxic zone to minimize nitrate effects on the anaerobic zone. Typically, anaerobic recycle is 1 to 2 times the influent flow. The fundamental difference between the VIP and UCT process is that shorter aeration times are used in the VIP.

In the modified UCT process (Figure 8.14), a second anoxic zone is added, with the first anoxic zone being used for anaerobic recycle. Because the first anoxic zone is used to denitrify only RAS rather than the full nitrate load in the anoxic recycle, this decreases the chance of nitrate being recycled to the anaerobic zone.

The Johannesburg process (Figure 8.15) is basically an A^2O process with an anoxic zone at the head of the bioreactor to denitrify RAS before entering the anaerobic zone. The modified Johannesburg process provides a recycle from the end of the anaerobic zone to the RAS denitrification zone to provide additional carbon to enhance denitrification of RAS.

Sequencing batch reactors (Figure 8.16) can achieve a moderate degree of both biological phosphorus and nitrogen removal by programming extended air-off periods to achieve anaerobic conditions during each cycle but, when reliable EBPR is required, a dedicated anaerobic zone and internal recycle are provided typically.

Membrane bioreactors (MBRs) (Figure 8.17) are a special version of the activated sludge process that, typically, is operated at high MLSS (6000 to 8000 mg/L) and use submerged membranes as the clarification mechanism instead of gravity settling clarifiers. A key difference between MBR and conventional activated sludge is that, typically, RAS in an MBR is 4 times

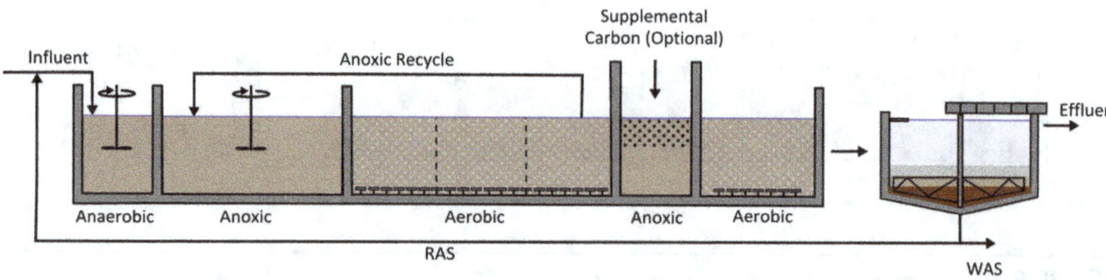

FIGURE 8.12 Five-Stage Bardenpho Process

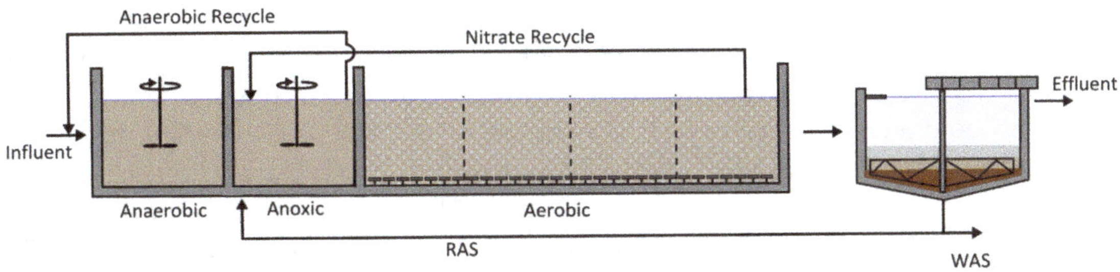

FIGURE 8.13 University of Cape Town Process and VIP

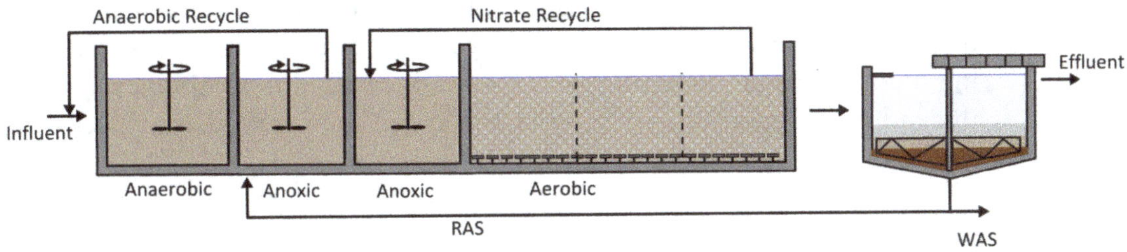

FIGURE 8.14 Modified UCT Process

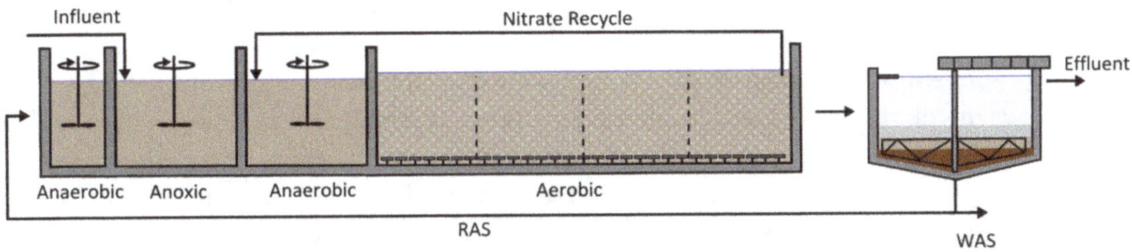

FIGURE 8.15 Johannesburg Process

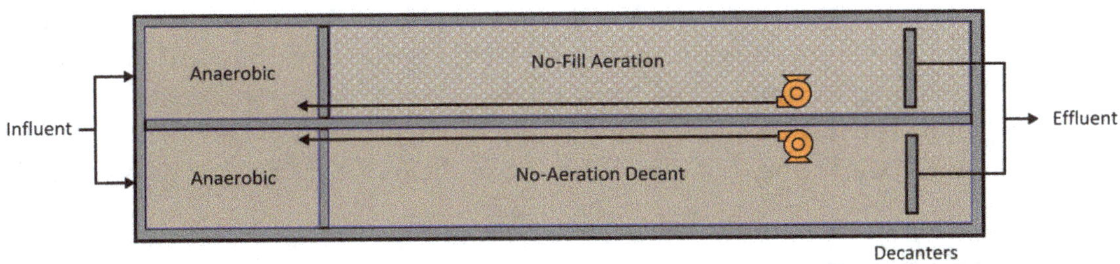

FIGURE 8.16 Biological Nutrient Removal SBR Configuration

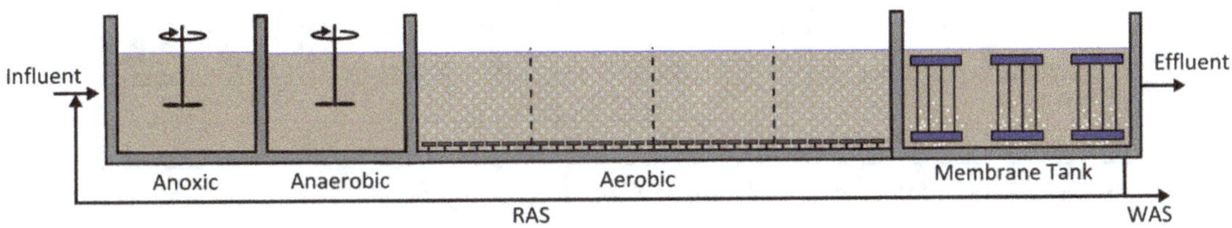

FIGURE 8.17 Basic BNR MBR Configuration

the influent flow to prevent excessive solids concentrations in the membrane tank (as opposed to conventional 0.5-to-1 RAS ratio). Because this high RAS rate results in high mass loading of nitrate in the first zone, anoxic and anaerobic zones can be reversed in an MBR BNR configuration. The disadvantage of this is that VFAs contained in raw wastewater may not be available in the anaerobic zone and EBPR is dependent on fermentation of mixed liquor in the anaerobic zone. Typically, this configuration is capable of meeting total nitrogen limits of 8 to 10 mg/L and total phosphorus limits of 1 to 2 mg/L.

To achieve lower total nitrogen and total phosphorus limits in an MBR, up to two additional mixed liquor recycles are added (Figure 8.18), allowing a conventional anaerobic–anoxic configuration. A single recycle from the end of the aeration zone to the anaerobic zone may also be adequate, although this is returning nitrate to the anaerobic zone.

4.0 CHEMICAL PHOSPHORUS REMOVAL BASICS

Removal of phosphorus via chemical precipitation reactions has been a common practice at utilities for decades. Much of the reason for its popularity is its simplicity from both a process and an operational standpoint, resulting in reliable and predictable performance. Chemical phosphorus removal is conventionally done by applying common metal salts, sometimes referred to as *coagulants*. When these metal salts dissociate in solution, resulting positively charged metal ions are free to bind with other negatively charged ions in solution. Because of the strong "electropotential" attraction, metals have a strong affinity and preference to rapidly bond with phosphate. Immediately, the newly formed compound becomes insoluble in water and comes "out of solution" as a solid particle. The sheer physical mass of the resulting precipitated particle allows it to be easily separated from solution. The most common method for this separation is gravimetric settling in a clarifier unit. Because of the advancement in membrane technology in

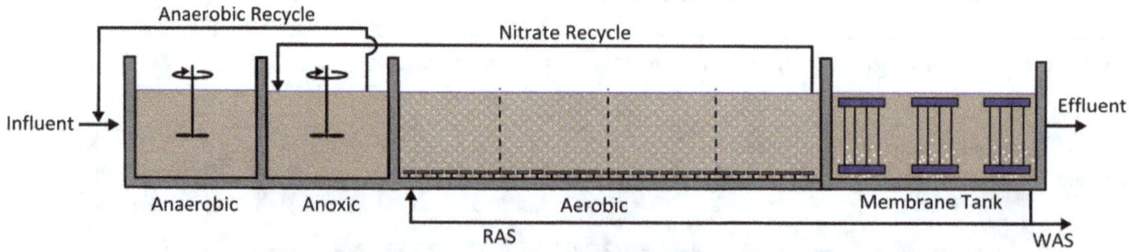

FIGURE 8.18 Biological Nutrient Removal MBR Configuration for Achieving Low Total Phosphorus

recent years, membranes are also extremely effective at separating these metal-phosphate complex particles from solution.

4.1 Chemical Precipitation Reactions

To achieve effective chemical phosphorus removal, it is important to understand basic chemical reactions that drive the process. The most widely used metals for chemical phosphorus removal are iron (as ferrous sulfate or ferric chloride) and aluminum (as aluminum sulfate [i.e., alum, polyaluminum chloride, or sodium aluminate]). There are also cases in which utilities apply a calcium salt (typically as lime) for phosphorus removal. However, because of unfavorable conditions surrounding operational handling of lime and significant amounts of lime sludges that are produced from these reactions, use of lime for phosphorus removal is in decline across the industry. Therefore, no further discussion on lime for phosphorus removal will be provided.

The predicted doses required from metal salts can be estimated by their equilibrium chemical reactions. These are listed below for iron and aluminum, respectively:

$$Fe^{3+} + H_2PO_4^- \rightarrow FePO_{4(s)} + 2H^+ \tag{8.12}$$

$$Al^{3+} + H_2PO_4^- \rightarrow AlPO_{4(s)} + 2H^+ \tag{8.13}$$

For both iron and aluminum, the reaction above for binding the phosphate is not the only reaction that is present. In each case, a portion of the dosed metal also reacts with water and forms a hydroxide, which also becomes immediately insoluble and precipitates out of solution. These "companion" hydroxide reactions are given as follows:

$$Fe^{3+} + 3H_2O \rightarrow Fe(OH)_{3(s)} + 3H^+ \tag{8.14}$$

$$Al^{3+} + 3H_2O \rightarrow Al(OH)_{3(s)} + 3H^+ \tag{8.15}$$

This means that the operator must account for these hydroxide reactions by providing for more than the stoichiometric dose required by Equations 8.14 and 8.15. These hydroxides add to the total mass of sludge that affects the operational capacity of solids-handling equipment and systems for ultimate solids disposal required from the facility. An operator can quickly approximate sludge production resulting from these reactions as follows: for aluminum, 2.9 mg of solids produced/mg of Al^{3+} applied, and, for iron, 2.7 mg of solids produced/mg of Fe^{3+} applied.

Another important point that must be recognized from these chemical reactions is the production of the hydrogen ion (H^+) in each case. This means that "acidity" is being produced by these reactions. This acidity consumes available alkalinity in the solution (primarily bicarbonates that predominate at pH levels typical of municipal wastewater), removing inorganic carbon and reducing the solution's buffering capacity, which are both necessary requirements for sustaining acceptable performance of nitrifiers when nitrification is required. When aluminum is used, alkalinity consumed is 5.8 mg as $CaCO_3$/mg of Al^{3+} applied, and, where iron is applied, it is 2.7 mg as $CaCO_3$/mg of Fe^{3+} applied. Sodium aluminate can be used if it is expected that conventional alum will lower pH excessively.

Operationally, it is critical to be accurate about doses of metal salts needed to achieve the level of phosphorus removal desired. In most National Pollutant Discharge Elimination System permits today that mandate effluent phosphorus limits, the typical requirement is to achieve at least 1 mg/L as total phosphorus; for especially sensitive receiving waters, levels well below 1 mg/L as total phosphorus are required. Unfortunately, the stoichiometry given in Equations 8.14 and 8.15 is accurately predictable down to a soluble phosphate level of approximately 1 mg/L. Below this level, other reactions begin to compete for reactants. Part of this has to do with the fact that some of the phosphorus is actually complexed with organic matter, and it is these compounds that are involved in competing reactions. This means that the stoichiometry, in and of itself, can no longer accurately predict the dose required. Thus, the field dose, or actual dose provided by the operator at the WRRF, must be more than what stoichiometric requirements of these equations indicate. The field dose can be estimated using published empirical data on molar ratio (metal/phosphorus) required to achieve a specific inorganic phosphorus concentration level. However, because all wastewaters are unique, it is preferable to conduct a set of simple jar tests in the laboratory testing a range of doses and measuring the removal performance. Table 8.6 provides a summary of these empirical relationships along with the corresponding minimum doses that would likely be required.

4.2 Chemical Feed Points

There are several locations within the treatment process train where metal salt additions for chemical phosphorus removal are acceptable. With each of these, the removal performance will be approximately equivalent, but, because each facility is unique, there will be operational differences and preferences that apply to the individual operator. These locations are indicated in Figure 8.19 by the red arrows labeled as "Me^{+}" and described as follows. Care must be taken against overdosing of iron and aluminum salts to avoid excessive consumption of alkalinity. It should be noted where chemical addition is applied to influent during primary treatment, frequently,

TABLE 8.6 Metal Salt Additions for Chemical Phosphorus Removal

Residual Ortho-P, mg/L	Estimated* aluminium dose, mg Al/mg Ortho-P_{INF}	Estimated* iron dose, mg Fe/mg Ortho-P_{INF}
1.00	1.4	4.5
0.80	1.7	4.5
0.50	1.9	5.4
0.10	3.0	27.0
0.05	5.0	Unknown

*Operators are encouraged to conduct jar tests to confirm site-specific doses.

this is done to enhance removal of BOD and TSS through primary sludge to increase carbon loading to the anaerobic digestion process and increase energy generation from the burning of digester gas. It can also be of benefit for facilities that experience overloaded conditions or facilities that struggle with periodic nitrification performance or that desire to nitrify year-round. The downside of this practice is that to obtain the removal performance desired, typically an order of magnitude higher dose is required, attributable primarily to competing reactions with organophosphates present in raw wastewater. Also, this practice can excessively deplete phosphorus, such that downstream biological processes that need phosphorus for bacterial growth suffer, and reduce available carbon necessary for BNR processes.

Table 8.7 provides the "typical" feed point locations in the process train and advantages and disadvantages for each dosing location.

4.3 Filtration

When considering methods for removal of phosphorus to achieve low effluent limits (i.e., <1.0 mg/L) and the fact that limits, typically, are written around total phosphorus as opposed to just inorganic soluble phosphorus (which is the focus of both biological and chemical phosphorus-removal

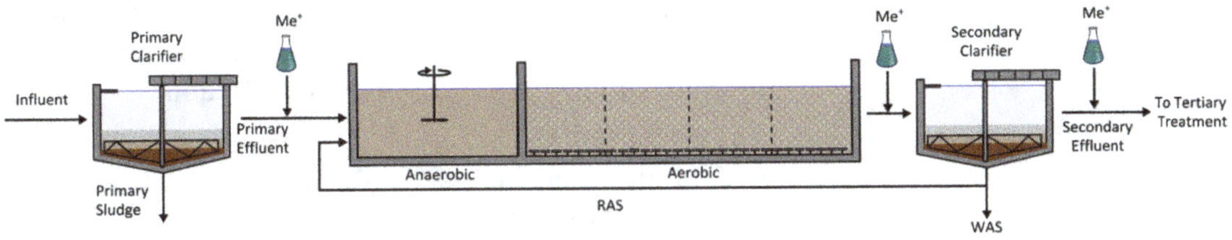

FIGURE 8.19 Typical Metal Salt Addition Points in BNR Process for Chemical Phosphorus Removal

TABLE 8.7 Advantages and Disadvantage of Metal Salt Addition Point Locations

Feed point location	Advantage	Disadvantage
Preprimary treatment (primary influent)	• Increases removal of total BOD and TSS to reduce these loads on downstream biological process • Good removal of soluble phosphorus • Can reduce sulfide problems in digester • Increased energy production	• Requires high doses • Produces large quantities of sludge • Can affect downstream biological process if too much phosphorus is removed • Reduces alkalinity and can be problematic with low-alkalinity wastewaters
Postprimary treatment (primary effluent)	• Permits application of ferrous salts, which may be more cost effective in some cases • Material handling of ferrous salts may be favorable	• Increases mixed liquor concentration • Results in hydroxide precipitates present in mixed liquor • Ferrous salts are limited in phosphorus removal because removals to low phosphorus levels require high pH
Postbiological treatment (mixed liquor ahead of final solids separation)	• Dose requirements are much lower • Phosphates removed with WAS • May enhance settleability where gravimetric settling is used for solids separation	• Increases WAS quantity • Requires appropriate and sufficient mixing at point of chemical feed that is not always "easy" to retrofit to existing infrastructure
Postsecondary treatment (secondary effluent)	• Lowest dose requirement of all locations because of minimal competing reactions • Minimal production of chemical solids • Typically, results in the highest phosphorus removal performance	• May have an effect on downstream systems, such as specific types of membranes or UV disinfection

systems), there will always be some amount of phosphorus tied up in effluent suspended solids that must be "counted" toward the compliance requirement. Therefore, regardless of whether the process uses biological phosphorus-removal methods or chemical removal methods, achieving low limits requires use of high-performance solids separation before discharge. Typically, this is achieved with the use of tertiary filters or membranes. Operation of these solids-separation systems is covered elsewhere in this manual. In the case of filters, polymers are used to aid in filtering performance. The decision to include a polymer, typically, is made following on-site jar tests using the secondary effluent directly. It should be noted that if polymer is to be added, appropriate mixing regimes immediately downstream of the point of addition are critical.

4.4 Operational Considerations

4.4.1 Online Phosphorus Analyzers

Understanding changes in phosphorus loading is critical for adjusting chemical dosage to remove the necessary amount of orthophosphate and avoid overfeeding conditions. Online phosphorus analyzers can play a significant role in optimizing chemical feed, whether dosage is controlled manually or automatically. Depending on the level of consistency of orthophosphate loads at the point of application at a facility, automation of chemical feed using online instrumentation may be determined to be a cost-effective approach.

4.4.2 Temperature and Mixing

In the normal operating temperature range of WRRFs, temperature changes should not have a significant effect on chemical reaction rates for orthophosphate removal. However, under all temperature conditions, the application point must be selected to enable thorough mixing and contact time for ortho phosphate removal reactions to occur. Each WRRF is unique, so feed systems and application points must be tailored for the application.

4.4.3 Alkalinity Addition

Iron salts are acidic and their use consumes alkalinity. If inadequate alkalinity is available, pH of the process will decline. The alkalinity consumed is based on the character of process loads and chemical dosage required to remove phosphorus to target levels; therefore, the potential effects to pH and the degree of alkalinity supplementation and pH control must be developed for each application.

4.4.4 Effect of Chemical Feed on Enhanced Biological Phosphorus Removal

Excessive reduction of ortho phosphate can be detrimental to the EBPR process because ortho phosphate is required for selection of PAOs. Overfeeding of chemicals in primary or secondary clarifier feed applications will reduce available ortho phosphate and the PAO population will adjust to that limiting condition; with reduced EBPR capacity, the system becomes more dependent on chemical phosphorus removal.

4.4.5 Secondary Release

Depending on the chemical used and the removal pathway of phosphorus (precipitate or adsorbed by hydroxyl complexes), return of soluble phosphorus in sidestreams from sludge digestion/biosolids-dewatering operations is possible. At normal anaerobic digester pH conditions, precipitates of phosphorus with alum and iron will not readily return to solution. However, with dynamic conditions occurring in anaerobic digesters, dissolution of precipitates, release of phosphorus from sludge complexes, and/or release because of microbial cell lysis of metabolic phosphorus can contribute to soluble phosphorus return to the EBPR system in these sidestreams. Sidestream monitoring is important to understanding release conditions that are characteristic to each WRRF.

5.0 PRACTICE QUESTIONS

5.1 Calculation of Wasting Rate to Achieve Desired Solids Retention Time

Given data:

Volume of aerated zone = 75 708 m^3 (20 mil. gal)

Volume of anaerobic zone = 18 927 m^3 (5 mil. gal)

Volume of anoxic zone = 37 854 m^3 (10 mil. gal)

Target MLSS in bioreactor = 4000 mg/L

Assume RAS/ WAS TSS = 8000 mg/L

Effluent TSS = 10 mg/L

Effluent flow = 227 ML/d (60 mgd)

Minimum temperature = 14 °C

Required nitrification safety factor = 2

Calculation:

Use Figure 8.3 to determine required aerobic SRT:

At 14 °C and safety factor of 2, aerobic SRT = 10 days

Using aerated volume only, calculate wasting rate:

Using Equation 8.6 in International Standard units:

$$\text{Solids Residence Time}_{\text{days}} = \frac{\text{Volume}_{\text{aerobic}} \times \text{MLSS}}{(\text{Flow}_{\text{WAS}} \times \text{TSS}_{\text{WAS}}) + (\text{Flow}_{\text{EFF}} \times \text{TSS}_{\text{EFF}})}$$

$$10\ \text{days} = \frac{75\ 708\ \text{m}^3 \times 4000 \frac{\text{mg}}{\text{L}}}{\left(\text{Wasting rate} \times 8000 \frac{\text{mg}}{\text{L}}\right) + \left(227 \frac{\text{ML}}{\text{d}} \times 10 \frac{\text{mg}}{\text{L}}\right)}$$

$$10\ \text{days} = \frac{75\ 708\ \text{m}^3 \times 4000 \frac{\text{mg}}{\text{L}} \times \frac{1000\ \text{L}}{\text{m}^3}}{\left(\text{Wasting rate} \times 8000 \frac{\text{mg}}{\text{L}}\right) + \left(227 \frac{\text{ML}}{\text{d}} \times 10 \frac{\text{mg}}{\text{L}} \times \frac{10^6}{\text{ML}}\right)}$$

$$10\ \text{days} = \frac{30.28 \times 10^{10}\ \text{mg}}{\left(\text{Wasting rate} \times 8000 \frac{\text{mg}}{\text{L}}\right) + \left(0.23 \times 10^{10} \frac{\text{mg}}{\text{d}}\right)}$$

Wasting rate × 8000 mg/L = $(3.03 \times 10^{10} - 0.23 \times 10^{10})$ mg/d

Wasting rate = 3.5×10^6 L/d or 3.5 ML/d

In U.S customary units:

$$\text{Solids Residence Time}_{\text{days}} = \frac{\text{Volume}_{\text{aerobic}} \times \text{MLSS}}{(\text{Flow}_{\text{WAS}} \times \text{TSS}_{\text{WAS}}) + (\text{Flow}_{\text{EFF}} \times \text{TSS}_{\text{EFF}})}$$

$$10\ \text{days} = \frac{20\ \text{mil. gal} \times 4000 \frac{\text{mg}}{\text{L}}}{\left(\text{Wasting rate} \times 8000 \frac{\text{mg}}{\text{L}}\right) + \left(60\ \text{mgd} \times 10 \frac{\text{mg}}{\text{L}}\right)}$$

$$10\ \text{days} = \frac{8 \times 10^4\ \text{mil. gal} \frac{\text{mg}}{\text{L}}}{\left(\text{Wasting rate} \times 8000 \frac{\text{mg}}{\text{L}}\right) + \left(600\ \text{mgd} \frac{\text{mg}}{\text{L}}\right)}$$

Divide both side by 10 days

Wasting rate $\times$ 8000 mg/L = 8 $\times$ 10^3 mgd $\times$ mg/L $-$ 0.6 $\times 10^3$ mgd $\times$ mg/L

Wasting rate = 0.925 mgd

5.2 Calculation of Acetic Acid Dose to Remove 5 mg/L of Nitrate

Given data:

Flow = 227.12 ML/d (60 mgd)

Acetic acid properties:

80% solution, COD = 0.913 g/L

5 g of COD required/g nitrate-nitrogen removed

Calculation in International Standard units:

$$\text{Nitrate-nitrogen to be removed} = \frac{227.12 \text{ ML}}{\text{d}} \times \frac{5 \text{ mg}}{\text{L}} \times \frac{10^6 \text{ L}}{1 \text{ ML}} = 1.14 \times 10^9 \text{ mg/d}$$

COD required = 5 g of COD required/g nitrate-nitrogen $\times$ 1.14 $\times$ 10^9 mg/d = 5.7 $\times$ 10^9 mg/d

Acetic acid required = (5.7 $\times$ 10^9 mg/d/0.913 g/L) $\times$ (10^{-3} g/mg) = 6.24 $\times$ 10^6 L/d or 6.24 ML/d

In U.S customary units:

Nitrate-nitrogen to be removed = 60 mgd $\times$ 8.34 $\times$ 5 mg/L = 2500 lb/d

COD required = 5 $\times$ 2500 = 12 500 lb/d

Acetic acid required = 12 500 lb/d / (0.913 g/L $\times$ Conversion factor)

Conversion factor = mg/L $\times$ 8.3454 = lb/mil. gal

0.913 $\times$ 10^3 mg/l $\times$ 8.3454 = 7.62 $\times$ 10^3 lb/mil. gal

Acetic acid required = 12 500 lb/d /7.62 $\times$ 10^3 lb/mil. gal = 1.64 mil. gal

5.3 Calculation of Anoxic Volume Required to Achieve 10 mg/L Nitrate-Nitrogen in Effluent in a Modified Ludzack–Ettinger Process

Given data:

Flow = 10 mgd

Influent BOD = 300 mg/L

Influent TKN = 50 mg/L

Temperature = 20 °C

Aeration basin volume = 10 mil. gal

IMLR rate = 4 × influent = 40 mgd

Aerobic SRT = 10 days

MLSS in aeration basin = 2600 mg/L

Waste activated sludge = 21 500 lb/d

Nonbiodegradable TKN in effluent = 1 mg/L

Calculation in U.S. customary units:

Target 9 mg/L nitrate-nitrogen in effluent for safety factor

Required nitrate as N in effluent = 10 mgd × 8.34 × 8 mg/L = 667 lb/d

Fraction not recycled by IMLR = 40

Nitrogen required for cell synthesis = 5% of WAS = 0.05 × (21 500/8.34/10) = 13 mg/L

Nitrate reduction required = 50 − 13 − 1 = 36 mg/L

Nitrate not returned = 36 × 8.34 × (1 − 0.8) × 10 mgd = 600 lb/d

Maximum allowable nitrate leaving anoxic zone = 667 − 600 = 67 lb/d

Nitrate recycled = 10 mgd × 8.34 × 36 × 0.8 = 2400 lb/d

Minimum nitrate reduction = 2400 − 67 = 2333 lb/d

Assume anoxic zone of 3 mil. gal to allow calculation of SNDR − iterative solution

F:M = (10 mgd × 300 mg/L BOD)/(2500 mg/L × 3 mil. gal) = 0.4

SNDR = (0.03 × F:M + 0.029) × $1.05^{(T-20)}$ = 0.043 lb N/lb MLSS

Nitrate reduced = 3 mil. gal × 0.043 × 8.34 × 2500 mg/L = 2690 lb/d

2690 lb/d exceeds nitrate recycled (2400 lb/d), so 3 mil. gal anoxic zone is adequately sized so some slight downsizing may be possible

Nitrate in effluent = Nitrate not recycled + Nonbiodegradable TKN = (36 mg/L × 0.2) + 1 mg/L = 8.2 mg/L

5.4 Biological Phosphorus Removal Example

Process design for biological phosphorus removal is complex. Typically, development of a reactor system to gauge performance is done using advanced biological process models. Nonetheless, there are simplistic methods for which to estimate the amount of phosphorus removal that can be expected from a biological phosphorus-removal process. Essentially, one can estimate phosphorus taken up by PAOs with that taken up by other heterotrophs and add the two. This results in an estimate of effluent phosphorus content. The following example illustrates this method.

Problem: A WRRF operates at an average daily flow of 20 mL/d (5.3 mgd) and uses an A/O process for biologically removing phosphorus. The amount of dissolved oxygen entering the anaerobic zone is assumed to be negligible from RAS. The following data are given:

Influent COD = 450 mg/L
Influent biodegradable COD (sCOD) = 320 mg/L
Influent readily biodegradable COD (rbCOD) = 50 mg/L
Influent phosphorus, as P = 7 mg/L
Endogenous specific decay rate, b = 0.08 mg VSS/mg VSS·d
Bacteria synthesis yield, Y = 0.45 mg VSS/mg COD used
Conversion of rbCOD to acetate in anaerobic zone = 85%
Phosphorus content of PAO biomass = 0.32 mg P/mg VSS
Phosphorus content of non-PAO biomass = 0.02 mg P/mg VSS
System SRT = 6 days
Effluent VSS concentration = 7 mg/L

Step 1: Determine phosphorus removed by the PAO biomass.

a. Determine the amount of acetate produced as COD

Acetate = (rbCOD converted, %)(rbCOD, mg/L)
Acetate = (0.85) (50 mg rbCOD/L) = 42.5 mg/L

b. Determine the amount of PAO biomass produced (neglect the cellular debris)

$$= \left[\frac{Y}{1+b(\text{SRT})}\right]\text{sCOD}$$

$$= \left\{\frac{\left(0.45\dfrac{\text{mg VSS}}{\text{mg COD}}\right)}{\left[\left[1+\left(\dfrac{0.08\text{ mg}}{\text{mg}-\text{d}}\right)(6\text{ d})\right]\right]}\right\}\left(42.5\frac{\text{mg}}{\text{L}}\right)$$

$$= 12.9\text{ mg VSS/L}$$

c. Determine the amount of phosphorus removed by the PAO biomass:

$$\begin{aligned} \text{P removed} &= \left(0.32 \frac{\text{g P}}{\text{g VSS}}\right)\left(12.9 \text{ mg} \frac{\text{VSS}}{\text{L}}\right) \\ &= 4.1 \text{ mg/L} \end{aligned}$$

Step 2: Determine phosphorus removed by other heterotrophs (this is the amount used for cell synthesis, and this synthesis occurs with the conversion of biodegradable COD):

a. Determine the amount of sCOD used by other heterotrophs:

$$\begin{aligned} &= 320 \frac{\text{mg}}{\text{L}} - 42.5 \frac{\text{mg}}{\text{L}} \\ \text{sCOD removed} &= 277.5 \text{ mg/L} \end{aligned}$$

b. Determine the amount of biomass of non-PAO heterotrophs produced:

$$\begin{aligned} \text{Non-PAO biomass} &= \left[\frac{Y}{1 + b(\text{SRT})}\right] \text{sCOD} \\ &= \left\{\frac{\left(0.45 \frac{\text{mg VSS}}{\text{mg COD}}\right)}{\left[1 + \left(\frac{0.08 \text{ mg}}{\text{mg} \cdot \text{d}}\right)(6 \text{ d})\right]}\right\}\left(277.5 \frac{\text{mg}}{\text{L}}\right) \\ &= 84.4 \text{ mg VSS/L} \end{aligned}$$

c. Determine the amount of phosphorus removed by this non-PAO synthesis:

$$\begin{aligned} \text{P removed} &= \left(0.02 \frac{\text{mg P}}{\text{mg VSS}}\right)\left(84.4 \text{ mg} \frac{\text{VSS}}{\text{L}}\right) \\ &= 1.7 \text{ mg/L} \end{aligned}$$

Step 3: Determine the total phosphorus removed by the biomass:

Total phosphorus removed by biomass = Phosphorus removed by PAO biomass + Phosphorus removed by non-PAO biomass

4.1 mg/L + 1.7 mg/L = 5.8 mg/L

Step 4: Determine effluent soluble phosphorus concentration:

Effluent soluble phosphorus = Influent phosphorus − Total phosphorus removed by biomass

Effluent soluble phosphorus = 7.0 mg/L − 5.8 mg/L = 1.2 mg/L

Step 5: Estimate phosphorus content in effluent VSS:

Average phosphorus content of effluent VSS:

$$\frac{\text{Phosphorus Removed by PAO Biomass} + \text{Phosphorus Removed by Non} - \text{PAO Biomass}}{\text{PAO Biomass Produced} + \text{Non} - \text{PAO Biomass Produced}}$$

$$= \frac{\left(0.32 \frac{\text{mg P}}{\text{mg VSS}}\right)\left(12.9 \ \frac{\text{mg VSS}}{\text{L}}\right) + \left(0.02 \frac{\text{mg P}}{\text{mg VSS}}\right)\left(84.4 \ \frac{\text{mg VSS}}{\text{L}}\right)}{(12.9 + 84.4) \text{ mg VSS/L}}$$

$$= 0.06 \frac{\text{mg P}}{\text{mg VSS}}$$

$$\text{P in effluent VSS} = \left(0.06 \frac{\text{mg P}}{\text{mg VSS}}\right)\left(7 \frac{\text{mg VSS}}{\text{L}}\right)$$

$$= 0.42 \frac{\text{mg P}}{\text{L}}$$

Total effluent phosphorus = Effluent soluble phosphorus + Average phosphorus content of effluent VSS = 1.2 + 0.42 = 1.62 mg phosphorus/L

5.5 Chemical Phosphorus Removal Example

The process design for sizing a chemical phosphorus removal system requires applying several simple relationships and empirical data gathered from research. Consider the following wastewater characteristics for a 20-ML/d (5.3-mgd) WRRF. From these data, compute the amount of aluminum sulfate (alum) dose that is required to meet an effluent total phosphorus limit of 0.6 mg/L if this chemical is added to both primary and secondary treatment.

Influent BOD_5 = 265 mg/L

Influent TSS = 280 mg/L

Influent total phosphorus = 7 mg/L

Influent orthophosphate = 5.3 mg/L

Influent alkalinity = 280 mg/L as $CaCO_3$

pH = 7.0

Effluent BOD_5 limit = ≤ 10 mg/L

Effluent TSS limit = ≤ 10 mg/L

Step 1: Determine the alum dose for primary treatment.

$$= \left(\frac{Al}{P}\right)(CP{,}in - CP{,}res)$$

Where

CP,in = influent Ortho-P,
CP,res = Ortho-P remaining after alum addition, and
$\left(\frac{Al}{P}\right)$ = molar ratio.

From Table 8.6, it is evident that for a CP,res < 0.1 mg/L, Al:P rapidly increases. It is appropriate, then, to keep Al:P as low as practical to minimize chemical costs. Therefore, select CP,res = 0.1 mg/L. This results in a ratio of 3. Using the equation above, we have the following:

$$\text{Alum dose (primary)} = \frac{(3)(5.3-0.1)\text{ mg}}{\text{L}} = 15.6\text{ mg/L}$$

Step 2: Determine alum dose required for the secondary treatment process.

For the secondary process, the influent phosphorus concentration to secondary includes all that which remains from the primary process, because the nonortho fraction of the phosphorus and the particulate fraction are converted to orthophosphate during the secondary biological process. Some of this converted phosphorus becomes available for assimilation into new biomass. Determining CP,in for this secondary treatment process is a conservative approach, as follows:

$$CP_{,in} = \text{Influent total phosphorus} - (CP_{,in} - CP_{,out})$$

$$CP_{,in} = 7\frac{\text{mg}}{\text{L}} - \frac{(5.3-0.1)\text{ mg}}{\text{L}} = 1.8\frac{\text{mg}}{\text{L}}$$

Again, using Table 8.6, we can determine the alum dose necessary for the secondary treatment process to obtain effluent soluble phosphorus. Again, we conservatively target 0.1 mg/L. This target is necessary because there will be phosphorus in the effluent TSS. This phosphorus content is assumed to be approximately 2% because this is typical of non-PAO biomass. Therefore, for effluent TSS = 10 mg/L, the phosphorus content in a particulate phosphorus is as follows:

$$\text{Effluent particulate P} = \left(10\frac{\text{mg}}{\text{L}}\right)(0.02) = 0.2\frac{\text{mg}}{\text{L}}$$

To achieve an effluent total phosphorus of ≤ 0.6 mg/L, the effluent soluble phosphorus fraction must be at least less than the following:

$$\text{Effluent soluble P} = \frac{(0.6 - 0.2)\ \text{mg}}{\text{L}} = 0.4\frac{\text{mg}}{\text{L}}$$

Now that we know the effluent soluble phosphorus (0.4 mg/L), we go to Figure 8.20 and select the appropriate Al:P, which is 2. From this, we can now compute the alum dose for secondary treatment, as follows:

$$\text{Al dose (secondary)} = \frac{(2)(1.8 - 0.4)\ \text{mg}}{\text{L}} = 2.8\frac{\text{mg}}{\text{L}}$$

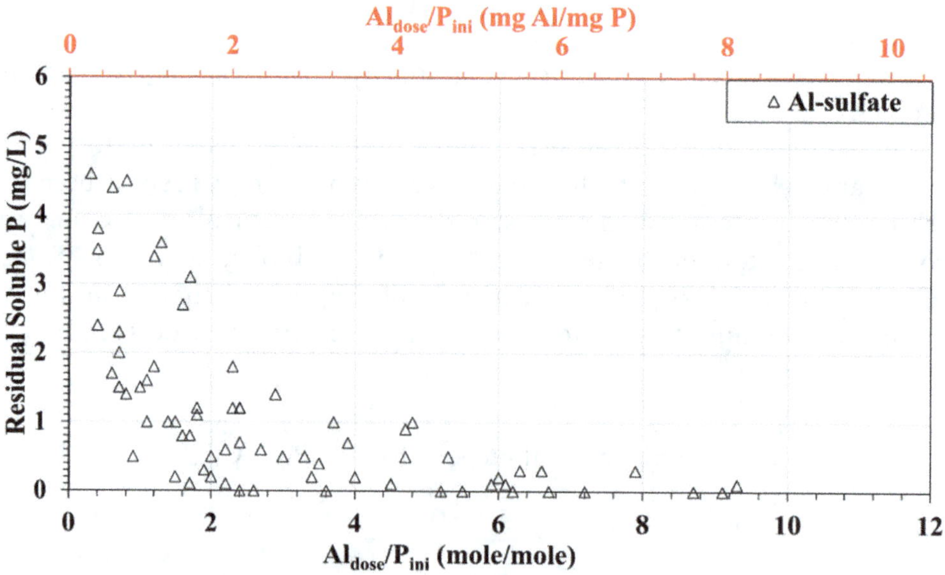

FIGURE 8.20 Aluminum Dosage, as a Ratio to the Initial Soluble Reactive Phosphorus, Versus the Residual Reactive Phosphorus Concentration (Modified from Szabo et al., 2008)

Step 3: Determine the total facility alum dose:

$$\text{Total alum dose} = \text{Primary} + \text{Secondary} = (15.6 + 2.8)\frac{\text{mg}}{\text{L}} = 18.4\frac{\text{mg}}{\text{L}}$$

It should be noted that a similar procedure is used if an iron salt, such as ferric chloride, is used as coagulant addition for the chemical phosphorus-removal computation.

6.0 REFERENCES

Barnard, J. L., Dunlap, P., & Steichen, M. (2017). Rethinking the mechanisms of biological phosphorus removal. *Water Environment Research*, *89*(11), 2043–2054.

Chen, G., van Loosdrecht, M. C. M., Ekama, G. A., & Brdjanovic, D. (2023). *Biological wastewater treatment: Principles, modelling and design* (2nd ed.). IWA Publishing.

Eckenfelder, W. W., Jr., Ford, D. L., & Englande, A. J., Jr. (2008). *Industrial water quality* (4th ed.). McGraw-Hill.

Filipe, C. D. M., Daigger, G. T., & Grady, C. P. L. (2001). pH as a key factor in the competition between glycogen-accumulating organisms and phosphorus-accumulating organisms. *Water Environment Research*, *73*(2), 223–232.

Lopez-Vazquez, C. M., Hooijmans, C. M., Brdjanovic, D., Gijzen, H. J., & van Loosdrecht, M. C. M. (2008). Factors affecting the microbial populations at full-scale enhanced biological phosphorus removal (EBPR) wastewater treatment plants in The Netherlands. *Water Research*, *42*(10–11), 2349–2360.

Mamais, D., & Jenkins, D. (1992, September 1). The effects of MCRT and temperature on enhanced biological phosphorus removal. *Water Science & Technology*, *26*(5–6): 955–965.

Sedlak, R. (1991) *Phosphorus and Nitrogen Removal from Municipal Wastewater: Principles and Practice* (2nd ed.). Lewis Publisher.

Szabo, A., Takács, I., Murthy, S., Daigger, G. T., Licsko, I., & Smith, S. (2008). Significance of design and operational variables in chemical phosphorus removal. *Water Environment Research*, *80*, 407–416.

Vollertsen, J., Petersen, G., & Borregaard, V. R. (2006). Hydrolysis and fermentation of activated sludge to enhance biological phosphorus removal. *Water Science & Technology*, *53*(12), 55–64.

Water Environment Federation; American Society of Civil Engineers; Environmental and Water Resources Institute (2018) *Design of Water Resource Recovery Facilities* (6th ed.); WEF Manual of Practice No. 8; ASCE Manuals and Reports on Engineering Practice No. 76. Water Environment Federation. McGraw-Hill.

Water Environment Federation. (2023). *Standards for design of water resource recovery facilities* (7th ed., WEF 8).

7.0 SUGGESTED READINGS

Grady, C. P. L., Jr., Daigger, G. T., Love, N. G., & Filipe, C. D. M. (2011). *Biological wastewater treatment* (3rd ed.). CRC Press.

Qasim, S. R. (1999). *Wastewater treatment plants: Planning, design, and operation* (2nd ed.). CRC Press.

Metcalf & Eddy, Inc./AECOM. (2014). *Wastewater engineering: Treatment and resource recovery* (5th ed., G. Tchobanoglous, H. D. Stensel, T. Tsuchihashi, F. Burton, Eds.). McGraw-Hill.

Wang, L. K., Shammas, N. K., & Hung, Y.-T. (2009). *Advanced biological treatment processes* (Vol. 9). Humana Press.

9

Process Simulators

1.0 INTRODUCTION

Process models refer to mathematical representations that describe fundamental mechanisms within a process. Simulators are software packages that execute these models. These process models can be linked together in the graphical interface of a simulator to produce a larger model that represents any water resource recovery facility (WRRF) configuration. Most simulators include models that can simulate biological, physicochemical, and separation processes as well as process control schemes. There are several process simulators that are commercially available and commonly used by consulting and utility engineers. The discussions in this chapter are directed at the use of those commercially available process simulators. The commercial names of specific simulators will not be referred to directly.

Since their early days, models have been used for design because they were shown to accurately estimate sludge production, oxygen demand, and denitrification capacity. Later, process simulators were applied more toward effluent quality prediction, operational optimization, and process control development. Today, process models are used for numerous purposes in the water resource recovery industry because they have accumulated all relevant knowledge about process mechanisms in a single place. These models include all the fundamentals that are in the "old" design and operations formulas and also consider additional mechanisms that are either more recently understood or that were too complex to consider in previous guidelines.

Despite their apparent complexity, models are still simplified representations of real systems and, as such, they represent only the key characteristics or behaviors of the process. Because modeling needs of the wastewater community are increasing, more processes are being included that are capturing the behavior of treatment processes more accurately. However, there continues to be significant gaps in what models can and cannot predict. For example, standard activated sludge models do not describe foaming, bulking, and deflocculation. Models must also be calibrated before they can be applied. Calibration is the process of modifying model parameters until simulation results match an observed set of data within a defined, acceptable error to the measured data (IWA Task Group on Good Modelling Practice, 2012).

Despite aforementioned shortcomings, models are well suited for many applications. They pull all of the different biological processes together—biochemical oxygen demand (BOD) removal, nitrification, denitrification, and phosphorus removal—in one place, which allows us to understand how these processes are interacting. For example, good denitrification is beneficial to biological phosphorus removal and the model will account for the effect of nitrogen removal and residual nitrate in determining biological phosphorus removal performance.

Simulators add the ability to look at diurnal flow effects as well as intra- and inter-annual dynamics; that is, simulators can reproduce changing conditions that WRRFs are subjected to typically. Dynamic simulations consider the effect of changing wastewater flow, characteristics, and temperature, and model results capture the dynamic response of the facility to these changing conditions. Dynamic simulations can be run to investigate the effect of short-term or long-term events. Short-term events include diurnal influent variations, storm events, taking equipment out of service, and many others. Long-term dynamic simulations can investigate the effect of weekday versus weekend loading or seasonal effects (IWA Task Group on Good Modelling Practice, 2012).

Process simulators can be used to run a "whole facility" model to investigate interactions between liquid and solid streams. Facility-wide wastewater treatment modeling combines modeling of liquid and solids streams into one overall model, which allows facilities to evaluate the effect of, for example, recycle streams generated by dewatering of digested sludge on facility performance and energy consumption. Most commercial simulators can be used for facility-wide modeling. They provide different methods to allow linking different process streams, such as activated sludge, anaerobic digestion, and sidestream treatment, in one layout.

Finally, process modeling can be helpful in evaluating control strategies and process changes before implementation. The development of a control strategy is intimately related to the dynamic behavior of a process, and simulators are ideal tools to study WRRF dynamics. They include a wide variety of control algorithms (on–off, proportional–integral–derivative, ratio, cascade, feedforward, etc.) that can be used to test control actions to be implemented. This chapter will review these applications of process simulators/models and discuss some common misconceptions and pitfalls of process model application. First, the data required for modeling projects will be discussed.

2.0 DATA REQUIREMENTS FOR MODEL CALIBRATION

2.1 Influent Characterization

Influent parameters vary considerably, both between facilities and under different environmental conditions at a given facility. As such, estimating influent parameters is one of the most important tasks that needs be completed before calibrating a model. Regular measurements of influent and effluent parameters are often taken to determine flows and loads for future rating and planning purposes and for demonstrating permit compliance. However, calibration and use of process models require extra information beyond what is necessary typically for these other purposes. To meet this additional need, regular facility sampling protocols can be adjusted, and periodic special sampling efforts can be undertaken as part of an ongoing model improvement plan. Accurate characterization of influent could require samples at other locations, such as some effluent and sidestream measurements.

Exactly which sample points are critical will vary with what questions one is trying to answer with the model and how accurate those answers need to be. Resources available and facilities present will affect the feasibility of measuring some parameters and will, therefore, be important considerations in developing a sampling plan for model improvement. If data are unavailable, then assumptions can be made, though the accuracy and predictive power of the model will be diminished. So, not all data discussed in this section are needed in all instances. Instead, careful consideration of each facility's resources and drivers is needed in developing a sampling plan for model development.

2.1.1 Carbonaceous Oxygen Demand Fractions

A variety of soluble, colloidal, and particulate organic components make up the total carbonaceous oxygen demand (COD) measurement. These components can further be divided on the basis of biodegradability; that is, biodegradable and nonbiodegradable COD, resulting in six bulk COD measurements. Specific COD components may also be important to measure; for example, the total volatile fatty acid (VFA) concentration is important in determining biological phosphorus removal performance. How the total COD is split between these subcomponents for a particular sample of wastewater is referred to as *COD fractionation*. This COD fractionation will have a significant effect on sludge production, aeration requirements, energy recovery potential, and expected effluent quality for a given WRRF. As such, characterizing COD fractions is an important part of any modeling project. The COD and BOD measurements, which are necessary for a full fractionation, are summarized in *Wastewater Treatment Process Modeling* (Water Environment Federation, 2013) (Table 8.8 in that manual), while Table 9.1 in this chapter provides an example of what COD measurement may be requested as part of a special sampling run, depending on the facility process. Often, facilities will have a long running dataset of influent BOD while relatively little COD data are available. Historically monitored BOD loads can be used as indicators for model input COD after COD:BOD has been established. A more complete discussion of how these measurements can be used to estimate various fractions is available in *Methods for Wastewater Characterization for Activated Sludge Modelling* (Melcer et al., 2003).

2.1.2 Nutrient Fractions

Characteristics of the nitrogen and phosphorus load to the WRRF will also have an effect on nutrient removal performance. At a minimum, total nitrogen (measured as total Kjeldahl nitrogen [TKN]) and total phosphorus should be measured if the model is to predict effluent nutrient quality or if influent characteristics are abnormal, such those caused by high CODs. Beyond this, data regarding free soluble forms of nitrogen and phosphorus (NH_X and PO_4, respectively) are often available on a regular basis. Remaining influent nutrients are most likely associated with one of the forms of COD discussed earlier, or, in the case of nitrogen, may occur as nitrogen oxides (NO_X). Even though simulator defaults are often used for these fractions, they can be estimated with analytical methods. Measurements that are necessary for full nitrogen and phosphorus fractionation are summarized in *Wastewater Treatment Process Modeling* (Water Environment Federation, 2013) (Table 8.8 in that manual), while Table 9.1 in this chapter provides an example of what nutrient measurement may be requested as part of a special sampling run, depending on the facility process. A more complete discussion of how these measurements can be used to estimate various fractions is available in *Methods for Wastewater Characterization for Activated Sludge Modelling* (Melcer et al., 2003).

2.1.3 Solids Parameter

To estimate volatile suspended solids (VSS) and total suspended solids (TSS) concentrations from COD-based variables used by all models, it is necessary to measure inert suspended solids (ISS) and the COD-to-VSS ratio at the influent and potentially at other solids streams throughout the process. The ISS can be calculated as the difference between TSS and VSS, data that are often measured regularly. The COD:VSS can be estimated to match VSS in the influent. Alternately, COD:VSS can be estimated by the following

TABLE 9.1 Example Sampling Protocol for Model Development Depending on Activated Sludge Process Type

Location	Component	Full biological nutrient removal	Nitrification	Biochemical oxygen demand removal	Sample type	Reasons/ remarks
Raw influent	Flow	Daily	Daily	Daily	Online	
	COD	Daily	Daily	Daily	Composite	Establish organic loading
	Filtered COD	3 per week	3 per week	3 per week	Composite	Determine COD : VSS
	ffCOD	3 per week	3 per week	3 per week	Composite	Flocculated/ filtered to determine readily biodegradable COD (rbCOD)
	TSS	Daily	Daily	Daily	Composite	Determine ISS
	VSS	Daily	Daily	Daily	Composite	Determine ISS, COD : VSS
	BOD_5[a]	Daily	Daily	Daily	Composite	Determine COD : BOD
	VFA	3 per week			Grab	3 grab samples/ day (composited)
	TKN	Daily	Daily	3 per week	Composite	Establish nitrogen loading
	Ammonia	Daily	Daily		Composite	Determine NH_X-N/TKN
	NO_X-N	3 per week	3 per week	3 per week	Composite	Typically, no nitrites or nitrates are expected
	Total phosphorus	Daily	3 per week	3 per week	Composite	Establish phosphorus loading
	Orthophosphate	3 per week			Composite	Determine orthophosphate/ phosphorus
	Alkalinity	Daily	Daily		Composite	Establish influent alkalinity/need for pH adjustment
	pH	Daily	Daily	Daily	Grab	Establish influent alkalinity/need for pH adjustment
	Magnesium	3 per week			Composite	Potential for struvite precipitation

(continued)

TABLE 9.1 Example Sampling Protocol for Model Development Depending on Activated Sludge Process Type (*Continued*)

Location	Component	Full biological nutrient removal	Nitrification	Biochemical oxygen demand removal	Sample type	Reasons/ remarks
Raw influent (*continued*)	Calcium	3 per week			Composite	Potential for formation of other phosphorus precipitates
	Diurnal flow	Every 2 hours	Every 2 hours	Every 2 hours	Online	Establish diurnal flow pattern, typically, 3 dry weather flow days
	Diurnal concentration	Every 2 hours	Every 2 hours	Every 2 hours	Grab	Establish diurnal load pattern, (COD, TSS, NH_3, orthophosphate), same 3 days as diurnal flow
Side-streams	Flow	1 per week	1 per week	1 per week	Online	
	COD	1 per week	1 per week	1 per week	Grab	3 grab samples/ day (composited)
	TSS	1 per week	1 per week	1 per week	Grab	3 grab samples/ day (composited)
	VSS	1 per week	1 per week	1 per week	Grab	3 grab samples/ day (composited)
	BOD_5	1 per week	1 per week	1 per week	Grab	3 grab samples/ day (composited)
	TKN	1 per week	1 per week	1 per week	Grab	3 grab samples/ day (composited)
	Total phosphorus	1 per week	1 per week	1 per week	Grab	3 grab samples/ day (composited)
	Alkalinity	1 per week	1 per week		Grab	3 grab samples/ day (composited)
Primary effluent	Flow	Daily	Daily	Daily	Online	
	COD	Daily	Daily	Daily	Composite	Bases for simulator, determine PST COD removal
	ffCOD	3 per week	3 per week	3 per week	Composite	Flocculated/ filtered to determine rbCOD
	TSS	Daily	Daily	Daily	Composite	Determine ISS

(*continued*)

TABLE 9.1 Example Sampling Protocol for Model Development Depending on Activated Sludge Process Type (*Continued*)

Location	Component	Full biological nutrient removal	Nitrification	Biochemical oxygen demand removal	Sample type	Reasons/ remarks
Primary effluent (*continued*)	VSS	Daily	Daily	Daily	Composite	Determine ISS
	BOD_5	Daily	Daily	Daily	Composite	Determine COD:BOD
	VFA	3 per week			Grab	3 grab samples/ day (composited)
	TKN	Daily	Daily		Composite	Determine PST TKN removal
	Ammonia	Daily	Daily		Composite	Determine if there is NH_X-N production across PSTs
	NO_X-N	3 per week	3 per week		Composite	No nitrites or nitrates are expected
	Orthophosphate	3 per week			Composite	Determine orthophosphate
	Alkalinity	3 per week	3 per week		Composite	Establish influent alkalinity/need for pH adjustment
	pH	Daily	Daily	Daily	Grab	Potential for struvite precipitation
Secondary effluent	Flow	Daily	Daily	Daily	Online	
	ffCOD	3 per week	3 per week	3 per week	Composite	Determine soluble nonbiodegradable COD
	TSS	Daily	Daily	Daily	Composite	Determine secondary clarifier performance
	$CBOD_5$	Daily	Daily	Daily	Composite	Determine COD:BOD
	ffTKN	3 per week	3 per week		Composite	Determine soluble nonbiodegradable TKN
	Ammonia	Daily	Daily	3 per week	Composite	Determine nitrification performance

(*continued*)

TABLE 9.1 Example Sampling Protocol for Model Development Depending on Activated Sludge Process Type (*Continued*)

Location	Component	Full biological nutrient removal	Nitrification	Biochemical oxygen demand removal	Sample type	Reasons/ remarks
Secondary effluent (*continued*)	Orthophosphate	3 per week			Composite	Determine orthophosphate/ phosphorus and EBPR performance
	pH	Daily	Daily	Daily	Grab	Establish need for pH adjustments
Mixed liquor	TSS	Daily	Daily	Daily	Grab	Determine SRT, MLSS
	VSS	3 per week	3 per week	3 per week	Grab	Determine mixed liquor ML ISS and COD:VSS
	COD	1 per week	1 per week	1 per week	Grab	Determine ML COD:VSS
	Filtered COD	1 per week	1 per week	1 per week	Grab	Determine ML COD:VSS
	Temperature	Daily	Daily	Daily	Online	Develop averages for model input
	Dissolved oxygen	Daily	Daily	Daily	Online	Develop averages for model input
WAS	Flow	Daily	Daily	Daily	Online	Determine SRT
	TSS	Daily	Daily	Daily	Grab	Determine SRT
	Total phosphorus	3 per week	3 per week	3 per week	Grab	Determine SRT through phosphorus mass balance
Miscella-neous process	Flow	Daily	Daily	Daily	As available	Mass balances around significant unit processes
	TSS	Daily	Daily	Daily	Grab	Mass balances around significant unit processes

Note. VSS = volatile suspended solids; ffCOD = filtered and flocculated COD; TSS = total suspended solids; ISS = inert suspended solids; NH_X-N = ammonia and ammonium nitrogen; NO_X-N = nitrate and nitrite-nitrogen; NH_3 = anhydrous ammonia; PST = primary settling tank; $CBOD_5$ = 5-day carbonaceous biochemical oxygen demand; ffTKN = filtered and flocculated TKN; EBPR = enhanced biological phosphorus removal; SRT = solids retention time; MLSS = mixed liquor suspended solids
[a]Parallel analysis of inhibited BOD (i.e., carbonaceous BOD [CBOD]) and uninhibited BOD should be performed intermittently to identify inhibitor effect on heterotrophs in the BOD test.

formula, as described in *Methods for Wastewater Characterization for Activated Sludge Modeling* (Melcer et al., 2003):

$$\text{COD:VSS} = \frac{\text{COD}_{\text{UNFILTERED}} - \text{COD}_{\text{GLASS-FIBER FILTERED}}}{\text{VSS}}$$

Table 9.1 includes sampling parameters necessary to estimate COD:VSS and ISS for influent solids and secondary sludge. Nutrients that are associated with solids can be estimated in a similar manner. Sampling required for this full fractionation is available in *Wastewater Treatment Process Modeling* (Water Environment Federation, 2013) (Table 8.8 in that manual), while a description of the method for determining this can be found in *Methods for Wastewater Characterization for Activated Sludge Modeling* (Melcer et al., 2003).

2.1.4 Translating Characterization Data Available into Simulator Inputs

Simulators differ slightly about how influent COD and nutrient fractions and solids characteristics are input and how resulting variables are calculated. Manuals for a specific simulator package should be referred to understand these calculations. Typically, simulators offer an influent advisor program or spreadsheet, which will assist in conversion of influent characterization data to appropriate fractions and help estimate COD:VSS. If specific fractions or ratios cannot be measured, they can be estimated from ranges of typical values available in different locations. The IWA Task Group on Good Modelling Practice (2012) provides a summary of typical influent and primary effluent ratios, complete with statistical information on the variability of those parameters (Table 5.2.3 in that report).

2.1.5 Diurnal Variations

Typically, 24-hour composite samples are used in influent characterization to determine average characteristics over the course of a day. Depending on the purpose of the model and the typical variation of the influent, a series of samples may be taken and analyzed over a 24-hour period to understand how influent loads and characteristics vary with flow over the course of a day. These could be taken on weekdays and weekends to determine the difference in patterns experienced throughout a week. This can be used to develop aeration control strategies through modeling or understand dynamics of an enhanced biological phosphorus removal system. Additionally, diurnal sampling over the course of a wet weather event can be helpful in determining how load varies with time and flow during a storm; this is information that can be especially useful if using models to develop operational strategies for wet weather events.

2.2 Operational Data

Operational data are important for developing and calibrating a model for two reasons. First, many of the operational controls at a facility are also model inputs. Second, measured effluent quality and operational parameters provide important points of reference that can be used to make comparisons between observations and model results during calibration. Similarly to influent characteristics, ensuring data quality and capturing the variability of operational measurements are important aspects of calibrating a model. These are discussed in more detail in the following subsection and are referenced in Table 9.1.

2.2.1 Operational Inputs

Operational inputs for which data are available are useful as inputs to the model. These primarily include the following:

- Significant process flows such as return activated sludge (RAS), waste activated sludge (WAS), and primary settling tank underflow;
- Internal process recycles and flow splits;
- Other flows needed to close flow balances around solids-separation processes;
- Chemical dose flows and concentrations; and
- Measured dissolved oxygen and airflow by zone in the aeration basin.

2.2.2 Effluent Measurements

Effluent monitoring for permit validation will provide much of the information to calibrate and validate the model. However, some additional monitoring will likely be necessary for model development, especially for modeling biological nutrient removal (BNR) processes. For example, effluent measurements can be used to estimate nonbiodegradable soluble influent fractions. These include the following:

- Total suspended solids,
- Biochemical oxygen demand,
- Carbonaceous oxygen demand,
- Filtered and flocculated COD,

- Total Kjeldahl nitrogen,
- Filtered and flocculated TKN,
- Ammonia and ammonium nitrogen (NH_X-N),
- Nitrate and nitrite nitrogen (NO_X-N),
- Total phosphorus, and
- Orthophosphates (PO_4-P).

2.2.3 Operational Monitoring

Any measurements within the process will be useful as reference points to compare model outputs against. Primarily, this would include TSS and flow data, which can be used to perform a solids mass balance around unit processes of the facility. If these data are not biased, it can be used to determine inputs of unit processes and estimate yields of biological processes. These include the following:

- Solids retention time (SRT) (and/or data necessary to calculate SRT),
- Mixed liquor suspended solids (MLSS) and mixed liquor volatile suspended solids (MLVSS),
- Waste activated sludge TSS and flowrate,
- Total suspended solids and flowrate of other significant sludge or recycle streams,
- Carbonaceous oxygen demand and phosphorus of MLSS and/or WAS for COD and phosphorus balancing,
- Basin profiles,
- Other TSS measurements needed to close solids mass balance around all solids-separation processes, and
- Significant concentrations of recycle streams.

Extensive discussions about operational data quality, typical influent fractions, and acceptable error as these relate to model calibration can be found in *Guidelines for Using Activated Sludge Models* (IWA Task Group on Good Modelling Practice, 2012).

3.0 APPLICATIONS OF MODELS

This section provides examples of how models can be used to improve facility operation. The first example discusses the use of models for operator training and process understanding and focuses on testing the effect of operational decisions on facility performance. The second example discusses the application of models for facility optimization and decision support. Finally, several case studies are given that show how process simulators have been used at WRRFs.

3.1 Models for Training

Simple models can be excellent tools for training operational staff on the effect of operational setpoints or control strategies on facility performance and behavior. When used solely for training about wastewater treatment concepts, models do not need to be calibrated. This section provides examples of how the effect of operational parameters can be demonstrated with process simulators for training.

3.1.1 Interaction between Mixed Liquor Suspended Solids, Solids Retention Time, Waste Activated Sludge Rate, and Ability to Nitrify

A simple model can be set up consisting of an influent object, an activated sludge tank, a clarifier, and RAS/WAS pumps. Change the wastage rate and plot wastage rate versus MLSS, SRT, and effluent ammonia, as shown in Figure 9.1. This simple modeling exercise can be used to demonstrate several

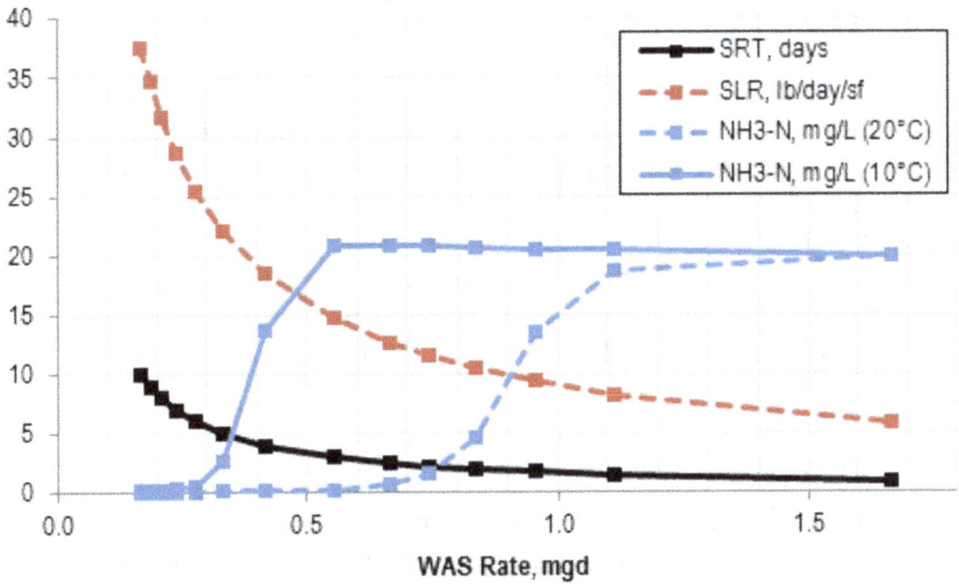

FIGURE 9.1 Training Model Output Illustrating Relationship Between Wasting and Effluent Ammonia (SLR = solids loading rate; NH_3-N = ammonia nitrogen; lb/ft^2·d × 4.89 = kg/m^2·d; mgd × 3.78 = ML/d)

concepts important to facility operators. As wastage rate increases, SRT and MLSS decrease and effluent ammonia increases. Above a specific wastage rate, no nitrification will be observed. This is the SRT at which all nitrifiers have been washed out of the system. The process water temperature can be varied and the exercise performed again to show that nitrifier growth rate, and, therefore, the minimum SRT to washout, is a function of temperature.

3.1.2 Effect of Dissolved Oxygen Setpoint

Set up a model with an influent object, a plug flow (three or more tanks in series), activated sludge tank, and a final clarifier with RAS and WAS pumps. For the first set of simulations, vary the dissolved oxygen setpoint and plot the predicted aeration required to meet the setpoint. For the second set of simulations, set up a dissolved oxygen controller to maintain a dissolved oxygen setpoint in one of the tanks in series at a time. Run the model with the same setpoint but change the tank to which the controller fixes dissolved oxygen to the selected setpoint. Repeat with a different dissolved oxygen setpoint. Plot effluent ammonia and airflow versus control location. As the dissolved oxygen control point moves from upstream to downstream in the tank, more air (higher dissolved oxygen setpoint) will be required to achieve the same amount of nitrification.

3.1.3 Effect of Varying Return Activated Sludge Rate on Return Activated Sludge Total Suspended Solids and Return Activated Sludge Control During Wet Weather Events

Set up a model with two influent objects (one influent will represent a typical domestic wastewater and the other will represent stormflow with no TSS, COD, nitrogen, or phosphorus). Also, include an activated sludge tank and a final clarifier with RAS and WAS pumps. Set the flow in the storm influent to 0. Vary the RAS rate and plot the RAS TSS and the MLSS in the activated sludge tank. Plot final effluent solids. As RAS rate increases, RAS concentration decreases and MLSS increases. Increase the flow of storm influent and increase RAS flow. Monitor the effect that RAS flow has on final clarifier performance.

3.1.4 Effect of Wet Weather Operational Strategies

Set up a model with an influent object, a plug flow (three or more tanks in series), activated sludge tank, and a final clarifier with RAS and WAS pumps. Set up the model so that influent can be delivered to any of the activated sludge tanks and such that the process total SRT is constant. Change the step-feed location and percentage and plot MLSS in each of the tanks as well as clarifier SRT to note how the distribution of solids shifts with more step-feed flow or step feeding to deeper within the aeration basin.

3.1.5 Effect of Recycle Streams

Set up a simple whole facility model using a mainstream BNR process and with solids treatment using anaerobic digestion and solids dewatering. Divert anaerobic digester supernatant to an effluent element to simulate off-site solids management at a regional facility and note effluent nitrogen and phosphorus concentrations. Now, tie in the recycle stream to a flow combiner at the facility influent and adjust the solids-dewatering solids-separation settings to achieve approximately 5% total solids to simulate a facility with liquid land application and note the effect of effluent nitrogen and phosphorus. Lastly, increase the dewatering performance such that the process achieves 20% total solids and note the further effect it has on facility effluent nitrogen and phosphorus. In this exercise, the influent biodegradable soluble COD fraction may need to be adjusted such that the facility is slightly carbon limited to achieve effective BNR performance.

3.2 Model Applications for Facility Optimization/Troubleshooting

In addition to training, operational engineers and staff can use calibrated models of their facilities to find data errors, optimize processes, and develop operational strategies. What follows is a brief explanation of simulator applications of some processes; in the next section, some case studies are provided that show how some WRRFs have used process modeling for operational improvements.

3.2.1 Mass Balance Investigation

Process simulators can be excellent tools for checking facility mass balances and thereby determine sample and flow measurement issues. Once a baseline model is constructed and calibrated, it can be used as a tool to check various facility mass balances. By simulating mass balance around thickeners and dewatering units and matching inputs and outputs to historical data, any discrepancies in solids capture rates can be determined. When this is performed, it can become clear where flow errors or bias solids samples have been taken. This knowledge can then be used to improve operational sampling protocols to fill in any data gaps that become apparent. This can be a valuable exercise that can lead to a better understanding of how solids are moving through the facility and where they are being stored. This understanding can also be used to reconcile data and any discrepancies in historical data measurements as well as for process optimization to improve facility performance. An example of how good modeling practices were applied to better understand the facility solids mass balance can be found in literature by Third et al. (2007).

Similarly, simulators can be used to detect and troubleshoot uneven flow splits between treatment trains. This can be done by modeling treatment trains separately and varying the flow split, which is least certain, whether that is secondary influent or RAS. Available data between the different trains, such as MLSS or, potentially, WAS production, can be matched to estimate the unknown flow split. This knowledge could inform revisions to flow control procedures, aid in more accurate rating of facility capacity, and help to develop a case for improved flow-splitting facilities.

3.2.2 Taking Tanks Out of Service

Process models can also be used to optimize the number of tanks in operation. The number of tanks in service can be adjusted by proportionally changing the volume of the appropriate reactor elements in a process model. The resulting effect in the model on capacity and effluent quality can be estimated in this way before trying the operation at full scale, which is much more operator intensive. The same process can be followed to investigate the effect of taking a tank out of service for planned maintenance under different seasonal conditions to decide if this is possible and, if so, when the best season is for taking it out of service (Phillips et al., 2010).

3.2.3 Dissolved Oxygen Control

Dissolved oxygen control strategies can be easily implemented in a process model to optimize aeration requirements and determine energy savings and effects on effluent quality. For example, in one scenario the model could be run at historical residual dissolved oxygen (say 4 mg/L) and a second scenario model could be run at residual dissolved oxygen of 2 mg/L. The difference in airflows required for the two scenarios could be used to calculate total energy savings while effects to nitrification and denitrification performance could be determined.

3.2.4 Strategies for Managing Extreme Flow/Load Events

Treatment of peak wet weather flows often requires facilities to take operational steps to temporarily enhance facility capacity, although these steps may have detrimental effects to effluent quality if they are overused. Simulators can be used to model a wet weather event and wet weather treatment alternatives to understand and evaluate the effects.

One option is step-feed operation, which involves feeding a portion of the secondary influent to a point further down the aeration basin, resulting in a reduction in MLSS at the end of the basin and, therefore, a reduction in final clarifier solids loading rate (SLR). The specific degree of MLSS reduction is dependent on the amount of flow diverted and the position within the aerobic zone where the flow is diverted. In this case, total solids inventory remains the same, but more of it is stored at the head of the aeration basin (higher MLSS) and less is stored at the end of the aeration basin (lower MLSS). A related strategy is offline sludge storage in which a portion of the RAS is diverted to a separate storage basin to again reduce MLSS and, therefore, the SLR on the secondary clarifiers. In both cases, the active biomass concentration and potentially the hydraulic retention time (HRT) will be reduced for the main flow through the facility. This could result in inadequate kinetics or inadequate aerobic HRT to achieve full nitrification. In general, the greater the reduction in SLR because of these strategies, the greater will be the effect on nitrification. This is not a simple relationship, and options are highly dependent on facility specifics such as basin volume and layout, clarifier area, wet weather peaking factors, and nutrient limits, making this is an excellent use of simulators. Modeling these alternatives can help to determine the amount of step-feed flow and/or the location of step-feed flow (or, similarly, the amount of sludge stored offline), which is acceptable without risking nitrification breakthroughs that risk permit violations, as shown in Figure 9.2 from an investigation of offline sludge storage (Sabherwal et al., 2006).

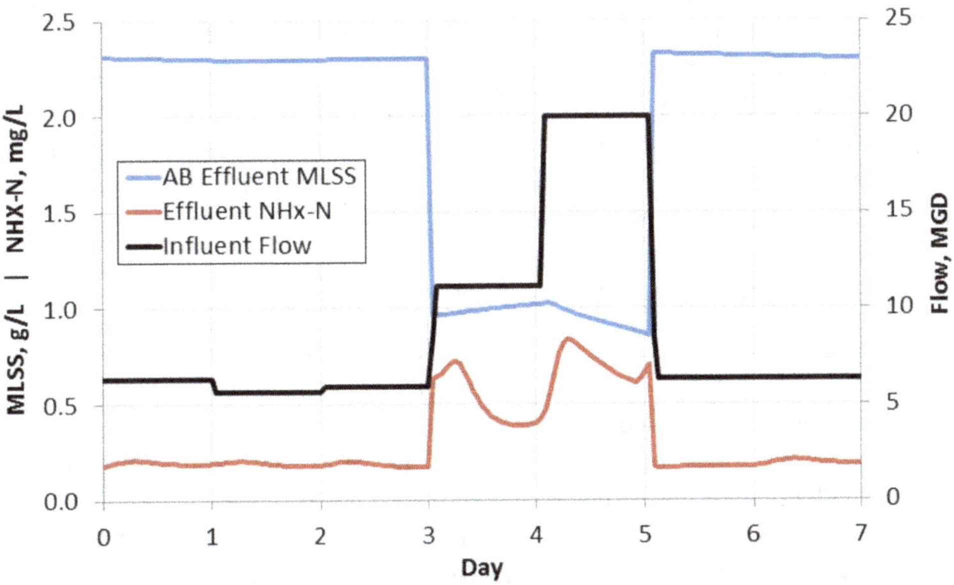

FIGURE 9.2 Dynamic Response of Clarifier Loading and Effluent NH_X-N to Wet Weather Event When Using Sidestream Sludge Storage (Adapted from Sabherwal et al. [2006]) (AB = activated sludge basin; mgd × 3.78 = ML/d)

Other wet weather treatment strategies will result in a similar tradeoff between short-term capacity enhancements and nutrient-removal performance. Again, dynamic simulators can be an effective way to evaluate this tradeoff. Enhanced primary treatment can be used to reduce loading to secondary treatment trains, which can limit MLSS and, therefore, SLR over a longer time scale. This diversion of COD to primary sludge will result in lower carbon-to-nitrogen and carbon-to-phosphorus ratios in the secondary influent affecting BNR performance. Steady-state models can be used to understand how significant this effect will be over the longer term, whereas dynamic models can be used to see how quickly effects on MLSS, SLR, and nutrient-removal performance will occur.

3.2.5 Advanced Controls

With dynamic models, more complex aeration strategies like advanced ammonia-based control can be simulated for a particular WRRF. If properly configured and calibrated, the model will take account of limitations of implementing ammonia-based control at a particular facility. This will allow the facility to investigate the effect on aeration energy requirements and effluent ammonia breakthrough at peak loads (Rieger et al., 2012).

Additionally, sensor placement can be investigated by using dynamic process models. Sensors should be located where they give the best information for operation and control. This should be a point where values are expected to change and/or need to be controlled to a setpoint. To guide the selection of the location, a dynamic process model can be used to investigate how and where an online measurement is expected to change. The model can also be used to investigate how moving a sensor location will change the performance of a control system (e.g., Should a dissolved oxygen probe be positioned at the front, middle, or back of a long basin to give the best control?).

3.2.6 Chemical Feed Needs and Effects

Models can be used to determine chemical feed rates and optimize dosage location or control for a treatment process. This could be for alkalinity supplementation, carbon supplementation, or metal salt addition for chemical phosphorus removal. If a model indicates that significantly less of a certain chemical is required than is currently used at the facility, this may result in the investigation of the feed location or dose control philosophy. Optimizations in this area can lead to significant reductions in chemical costs. Metal coagulants could be added to an area with low mixing energy or carbon could be added to an area with some residual dissolved oxygen, both of which would lead to wasted chemical dosing. If a dynamic model is available, then dosage control philosophies can also be investigated that can lead to more efficient use of the chemicals toward stated goals.

3.2.7 Energy and Cost Calculations

Finally, models can be used as effective tools to track energy consumption throughout a facility to estimate future operating costs, and, ultimately, develop plans to reduce operating costs. As with any prediction by a model, the accuracy of energy and cost estimates is governed by the quality of data, accuracy of the calibration, and cost assumptions. Further, accuracy will largely be governed by what scenarios are run as the basis for this estimation. For example, an annual average steady-state model output will provide a poor estimation of energy and costs compared to a full-year dynamic simulation with facility controls properly described. More accurate energy cost inputs are also important in some markets and simulators today allow the use of more complex cost calculations such as peak and off-peak electricity tariffs, which will allow for more realistic cost calculations.

3.3 Case Studies

3.3.1 Denver Metro Wastewater Reclamation District

The Denver Metro Wastewater Reclamation District's (the District's) Robert W. Hite Treatment Facility has a current annual average flowrate of 530 000 m^3/d (140 mgd). Current discharge limits include monthly average ammonia concentrations that vary by month (9.7 to 15 mg N/L), and a 7-day nitrate-nitrite (NO_X-N) concentration of 8.68 mg N/L. The facility consists of the North Secondary Complex (air activated sludge in the Modified Ludzack–Ettinger configuration) and the South Secondary Complex (high-purity oxygen activated sludge). Future upgrades at the North Secondary and an expansion of the South Secondary will both include nitrification, denitrification, centrate, and RAS reaeration basins and provisions for future biological phosphorus removal.

The Metro Wastewater Reclamation District owns and regularly calibrates and validates a whole facility model, which it uses for a variety of applications, including day-to-day operations, forecasting constructability challenges, and verifying designs developed using other models. The model encompasses all liquid processes, with the exception of disinfection, and all solids processes. It is dynamic in the sense that it includes diurnal flows and loads and user-specified pumping rates, and because it is routinely validated and updated. Several special sampling campaigns have been carried out to better characterize influent in the model and determine the effect of seasonal variations.

The model has been useful to the District in simulating construction interruptions, experimenting with performance, sizing future facilities, and providing many answers to "what if?" questions. The whole facility model was applied to evaluate the effect of new industrial nitrogen load on performance during construction. The primary objective here was to quantify costs associated with additional aeration as well as chemicals. At another time, the District experienced problems with its new submersible mixed liquor recycle pumps, so a sensitivity analysis was conducted to assess the effect of pump failure on the ability to maintain compliance with the effluent NO_X-N limit. Modeling results enabled staff to efficiently identify critical months and loading conditions so that maintenance could be scheduled to reduce the risk of permit violations. The facility model has helped the District understand the effects of temporary decreases in capacity when basins are taken offline and it has helped gauge risk associated with making process changes. Modeling has facilitated the proactive development of operating procedures that properly consider and balance effluent quality, risk of permit noncompliance, and cost effects.

More than $1 million worth of time and effort had been invested in the District's facility-wide model, but this is just a fraction of the savings anticipated by refinement of nearly $800 million of capital improvements and the associated operating costs. Dynamic modeling of the South Secondary BNR expansion enabled the design team to select an SRT that resulted in saving 3790 m^3 (1 mil. gal) of aeration basin volume, or $2 million, from planning estimates. Modeling also identified construction scenarios where adding acetic acid could increase operating budgets by hundreds of thousands of dollars. The model provides the District a tool to be able to simulate changes to the facility and develop ways to reduce expenses before spending the money and time to implement at full scale (Phillips et al., 2010).

3.3.2 Northeast Ohio Regional Sewer District

Northeast Ohio Regional Sewer District's (NEORSD) Southerly Wastewater Treatment Center (WWTC) is designed to treat an average daily flow of 663 ML/d (175 mgd) and a peak wet weather flow of 1512 ML/d (400 mgd). The Southerly facility provides tertiary treatment by using a two-stage activated sludge system with effluent filtration and chlorination/dechlorination. The first-stage activated sludge system is devoted to BOD removal and the second stage activated sludge system provides BOD polishing and nitrification. Residual solids are processed by gravity thickening or gravity belt thickening, thermal sludge conditioning, centrifuge dewatering, and incineration.

The NEORSD decided to include a model and a customized interface of the facility as part of the Southerly WWTC Comprehensive Facilities Plan.

The objective for the customized simulator was to provide an easy-to-use model of the Southerly facility that could be used by NEORSD staff for training, conducting "what if" scenarios and planning studies, troubleshooting, and investigating operational improvements. Specific uses planned for the tool included the following:

- Prediction of sludge production, storage, and feed rates to the incinerators for selected "what if" scenarios such as wet weather events (which are predefined in the software) and
- Optimization of current facility operational controls such as flow split between first and second activated sludge stages, number of tanks and clarifiers in operation, sludge return rates, sludge wasting rates, and dissolved oxygen concentration or air supply.

Project benefits included increased accessibility of models within the organization, reduced training requirements for process modeling, and the availability of a tool to support operational and planning decisions. In particular, high-level operations staff at the NEORSD have been using the Southerly WWTC simulator as a decision support tool. The simplified interfaces help ensure access of the underlying complex facility models to operations staff.

Despite the benefits achieved, certain obstacles for the use of models by operations staff remain. The most significant is the lack of time available for modeling by operations staff and the need to have dedicated personnel responsible for the implementation of modeling within the organization. An important requirement is managing expectations in terms of model predictive accuracy.

For continued use of models in the future, it is important to have a champion of the model within the organization and a plan for maintenance of the model. Spreading the modeling knowledge among multiple users within the organization helps minimize the effect of staffing changes.

3.3.3 Pima County Regional Wastewater Reclamation Department

Pima County Regional Wastewater Reclamation Department recently upgraded its Tres Rios WRRF to increase its design capacity from (37.5 to 50.0 mgd) to meet projected organic loads for the year 2030. The facility consists of a modified BNR activated sludge train and a new Bardenpho train.

The Tres Rios WRRF is equipped with a supervisory control and data acquisition system (SCADA) system that collects a substantial amount of online and offline data. A calibrated model of the facility was developed and integrated with SCADA data to help facilities undertake automated periodic reviews of operational data for performance reviews and benchmarking of facility performance against predefined baseline values. Benchmark indicators included performance variables directly related to operational and maintenance issues (e.g., air requirements, energy consumption, biogas production rate, and online data quality). The objective was to identify process improvement opportunities and formulate and prioritize improvement strategies through regular analysis of data collected at the facility. Additionally, by doing these performance reviews, the model results are compared against facility monitoring data regularly, leading to further refinement of model calibration and/or overall improvement in data quality.

The hourly SCADA data include flows, online sensor water quality measurements such as dissolved oxygen, nitrate-nitrogen , and ammonium and operational variables such as airflows and chemical dosing were used to conduct dynamic simulations and analyze the response of the model to assess noticeable differences between the model and observed data. These differences were then used to evaluate model and data quality and define actions for improvement. The following are a few examples of the way that the SCADA integrated model was used:

- Troubleshooting MLSS sensor calibration issues in bioreactors;
- Predicting airflow rate values lower than actual airflows used, suggesting potential issues with oxygen-transfer efficiency, diffuser fouling, air leaks, blower control strategy, and so on; and
- Providing insights to operations of the digesters via biogas generation simulation, confirming expected biogas generation rate.

The full paper describes the use of a detailed operation model for performance reporting and benchmark application, including broad outlines of a protocol for collating, cleaning, and converting SCADA and laboratory records to a format suitable for running dynamic simulations (Goel et al., 2016).

3.3.4 Eastman Chemical Company

Eastman Chemical's wastewater treatment facility in Kingsport, Tennessee, is the largest (4400 m^3/h) of the Eastman wastewater facilities. It treats waste from a facility that manufactures chemicals, plastics, and synthetic fibers. Wastewater is highly variable in both flow and composition and is primarily comprised of short-chained alcohols and acids. The facility has a National Pollutant Discharge Elimination System permit with seasonal BOD, TSS, and total nitrogen limits (5-day BOD): 4000 to 6000 lb/d, TSS: 10 093 lb/d, and total nitrogen: 6664 lb/d or 30.5 mg/L. The facility is a step-feed activated sludge facility with conventional secondary clarification.

The aerobic reactor system consists of four parallel trains of three aeration basins in series. Influent flow to the aerobic reactor system is divided equally among the trains. Although, typically, the facility is operated in step-feed mode, individual trains can be operated throughout a range, from plug flow mode to complete-mix mode. The facility also has equalization and diversion tanks to manage loading to the activated sludge system and belt filter presses for WAS dewatering. The Kingsport WRRF is described in detail by Churn et al. (1990).

A calibrated facility model was used to conduct "what if" scenarios to improve understanding of the facility and assess the effect of operational changes such as SRT, toxic spills, and increased organic and nutrient loadings. Steady-state and dynamic sensitivity analysis studies were performed to determine the effect of changes to the facility. The model has been used to predict the effect of SRT on the aeration basin airflow rate required, the sludge produced, and the amount of dewatering polymer used. By adding operating and sludge disposal costs to the model, facility staff could use the sensitivity analysis tool to select an optimal SRT for the facility that minimizes costs while maintaining acceptable effluent quality. In addition, this type of analysis can be used by the facility to assess the effect of SRT on the solids loading to clarifiers.

The model provided additional valuable insights beyond predefined scenarios. For example, by separating combined waste streams into simpler individual waste streams in the model, facility staff were able to evaluate the effect of different product lines of the manufacturing facility. Problems with secondary sludge settleability are common in industrial WRRFs and settling aids such as polymer may be used. With the aid of the model, variable settling characteristics could be evaluated without having to experiment with the facility (Schraa et al., 2004).

4.0 REALISTIC EXPECTATIONS FOR PROCESS MODELS

Activated sludge models were originally designed to predict sludge production and aeration demand and develop concepts for early BNR processes. They were not intended to predict effluent quality or settler performance with a high degree of accuracy, especially at low levels. While these are still some of the primary uses of models, their applications have extended significantly beyond this, as highlighted in Section 3.0 of this chapter. As a result, understanding model limitations, which aspects of a process models cannot accurately predict, and which outputs, therefore, require interpretation based on engineering judgment and operational experience is essential.

4.1 Sources of Error in Biological Reactions

Biological models adjust kinetic rates because of substrate limitations based on the Monod equation (see Chapter 2). This is an empirical relationship with two constants (the maximum specific growth rate and the half-saturation coefficient) and it controls reaction rates and how these rates slow down as necessary constituents in those reactions approach low levels. In reality, these rate reductions could be caused by biological phenomena or diffusion limitations (likely a function of mixing energy, for example). This explains the variability in observed kinetic performance. In simulators, default values of kinetic rate constants have been derived from average performance data of typical WRRFs. However, at lower concentrations observed at the end of aeration basins, kinetic constant values may differ and the model will not necessarily be predictive without additional calibration.

The effect of this variability is not important for prediction of BOD or COD removal. Soluble biodegradable and nonbiodegradable COD fractions can be determined relatively easily as part of influent characterization. Soluble BOD degradation is fast, and this reaction is close to completion in aeration basin effluent, whereas removal of particulate BOD and COD is more closely tied to secondary clarifier performance and effluent solids concentrations. While the effluent solids concentrations will affect effluent total nitrogen and total phosphorus, soluble species (NH_X-N, NO_X-N, and PO_4-P) are harder to predict with the model and will have significant effects on total nitrogen and total phosphorus. For these reasons, accurate prediction of effluent nutrient concentrations has its limits below a certain concentration. The accuracy could potentially be refined with considerable calibration involving the adjustment of kinetic values, something which is beyond the scope of most projects and that may require more detailed data than are available typically.

The strength of models lies in their ability to investigate effects of trends on effluent quality and comparing relative differences in effluent quality among alternatives. For example, effects because of higher flows, higher concentrations resulting from water conservation, or increasing sidestream nutrient loads resulting from increased hauled waste acceptance will be represented in the model. Additionally, relative comparisons among different alternatives can be made, such as the improvement because of sidestream treatment or the difference between BNR process alternatives.

4.2 Sources of Error in Settlers

Most mainstream processes continue to use gravity separation for primary and secondary solids removal. For a settler model to predict the performance of these processes, settling characteristics of the sludge (such as those discussed in Chapter 5) must be known. Because of the variability of settling rates, models have difficulty predicting settler performance.

Secondary settling rates are largely the outcome of biological processes, and the fraction of solids made up of denser inert particles. For example, the selection of certain bacteria in aeration basins can result in sludge "bulking", which significantly reduces solids-settling rate and the degree of thickening possible. Variables and reactions that control these bulking phenomena within the aeration basin are not well understood and are not part of biological models to date. Inert solids can have a ballasting effect, that is, they will weigh down other particles and substantially increase settling rates of all solids. This could be affected by the presence or performance of primary clarifiers or the addition of metal flocculants to aeration basins for chemical phosphorus removal, resulting in the formation of denser precipitates.

Primary settling rates and removal performance of various settleable constituents are characteristics of influent wastewater. Raw influent solids include inert solids, nonbiodegradable volatile solids, and biodegradable volatile solids, all of which may be associated with particles with different settling characteristics. This distribution of influent solids and their settling characteristics may change seasonally or with higher degrees of infiltration and inflow. Because of this complexity, primary clarifier models may also need to be calibrated against observable primary effluent characteristics to capture any selective removal effect.

The performance of clarifiers is also largely dependent on specifics of clarifier design. However, settling models used in commercially available simulators are one-dimensional models that do not account for these facility specifics. This can be overcome by using two-dimensional or three-dimensional computational fluid dynamics (CFD) models to evaluate clarifier performance. However, those evaluations will still, at best, be only as good as the settling characteristics input, which, again, is not predicted by any models. As a result of these shortcomings, percent solids removal based on historical performance is often input directly to the model (overriding the settling model) to yield a representative mass balance of the whole facility. Settling characteristics could be input based on settling rate tests for the facility or from historical sludge volume index (SVI) data and published SVI correlations. However, this is still an estimate, which may be appropriate for design but may not be a predictive model. As a result, operational modeling should not rely on clarifier performance predictions without extensive investigation.

4.3 Sources of Error in Biological Reactor Hydraulics

Hydraulic behavior within activated sludge basins can have a significant effect on performance because it affects the residence time available for reactions to occur. For example, short-circuiting will result in a shorter reaction time and incomplete reactions for a portion of flow through the basins. In contrast, more ideal hydraulic behavior will result in the reaction proceeding to a greater extent. All reactors behave somewhere between ideal completely mixed and ideal plug flow behavior, which will affect the extent of reactions and, thereby, process performance.

Despite its importance, the degree to which a basin exhibits plug flow or short-circuiting behavior is not predicted by models. Reactors must be modeled in the simulator as a series of reactor elements that best capture actual performance. All simulators currently use reactor elements, which represent one or multiple completely mixed tanks in series; however, the modeler must pick the number of tanks in series necessary to represent the basin being modeled and whether parallel reactors must be added to accurately represent short-circuiting behavior. Several methods can be used to estimate the number of reactors necessary to appropriately model a basin. A general rule can be used based on the aspect ratio of the basins, and there are simple formulas available that will take more factors into account, such as the empirical formula developed by Fujie et al. (1983). For more accurate assessments or to assess if short-circuiting is occurring, methods such as full-scale tracer tests or CFD modeling are necessary.

4.3.1 Garbage In = Garbage Out

Books on good modeling practice referenced in this chapter stress the need for high-quality data for model inputs. Models are only as good as the information used to build them. If data are of low quality, then the predictive ability of the model will be poor. This is why a lot of time and energy has to be spent on providing good data, as discussed in Section 2.0 of this chapter (IWA Task Group on Good Modelling Practice, 2012; Menniti et al., 2015).

The most important data for any process model are influent data. To get good estimates of sludge production, aeration requirements, and treatability of incoming wastewater, it is important to have a detailed understanding of influent waste concentrations and fractionations. An error in this part of the model can result in errors passing all the way through to digester performance and sludge production. The availability of the necessary influent and fractionation data for model calibration can be ensured by following guidelines outlined in Section 2.0 of this chapter. Additionally, models must be calibrated to actual facility data. Modern process simulators have user-friendly interfaces and are quite easy to use. Because of this, there is a temptation to put together a model quickly and run some simulations "straight out of the box" without paying attention to proper data inputs. Although out-of-the-box models may be adequate when used for training on wastewater treatment concepts, the setup and calibration of the model are important if simulations are expected to give meaningful and useful results.

4.3.2 Unrealistic Accuracy Expectations

It is important to understand that no model is perfect and, in the world of wastewater treatment, where everything is constantly changing and accurate measurements are difficult to take, it should not be expected that models will produce extremely accurate predictions. Some of the significant reasons for this already have been discussed throughout this chapter, but inaccuracies could also be caused by inaccurate influent or process data or simply the natural variability of model parameters. For the most part, accuracy of the model depends on the accuracy of the data that are put into it. Typically, however, it is too costly to gather enough data to approach 100% accuracy. Therefore, the modeler has to strike a balance between the effort of data gathering and calibration with practicalities of cost and time to produce a model that is "good enough" for the purpose for which it is designed.

Accuracy expectations should depend on the level of effort that was put into the model and data gathering. In turn, levels of effort should be tied to the question that is being asked and how the model will ultimately be used. These are important scope items to define at the beginning of a modeling project so that time spent modeling will not be wasted. Goals and expected accuracy levels of the model should be defined in line with *Guidelines for Using Activated Sludge Models* (IWA Task Group on Good Modelling Practice, 2012).

For example, if a model was calibrated originally to estimate future sludge production, then it is not realistic to use the same model to estimate current effluent ammonia concentrations. If a model is needed to predict more accurately effluent ammonia, then more data and a more rigorous calibration are needed, most likely using a dynamic model. To give some idea of the level of accuracy that can be expected from a well-calibrated model, Table 9.2 presents common target variables and stop criteria for model calibration. This is adapted from Table 8.11 in *Wastewater Treatment Process Modeling* (Water Environment Federation, 2013). These "stop criteria" define the level of error that is acceptable (based on the expert opinion of the IWA Task Group on Good Modelling Practice) during a model calibration for a series of model purposes. Note that, in general, even for detailed modeling efforts, it is unrealistic to expect that effluent outputs will be within approximately 1 mg/L for nitrogen compounds and 0.5 mg/L for phosphorus compounds. Likewise, most criteria for airflow requirements and sludge production are approximately 10% of measured values. When using a model for predictive purposes, it is reasonable to assume that the

TABLE 9.2 Common Target Variables and Stop Criteria for Model Calibration (WEF, 2013)

Model purposes	Averaging period	Target variable	Acceptable error range
Calculate sludge production	Monthly average	MLSS MLVSS/MLSS WAS mass load Effluent TSS SRT	10% 5% 5% 5 mg/L 1 d or 15% for SRT < 5 days
Design aeration system	Monthly average daily average hourly peaks	Airflow rate Dissolved oxygen (profile) Oxygen uptake rate	10% 0.3 mg/L 10 mg/L·h
Develop a process configuration for nitrogen removal	Monthly or annual average	NH_X-N NO_X-N Total nitrogen	1.0 mg/L 1.0 mg/L 1.0 mg/L
Design a treatment system to meet peak effluent nitrogen limits	Instantaneous values	NH_4-N NO_X-N Total nitrogen	0.5 mg/L 0.5 mg/L 0.5 mg/L
Develop a process configuration for phosphorus removal	Monthly or annual average	PO_4-P Total phosphorus	0.5 mg/L 0.5 mg/L
Assess facility capacity for nitrogen removal	Monthly or annual average	NH_X-N NO_X-N Total nitrogen	1.0 mg/L 1.0 mg/L 1.0 mg/L
Make the most of aeration control	Hourly average	Airflow rate Dissolved oxygen Effluent NH_X-N	10% 0.5 mg/L 0.5 mg/L
Test effect of taking tanks out of service	Monthly average	NH_X-N NO_X-N Total nitrogen PO_4-P Total phosphorus	1.0 mg/L 1.0 mg/L 1.0 mg/L 0.5 mg/L 0.5 mg/L
Use model to develop sludge-wastage strategy	Weekly average Daily average	WAS load SRT NH_X-N PO_4-P	10% 1 d 1.0 mg/L 0.5 mg/L

(*continued*)

TABLE 9.2 Common Target Variables and Stop Criteria for Model Calibration (WEF, 2013) (*Continued*)

Model purposes	Averaging period	Target variable	Acceptable error range
Develop a strategy to handle stormflows	Hourly average	MLSS	10%
		Effluent TSS	10 mg/L
		NH_X-N	1.0 mg/L
		NO_X-N	1.0 mg/L
		PO_4-P	0.5 mg/L
Develop a site-specific model for operator training	Monthly average	MLSS	10%
		WAS load	5%
		Effluent TSS	5.0 mg/L
		NH_X-N	1.0 mg/L
		NO_X-N	1.0 mg/L
		PO_4-P	1.0 mg/L
		Airflow rate	10%
		Dissolved oxygen	0.5 mg/L

model will be less accurate than during calibration because of additional uncertainties.

5.0 PRACTICE QUESTIONS

1. Total COD is divided into fractions that are used by the model on the basis of what?
   a. Chemical formula and particle size distribution
   b. Nitrogen and phosphorus content
   c. Particulate/colloidal/soluble form and biodegradability
   d. Loss of mass in a furnace and settling rate

2. Typically, what are the most important adjustments in model calibration?
   a. Tuning of maximum growth rates and half-saturation coefficients
   b. Influent characterization and understanding data quality
   c. Adjusting stoichiometric ratios and biological process yields
   d. Estimating settling rate characteristics and determining the causes of bulking sludge

3. Which of these is it not reasonable to expect a well-calibrated process model to predict?

a. Ammonia breakthrough within ~2.0 mg N/L because of a wet weather event
b. Waste activated sludge production within 10% of the observed value
c. Clarifier failure because of bulking sludge brought on by high food-to-microorganism ratio
d. Relative effect of sidestream flow equalization on dynamic BNR performance

4. Which of the following describes the difference between a process model and a process simulator?
   a. Simulators are used for estimating solids removal, whereas models calculate biological processes
   b. A process model is a mathematical representation of reality that is implemented by a simulator
   c. Models were referred to as *simulators* before the introduction of Activated Sludge Model 1
   d. A simulator is a mathematical representation of reality that is implemented by a model

5. Which of the following is not a realistic step in using a model to answer an operational question?
   a. Precisely defining the operational question and the degree of accuracy expected from the model
   b. Gathering data that are relevant to the model needed to answer the question being posed
   c. Producing a model that is 100% accurate under all scenarios experienced at the facility
   d. Interpreting the results of the model and using that knowledge to inform real-world decisions

Answer key:

1. c, 2. b, 3. c, 4. b, and 5. c.

6.0 REFERENCES

Churn, C., Johnson, D., & Severin, B. (1990). Facility and process design for large scale activated sludge industrial wastewater plant. *Proceedings of the 44th Industrial Waste Conference*, West Lafayette, IN, United States.

Fujie, K., Sekizawa, T., & Kubota, H. (1983). Liquid mixing in activated sludge aeration tank. *Journal of Fermentation Technology, 61*(3), 295–304.

Goel, R., Nelson, E., Sherlock, J., Goel, R., & Fabiyi, M. (2016, September 24–28). Enhancing plant operations using SCADA data for model based plant performance reporting: An example from Tres-Rios WRRF, Pima County. *Proceedings of the 89th Annual Water Environment Federation Technical Exhibition and Conference* [CD-ROM], New Orleans, LA, United States. Water Environment Federation.

IWA Task Group on Good Modelling Practice. (2012). *Guidelines for using activated sludge models*. IWA Publishing.

Melcer, H., Dold, P. L., Jones, R. M., Bye, C. M., Takacs, I., Stensel, H. D., Wilson, A. W., Sun, P., & Bury, S. (2003). *Methods for wastewater characterization in activated sludge modelling*. Water Environment Research Foundation.

Menniti, A., Schauer, P., Szilágyi, N., Shanely, R., Cullen, N., & Takacs, I. (2015, September 26–30). The Clean Water Services approach to developing a comprehensive process modeling program. *Proceedings of the 88th Annual Water Environment Federation Technical Exhibition and Conference* [CD-ROM], Chicago, IL, United States. Water Environment Federation.

Phillips, H. M., Rogowski, S., & Anderson, W. (2010). Long-term operations modelling: When does validation become recalibration? IWA Wastewater Treatment Modeling Seminar, WWTmod 2010. International Water Association.

Rieger, L., Jones, R. M., Dold, P. L., & Bott, C. B. (2012, September 29–Oct 3). Myths about ammonia feedforward aeration control. *Proceedings of the 85th Annual Water Environment Federation Technical Exhibition and Conference* [CD-ROM], New Orleans, LA, United States. Water Environment Federation.

Sabherwal, B., Kobylinski, E., Keller, J., Lawrence, M., & Moore, R. (2006, October 22–26). Novel approach to storm water management for BNR facilities. *Proceedings of the 79th Annual Water Environment Federation Technical Exhibition and Conference* [CD-ROM], Dallas, TX, United States. Water Environment Federation.

Schraa, O. O., Belia, E., & Churn, C. (2004, October 2–6). The use of process modeling to investigate the performance of a large industrial wastewater treatment plant. *Proceedings of the 77th Annual Water Environment Federation Technical Exhibition and Conference* [CD-ROM], New Orleans, LA, United States. Water Environment Federation.

Third, K. A., Shaw, A. R., & Ng, L. (2007). Application of the good modelling practice unified protocol to a plant wide process model for Beenyup WWTP design upgrade. 10th Specialized Conference on Design, Operation, and Economics of Large Wastewater Treatment Plants, Vienna, Austria, pp. 251–258. International Water Association.

Water Environment Federation. (2013). *Wastewater treatment process modeling* (2nd ed., Manual of Practice No. 31).

7.0 SUGGESTED READINGS

Schraa, O. O., Tole, B., Belia, E., Luo, J., Froehlich, D., Jenkins, J., Becker, E., & Martinak P. (2008, October 18–22). Promoting On-Site Use of Wastewater Process Models with Customized Simulators. *Proceedings of the 81st Annual Water Environment Federation Technical Exhibition and Conference* [CD-ROM], Chicago, IL, United States. Water Environment Federation.

Water Environment Federation. (2016). Process modeling fact sheets. access water.org

Index

S

W

Y

Z

www.ingramcontent.com/pod-product-compliance
Lightning Source LLC
LaVergne TN
LVHW081044280526
839639LV00002B/2
* 9 7 8 1 5 7 2 7 8 4 8 3 3 *